STIRLING CONVERTOR REGENERATORS

STIRLING CONVERTOR REGENERATORS

Mounir B. Ibrahim

Roy C. Tew, Jr.

CRC Press
Taylor & Francis Group
Boca Raton London New York

CRC Press is an imprint of the
Taylor & Francis Group, an **informa** business

CRC Press
Taylor & Francis Group
6000 Broken Sound Parkway NW, Suite 300
Boca Raton, FL 33487-2742

First issued in paperback 2017

ISBN-13: 978-1-4398-3006-2 (hbk)
ISBN-13: 978-1-138-07559-7 (pbk)

Library of Congress Cataloging-in-Publication Data

Ibrahim, Mounir B.
    Stirling convertor regenerators / Mounir B. Ibrahim, Roy C. Tew, Jr.
        p. cm.
    "A CRC title."
    Includes bibliographical references and index.
    ISBN 978-1-4398-3006-2 (hardcover : alk. paper)
    1. Stirling engines--Research. 2. Space vehicles--Propulsion systems--Research. I. Tew, Roy C. II. Title.

TJ765.I27 2012
621.4'2--dc23   2011038431

Visit the Taylor & Francis Web site at
http://www.taylorandfrancis.com

and the CRC Press Web site at
http://www.crcpress.com

To my wife Nagwa (Basily) Ibrahim, and to my children and family: Joseph M. Ibrahim and wife Daniela, and their daughter Abigail Ibrahim; Victor M. Ibrahim, M.D. and wife Ereni, and their son, Luke Roess Ibrahim; Mary M. Soryal and husband Soryal Soryal, M.D., and their daughters, Julia Marie and Gabriella Sophie Soryal; and Andrew M. Ibrahim, M.D.

Mounir Ibrahim

To my wife Margaret (Hoel) Tew, and to my sons and their families: David Tew and wife Patia McGrath; Jonathan Tew and wife Melissa (Hila) Tew, and their daughters, Greta and Maci Tew; Robert Tew and wife Jennifer (Lamielle) Tew, and their sons, Evan and Colin Tew, and daughter, Rylee Tew.

Roy Tew

Contents

Preface

Stirling engines and coolers are both practically useful and theoretically significant devices. Since 1980, the National Aeronautics and Space Administration (NASA) has worked on development of Stirling engine technology, first for automotive and then for space applications. Development of Stirling engines for generation of space auxiliary power began in 1983 with Mechanical Technology Inc. (MTI) as the primary contractor. These Stirling space engines use helium as the working fluid, drive linear alternators to produce electricity, and are hermetically sealed. Since then, significant advancements have taken place in Stirling engine/alternator development for both terrestrial and space applications. NASA is expected to launch a Stirling engine/alternator (for power generation) for a deep-space application sometime around 2017. The high-efficiency Stirling radioisotope generator (SRG) for use on NASA space science missions is being developed by the Department of Energy (DOE), Lockheed Martin (Bethesda, Maryland), Sunpower Inc. (Athens, Ohio), and NASA Glenn Research Center (GRC). Potential missions include providing spacecraft onboard-electric power for deep-space missions and power for unmanned Mars rovers. Advanced Stirling convertors (i.e., engine/alternators) would provide substantial performance and mass benefits for these missions and could also provide power for electric propulsion. A combined Stirling convertor/cooler, based on a high-temperature heater head, may enable an extended-duration Venus surface mission. Therefore, GRC is also developing advanced technology for Stirling convertors aimed at improving specific power and efficiency of the convertor and the overall power system. Performance and mass improvement goals have been established for these next-generation Stirling radioisotope power systems. Efforts to achieve these goals have been conducted both in-house at GRC and via grants and contracts. These efforts include further development, validation, and practical use of a multidimensional (multi-D) Stirling computational fluid dynamics (CFD) model, and research and development on high-temperature materials, advanced controllers, low-vibration techniques, advanced regenerators, and reducing convertor weight.

In 1988, Mounir Ibrahim (author of this book) worked as a Summer Fellow with the Stirling Engine Branch at NASA GRC, then known as NASA Lewis Research Center, in close association with Roy Tew (coauthor). The Stirling group at NASA, at that time, hosted regular workshops for grantees and contractors with the goal of improving the understanding of the losses in Stirling machines. Several companies and universities participated in those activities. David Gedeon, sole proprietor of Gedeon Associates (Athens, Ohio), and author of the Sage code that is in use today for designing Stirling machines, was one of the participants in those NASA programs and workshops.

Terry Simon of the Mechanical Engineering Department, University of Minnesota, was one of the university participants. As a result of such interactions, in conjunction with Mounir Ibrahim of Cleveland State University (CSU), developed an analytical technique for modeling unsteady flow and heat transfer in Stirling engine heater and cooler tubes (Simon et al., 1992; Ibrahim et al., 1994); the technique is now a component of the Sage one-dimensional (1-D) system simulation code (at that time called GLIMPS). This successful collaboration brought together three components of the research activities: experiment (Simon), multi-D CFD modeling (Ibrahim), and 1-D engine design and modeling (Gedeon). Further collaborations among these participants and with Gary Wood (Sunpower Inc.) and Songgang Qiu (Infinia Corporation, Kennewick, Washington) led to several DOE and NASA grants/contracts for Stirling regenerator research and development over several years. Dean Guidry, Kevin Kelly, and Jeffrey McLean of International Mezzo Technologies (Baton Rouge, Louisiana) guided the final design, development, and fabrication of the successful segmented-involute-foil regenerators that are reported on in this book.

This book brings together the results of regenerator research and development done by the above participants, as well as others in the United States (David Berchowitz, Matthew Mitchell, Scott Backhous) and outside of the United States (Noboro Kagawa, Japan) to benefit researchers in this field.

The authors would like to acknowledge the following persons who made major contributions in helping to generate much of the content of this book under various DOE- and NASA-funded contracts and grants: David Gedeon of Gedeon Associates, did many supporting Sage simulations, made us aware of several of the microfabrication geometries considered for regenerator development, worked out the involute mathematics needed to design the involute foils, and wrote many memorandums summarizing his analysis and other work for the rest of the regenerator research and development team. Terry Simon of the University of Minnesota supported the effort via large-scale regenerator testing; he worked closely with Mounir Ibrahim in coordinating his testing with the CFD simulations. Gary Wood of Sunpower Inc. provided insight from the Stirling designer's point of view; he planned modifications of the Sunpower frequency test bed (FTB) to prepare it for testing of an involute-foil regenerator, and coordinated various tests of random fibers and involute foils in the NASA/Sunpower oscillating-flow test rig. In earlier years, Gary Wood designed and developed the oscillating flow rig; and David Gedeon developed the techniques for analyzing the data with the help of his Sage code; Gedeon still does the oscillating-flow test rig data analysis. The above were asked to be coauthors of this book but declined, due to other commitments—we like to think.

Susan Mantell of the University of Minnesota Mechanical Engineering Department provided much valuable support in guiding investigations of microfabrication techniques and possible manufacturing vendors during Phase I of a NASA Research Award (NRA) Regenerator Microfabrication contract; Songgang Qiu of Infinia (a former doctoral student of Simon's) provided structural analysis support during Phase I and Phase II of that NRA contract. Infinia's structural analysis results are a part of this book. After International Mezzo Technologies was chosen to be the regenerator manufacturing vendor for the microfab regenerator development effort, Dean Guidry, Kevin Kelly, and Jeffrey McLean of Mezzo were responsible for guiding the development of the segmented-involute-foils and fabricating them for testing, and reporting on their work.

Others attended regenerator research and development meetings and provided support due to their interest even though they may not have been funded by DOE or NASA. An example is David Berchowitz of Global Cooling (Athens, Ohio). He also provided polyester material, such as that used in Stirling coolers, for test purposes.

Several University of Minnesota graduate students assisted Simon in his large-scale regenerator tests, and their research contributed to the contents of this book. Some of them are Yi Niu, Nan Jiang, Liyong Sun, David Adolfson, and Greg McFadden. Joerge Seume, in earlier years, designed the scotch yoke drive mechanism that provided oscillating flow for the large-scale regenerator tests.

Several Cleveland State University students assisted Ibrahim in development of the CFD techniques and simulations that supported the effort. Among them are Zhiguo Zhang, Wei Rong, Daniel Danilla, Ashvin V. Mudalier, S. V. Veluri, Miyank Mittal, and Mandeep Sahota.

Among the NASA employees who supported the regenerator research and development (R&D) efforts were Branch Chief Richard Shaltens, Lanny Thieme, Jim Cairelli, Scott Wilson, Rodger Dyson, and Rikkako Demko. In earlier years, Branch Chief Jim Dudenhoefer was an enthusiastic supporter of the Stirling Loss Understanding Workshops and Stirling loss research. Steven Geng helped organize and execute those workshops. Randy Bowman of the Materials Division took scanning electron microscope (SEM) photographs for various matrices.

The authors appreciate the permission of Noboro Kagawa of the National Defense Academy of Japan to include material from the many Mesh Sheet Regenerator papers written by him, his students, and his staff; he also kindly reviewed and commented on drafts of Chapter 10, reporting on his work.

The authors appreciate the contributions of the above, and other contributors/participants who we may have failed to mention.

The Authors

Mounir Ibrahim is professor of mechanical engineering at Cleveland State University (CSU), Ohio. Ibrahim has been involved in fluid flow and heat transfer research in different applications such as heat transfer in gas turbines, gas turbine combustors, Stirling engines, Stirling regenerator design using microfacbrication techniques, among others. He has more than 35 years of administrative, academic, research, and industrial experience. He is a Fellow of the American Society of Mechanical Engineers (ASME) and Associate Fellow of the American Institute of Aeronautics and Astronautics (AIAA). Ibrahim has been the chair of the ASME K-14 (Heat Transfer in Gas Turbines) Committee (July 2006 to June 2008). He was the chair of the Mechanical Engineering Department at CSU from March 1998 to June 2002. He was a visiting scholar at the University of Oxford, United Kingdom, in 2008, and at the University of Minnesota, Minneapolis, in 2002. He was awarded over $5 million externally funded research. Ibrahim has supervised more than 60 master and doctoral students. He has published in more than 100 publications in prestigious journals and conference proceedings. Ibrahim has two patents: "High-Temperature, Non-Catalytic, Infrared Heater," U.S. Patent #6368102 and U.S. Patent #6612835.

Roy Tew worked as an analytical research engineer for over 46 years at the National Aeronautics and Space Administration (NASA) Glenn Research Center on space-power projects, with particular emphasis on Stirling power-convertor analysis, until his retirement in January 2009. In these areas, he also acted as grant and contract monitor for efforts including research into Stirling thermodynamic loss understanding, Stirling regenerator research and development, and development of Stirling multidimensional modeling codes. Tew was an author or coauthor on 29 NASA reports and other published papers, while employed at NASA. He earned degrees in physics (B.S. from the University of Alabama), engineering science (M.S. from Toledo University, Ohio), and mechanical engineering (Dr.Eng. from Cleveland State University). He is a member of the American Society of Mechanical Engineers (ASME) and the American Institute of Aeronautics and Astronautics (AIAA). He was an Ohio Registered Professional Engineer until he let his license expire after retirement. Since retirement, Tew has been working with Mounir Ibrahim of Cleveland State University on the preparation of this book. During the fall semester of 2010, he taught a graduate course in energy conversion at Cleveland State University (his first experience in teaching a course).

1

Introduction

The Stirling engine has been identified as an excellent candidate for conversion of solar thermal energy to electric power at the 500 W to 5 kW level. As for space applications, NASA has recently developed a system that will utilize Stirling engines for deep-space missions. Those engines are expected to run without the need for refueling or any maintenance for a period of 14 years, continuously. Attributes of the engine that make it a strong candidate for such terrestrial and space applications are its high efficiency and the fact that heat is added to the cycle externally. Critically important to the performance of the cycle is the regenerator, a component within which thermal energy is extracted from the working fluid as it flows from the hot end of the engine to the cold end and then returned to the working fluid when the working fluid proceeds back to the hot portion of the engine. Use of a regenerator greatly increases the Stirling efficiency. It is the regenerator that is considered by many designers to be the critical component to target for improvement in the next generation of Stirling engine (and cooler) systems. A survey of performance for small (<100 We) engines indicates that regenerator thermal inefficiency contributes 1.5% to engine thermal inefficiency while pressure drop losses contribute about 11% engine inefficiency.

On the other hand, cryocoolers are cooling devices that produce temperatures below 100°K. They are used for cooling infrared sensors and low-noise amplifiers used in a number of NASA missions. Stirling coolers have higher cold-end temperatures, of the level needed for such applications as food storage and electronic cooling, for example. Electrical power is supplied to drive the piston and displacer, which causes thermal energy to be absorbed at one end and rejected at the other. The regenerator component is even more critical to the performance of the Stirling cryocooler than for a power convertor. In the case of a Stirling cryocooler regenerator, thermal losses come directly off the cooling payload. Improved cryocooler regenerators could boost overall efficiency (i.e., coefficient of performance, or COP) by 40% or more, based on some numerical simulations (using the Sage computer code, Gedeon, 2010) of a typical Stirling cryocooler operating at about 80°K.

The Stirling-engine regenerator has been called "the crucial component," Organ (2000), in the Stirling-cycle engine. The regenerator, which obtains heat from the hot working fluid and releases heat to the cold working fluid, recycles the energy internally, allowing the Stirling cycle to achieve high efficiency. The location of the regenerator within a radioisotope powered Stirling convertor is shown in Figure 1.1 .

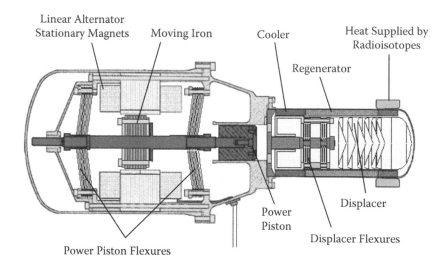

FIGURE 1.1
A radioisotope-powered Stirling convertor showing the location of the regenerator.

Currently, regenerators are usually made of woven screens or random fibers. Woven-screen regenerators have relatively high flow friction. They also require long assembly times that tend to increase their cost. Random fiber regenerators also have high flow friction but are easy to fabricate and therefore are inexpensive. Figure 1.2 shows a typical random-fiber

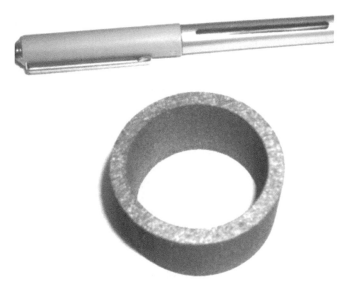

FIGURE 1.2
Random-fiber regenerator.

Bek-12_01 15.0 kV 12.0 mm ×500 SE(L) 05/13/2003 100 um   Bek30 F 6.0 kV 13.1 mm ×1.00 k SE(L) 05/13/2003 50.0 um

FIGURE 1.3
Electron micrography of a random-fiber regenerator matrix. (Courtesy of NASA Glenn Research Center.)

regenerator, and Figure 1.3 shows a close-up of the fibers. Due to the method of fabrication, the fibers are random primarily in a plane perpendicular to the main flow path. Thus, both woven screens and random fibers experience flow primarily across the wires (cylinders in cross-flow). Cylinders in cross-flow tend to cause flow separation resulting in high flow friction and considerable thermal dispersion, a thermal loss mechanism that causes an increase in apparent axial thermal conduction. For space engines, there must be assurance that no fibers of this matrix will eventually work loose and damage vital convertor parts during the mission. It is also important that local variations in porosity inherent to random fiber regenerators not result in local mismatches in flow channels, which would contribute to axial thermal transport. Wire screens have some randomness associated with their stacking and thus may have locally nonuniform flow. Random-fiber felts are much cheaper to manufacture than woven screens and perform about as well. And, they generally outperform wrapped foils. According to idealized one-dimensional theories, it should be the other way around, with foil regenerators providing the best performance (measured by heat transfer per unit flow resistance). But, in practice, wrapped-foil regenerators are difficult to manufacture in such a way that the theoretically good performance can be achieved. In particular, it is difficult to maintain the uniform spacing between the wrapped-foil layers required to achieve uniform flow. Research efforts have shown that attractive features for effecting high fluid-to-matrix heat transfer with low pressure drop are a matrix in which (1) the heat transfer surface is smooth, (2) the flow acceleration rates are controlled, (3) flow separation is minimized, and (4) passages are provided to allow radial mass flow for a more uniform distribution when the inlet flow or the in-channel characteristics are not radially uniform. It is thought that properly designed microfabricated regular geometries could not only reduce pressure drop, maintain high heat transfer, and allow some flow redistribution when needed, but could show improved regenerator durability for long missions.

A microfabricated segmented-involute-foil regenerator has been designed and built to achieve those goals; initial test results in an oscillating-flow test rig and also in an engine have been quite successful. This new design will be discussed in detail in Chapter 8.

This short introductory chapter has provided some background for this book. An outline of the scope of the rest of the book follows: Chapter 2 discusses unsteady fluid-flow and heat-transfer theory. Chapter 3 reviews correlations for steady and unsteady fluid flow and heat transfer. Fundamentals of operation and types and ranges of Stirling power convertors and coolers will be dealt with in Chapter 4, and different types of regenerators and recent development of their design will be discussed in Chapters 5 through 10. Chapter 5 will deal with different types of Stirling engine regenerators (SERs). Experiments, analysis, and computational fluid dynamics (CFD) will be presented for actual-scale random-fiber regenerators (Chapter 6), for large-scale random-fiber regenerators (Chapter 7), for actual-scale segmented-involute-foil regenerators (Chapter 8), and for large-scale segmented-involute-foil regenerators (Chapter 9). Chapter 10 will deal with other regenerator matrices such as "mesh sheet," flat plate, and so forth. The use of a compact porous-media device (similar to the Stirling regenerator) as an efficient heat exchanger and for thermal energy storage for different applications will be presented in Chapter 11. Finally, concluding remarks and discussion of future work will be presented in Chapters 12 and 13, respectively.

2

Unsteady Flow and Heat Transfer Theory

2.1 Governing Equations

Most real-life fluid flow phenomena are mathematically represented by the well-known Navier-Stokes (N-S) equations that are based on the continuum hypothesis. The N-S equations are a set of a nonlinear partial-differential equations arrived at by conservation of transport properties such as mass, momentum, and energy for an infinitesimal control volume. The N-S equations are based on the following universal laws of conservation: Conservation of Mass, Conservation of Momentum, and Conservation of Energy. The equation that results from applying the Conservation of Mass law to a fluid is called the continuity equation. The Conservation of Momentum law is nothing more than Newton's Second Law. When this law is applied to a fluid flow, it yields a vector equation known as the momentum equation. The Conservation of Energy law is identical to the First Law of Thermodynamics, and the resulting fluid dynamic equation is named the energy equation. Below is a description of each one of those governing equations. They are taken directly from Tannehill et al. (1997). In addition, a presentation is made for porous media models.

2.1.1 Continuity Equation

The Conservation of Mass law applied to a fluid passing through an infinitesimal, fixed control volume yields the following equation of continuity:

$$\frac{\partial \rho}{\partial t} + \nabla \bullet (\rho V) = 0 \tag{2.1}$$

where ρ is the fluid density, and V is the fluid velocity. The first term in this equation represents the rate of increase of the density in the control volume, and the second term represents the rate of mass flux passing out of the control surface (which surrounds the control volume) per unit volume. The substantial derivative notation:

$$\frac{D()}{Dt} \equiv \frac{\partial()}{\partial t} + V \bullet \ ()$$

(2.2)

can be used to change Equation (2.1) into the following form:

$$\frac{D\rho}{Dt} + \rho(\ \bullet V) = 0$$

(2.3)

For a Cartesian coordinate system, where u, v, w represent the x, y, z components of the velocity vector, Equation (2.1) becomes:

$$\frac{\partial \rho}{\partial t} + \frac{\partial}{\partial x}(\rho u) + \frac{\partial}{\partial y}(\rho v) + \frac{\partial}{\partial z}(\rho z) = 0$$

(2.4)

For incompressible flow, Equation (2.4) becomes:

$$\frac{\partial u}{\partial x} + \frac{\partial v}{\partial y} + \frac{\partial w}{\partial z} = 0$$

(2.5)

2.1.2 Momentum Equation

Newton's Second Law applied to a fluid passing through an infinitesimal, fixed control volume yields the following momentum equation:

$$\frac{\partial}{\partial t}(\rho V) + \ \bullet \rho VV = \rho f + \ \bullet \Pi_{ij}$$

(2.6)

The first term in this equation represents the rate of increase of momentum per unit volume in the control volume, the second term represents the rate of momentum lost by convection (per unit volume) through the control surface. ρVV is a tensor, and $\nabla \bullet \rho VV$ is not a simple divergence term and can be expanded as:

$$\bullet \rho VV = \rho V \bullet \ V + V(\ \bullet \rho V)$$

(2.7)

Substituting the above expression in Equation (2.6) and using the continuity equation, the momentum equation reduces to:

$$\rho \frac{DV}{Dt} = \rho f + \ \bullet \Pi_{ij}$$

(2.8)

The first term on the right-hand side of Equation (2.8) is the body force per unit volume, and the second term represents the surface forces per unit volume. These forces are applied by the external stresses on the fluid element. The stresses consist of normal stresses and shearing stresses and are represented by the components of the stress tensor Π_{ij}. For all gases that can be treated as a continuum, and most liquids, the stress at a point is linearly dependent on the rates of strain (deformation) of the fluid. A fluid that behaves in this manner is called a Newtonian fluid. With this assumption, a general deformation law can be derived that relates the stress tensor to the pressure and velocity components. In compact tensor notation, the following equation is obtained:

$$\Pi_{ij} = -p\delta_{ij} + \mu\left(\frac{\partial u_i}{\partial x_j} + \frac{\partial u_j}{\partial x_i}\right) + \delta_{ij}\mu'\frac{\partial u_k}{\partial x_k} \quad i,j,k = 1,2,3 \tag{2.9}$$

where δ_{ij} is the Kronecker delta function ($\delta_{ij} = 1$ if $i = j$ and $\delta_{ij} = 0$ if $i \neq j$); u_1, u_2, u_3, represent the three components of the velocity vector V; x_1, x_2, x_3 represent the three components of the position vector; μ is the coefficient of viscosity (dynamic viscosity); and μ' is the second coefficient of viscosity. The two coefficients of viscosity are related to the coefficient of bulk viscosity κ by the expression:

$$\kappa = \frac{2}{3}\mu + \mu' \tag{2.10}$$

In general, κ is negligible except in the study of the structure of shock waves and in the absorption and attenuation of acoustic waves; therefore, the second coefficient of viscosity becomes:

$$\mu' = -\frac{2}{3}\mu \tag{2.11}$$

and the stress tensor becomes:

$$\Pi_{ij} = -p\delta_{ij} + \mu\left[\left(\frac{\partial u_i}{\partial x_j} + \frac{\partial u_j}{\partial x_i}\right) - \frac{2}{3}\delta_{ij}\frac{\partial u_k}{\partial x_k}\right] \quad i,j,k = 1,2,3 \tag{2.12}$$

By substituting Equation (2.12) into Equation (2.8), the N-S equation is obtained as:

$$\rho\frac{DV}{Dt} = \rho f - p + \frac{\partial}{\partial x_j}\left[\mu\left(\frac{\partial u_i}{\partial x_j} + \frac{\partial u_j}{\partial x_i}\right) - \frac{2}{3}\delta_{ij}\mu\frac{\partial u_k}{\partial x_k}\right] \tag{2.13}$$

For a Cartesian coordinate system, Equation (2.13) can be separated into the following three scalar N-S equations:

$$\rho \frac{Du}{Dt} = \rho f_x - \frac{\partial p}{\partial x} + \frac{\partial}{\partial x}\left[\frac{2}{3}\mu\left(2\frac{\partial u}{\partial x} - \frac{\partial v}{\partial y} - \frac{\partial w}{\partial z}\right)\right] + \frac{\partial}{\partial y}\left[\mu\left(\frac{\partial u}{\partial y} + \frac{\partial v}{\partial x}\right)\right] + \frac{\partial}{\partial z}\left[\mu\left(\frac{\partial w}{\partial x} + \frac{\partial u}{\partial z}\right)\right]$$

$$\rho \frac{Dv}{Dt} = \rho f_y - \frac{\partial p}{\partial y} + \frac{\partial}{\partial x}\left[\mu\left(\frac{\partial u}{\partial y} + \frac{\partial v}{\partial x}\right)\right] + \frac{\partial}{\partial y}\left[\frac{2}{3}\mu\left(2\frac{\partial v}{\partial y} - \frac{\partial u}{\partial x} - \frac{\partial w}{\partial z}\right)\right] + \frac{\partial}{\partial z}\left[\mu\left(\frac{\partial w}{\partial y} + \frac{\partial v}{\partial z}\right)\right]$$

$$\rho \frac{Dz}{Dt} = \rho f_z - \frac{\partial p}{\partial z} + \frac{\partial}{\partial x}\left[\mu\left(\frac{\partial w}{\partial x} + \frac{\partial u}{\partial z}\right)\right] + \frac{\partial}{\partial y}\left[\mu\left(\frac{\partial w}{\partial y} + \frac{\partial v}{\partial z}\right)\right] + \frac{\partial}{\partial z}\left[\frac{2}{3}\mu\left(2\frac{\partial w}{\partial z} - \frac{\partial v}{\partial y} - \frac{\partial u}{\partial x}\right)\right]$$

$$(2.14)$$

The above equations can be written in conservation-law form as:

$$\frac{\partial \rho u}{\partial t} + \frac{\partial}{\partial x}(\rho u^2 + p - \tau_{xx}) + \frac{\partial}{\partial y}(\rho uv - \tau_{xy}) + \frac{\partial}{\partial z}(\rho uw - \tau_{xz}) = \rho f_x$$

$$\frac{\partial \rho v}{\partial t} + \frac{\partial}{\partial x}(\rho uv - \tau_{xy}) + \frac{\partial}{\partial y}(\rho v^2 + p - \tau_{yy}) + \frac{\partial}{\partial z}(\rho vw - \tau_{yz}) = \rho f_y \qquad (2.15)$$

$$\frac{\partial \rho w}{\partial t} + \frac{\partial}{\partial x}(\rho uw - \tau_{xz}) + \frac{\partial}{\partial y}(\rho vw - \tau_{yz}) + \frac{\partial}{\partial z}(\rho w^2 + p - \tau_{zz}) = \rho f_z$$

where the components of the viscous stress tensor τ_{ij} are given by:

$$\tau_{xx} = \frac{2}{3}\mu\left(2\frac{\partial u}{\partial x} - \frac{\partial v}{\partial y} - \frac{\partial w}{\partial z}\right)$$

$$\tau_{yy} = \frac{2}{3}\mu\left(2\frac{\partial v}{\partial y} - \frac{\partial u}{\partial x} - \frac{\partial w}{\partial z}\right)$$

$$\tau_{zz} = \frac{2}{3}\mu\left(2\frac{\partial w}{\partial z} - \frac{\partial v}{\partial y} - \frac{\partial u}{\partial x}\right)$$

$$\tau_{xy} = \mu\left(\frac{\partial u}{\partial y} + \frac{\partial v}{\partial x}\right) = \tau_{yx}$$

$$\tau_{xz} = \mu\left(\frac{\partial u}{\partial z} + \frac{\partial w}{\partial x}\right) = \tau_{zx}$$

$$\tau_{yz} = \mu\left(\frac{\partial w}{\partial y} + \frac{\partial v}{\partial z}\right) = \tau_{zy}$$

For incompressible flows and flows with constant coefficient of viscosity (μ), Equation (2.13) will reduce to the much simpler form:

$$\rho \frac{DV}{Dt} = \rho f - \nabla p + \mu \nabla^2 V \tag{2.16}$$

2.1.3 Energy Equation

The First Law of Thermodynamics applied to a fluid passing through an infinitesimal, fixed control volume yields the following energy equation:

$$\frac{\partial E_t}{\partial t} + \nabla \bullet E_t V = \frac{\partial Q}{\partial t} - \nabla \bullet q + \rho f \bullet V + \nabla \bullet (\Pi_{ij} \bullet V) \tag{2.17}$$

where E_t is the total energy per unit volume given by:

$$E_t = \rho \left(e + \frac{V^2}{2} + \text{potential energy} + \ldots \ldots \right) \tag{2.18}$$

and e is the internal energy per unit mass. The first term on the left-hand side of Equation (2.17) represents the rate of E_t in the control volume, and the second term represents the rate of total energy lost by convection (per unit volume) through the control surface. The first term on the right-hand side of Equation (2.17) is the rate of heat produced per unit volume by external agencies, and the second term is the rate of heat lost by conduction (per unit volume) through the control surface. By assuming the Fourier's law for heat transfer by conduction, the heat transfer q can be expressed as:

$$q = -k \nabla T \tag{2.19}$$

where k is the coefficient of thermal conductivity, and T is the temperature. The third term on the right-hand side of Equation (2.17) represents the work done on the control volume (per unit volume) by the body forces, and the fourth term represents the work done on the control volume (per unit volume) by the surface forces.

For a Cartesian coordinate system, Equation (2.17) becomes:

$$\frac{\partial E_t}{\partial t} - \frac{\partial Q}{\partial t} - \rho(f_x u + f_y v + f_z w) + \frac{\partial}{\partial x}(E_t u + pu - u\tau_{xx} - v\tau_{xy} - w\tau_{xz} + q_x)$$

$$+ \frac{\partial}{\partial y}(E_t v + pv - u\tau_{xy} - v\tau_{yy} - w\tau_{yz} + q_y) \tag{2.20}$$

$$+ \frac{\partial}{\partial z}(E_t w + pw - u\tau_{xz} - v\tau_{yz} - w\tau_{zz} + q_z) = 0$$

which is in conservation law form. Using the continuity equation, the left-hand side of Equation (2.17) can be replaced by the following expression:

$$\frac{\partial E_t}{\partial t} + \bullet E_t V = \rho \frac{D(E_t/\rho)}{Dt} \tag{2.21}$$

If only internal energy and kinetic energy are considered in Equation (2.18), then:

$$\rho \frac{D(E_t/\rho)}{Dt} = \rho \frac{De}{Dt} + \rho \frac{D(V^2/2)}{Dt} \tag{2.22}$$

Forming the scalar dot product of Equation (2.8) with the velocity vector V, the following equation is obtained:

$$\rho \frac{DV}{Dt} \bullet V = \rho f \bullet V - \ p \bullet V + (\ \bullet \tau_{ij}) \bullet V \tag{2.23}$$

By combining Equations (2.21), (2.22), and (2.23) and substituting into Equation (2.17), the energy equation is changed into:

$$\rho \frac{De}{Dt} + p(\ \bullet V) = \frac{\partial Q}{\partial t} - \ \bullet q + \ \bullet (\tau_{ij} \bullet V) - (\ \bullet \tau_{ij}) \bullet V \tag{2.24}$$

The last two terms in Equation (2.24) can be combined into a single term as:

$$\tau_{ij} \frac{\partial u_i}{\partial x_j} = \ \bullet (\tau_{ij} \bullet V) - (\ \bullet \tau_{ij}) \bullet V \tag{2.25}$$

This term is known as the dissipation function Φ, and it represents the rate at which mechanical energy is expended in the process of deformation of the fluid due to viscosity. After substituting the dissipation function in Equation (2.24), it becomes:

$$\rho \frac{De}{Dt} + p(\ \bullet V) = \frac{\partial Q}{\partial t} - \ \bullet q + \Phi \tag{2.26}$$

Using the definition of enthalpy:

$$h = e + \frac{p}{\rho} \tag{2.27}$$

and the continuity equation, Equation (2.26) can be rewritten as:

$$\rho \frac{Dh}{Dt} = \frac{Dp}{Dt} + \frac{\partial Q}{\partial t} - \ \bullet q + \Phi \tag{2.28}$$

For incompressible flows, and flows with constant coefficient of thermal conductivity, Equation (2.26) reduces to:

$$\rho \frac{De}{Dt} = \frac{\partial Q}{\partial t} + k \ ^2 T + \Phi \tag{2.29}$$

2.1.4 Turbulence Models

According to Hinze (1975), turbulent fluid motion is an irregular condition of flow in which the various quantities show a random variation with time and space coordinates so that statistically distinct average values can be discerned.

In this section, we provide a brief description for some turbulence models which has been applied in some sections of this book. These models are standard $k - \varepsilon$ model, standard $k - \omega$, and the $\overline{v^2} - f$ model. The Reynolds-averaged Navier-Stokes (RANS) equations were used as the transport equations for the mean flow. The RANS equation in a Cartesian tensor form can be written as (continuity, momentum, and energy, respectively):

$$\frac{\partial \rho}{\partial t} + \frac{\partial}{\partial x_i}(\partial u_i) = 0 \tag{2.30}$$

$$\frac{\partial}{\partial t}(\rho u_i) + \frac{\partial}{\partial x_j}(\rho u_i u_j) = -\frac{\partial p}{\partial x_i} + \frac{\partial}{\partial x_j}\left[\mu\left(\frac{\partial u_i}{\partial x_j} + \frac{\partial u_j}{\partial x_i} - \frac{2}{3}\delta_{ij}\frac{\partial u_k}{\partial x_k}\right)\right] + \frac{\partial}{\partial x_j}(-\overline{u_i' u_j'}) \tag{2.31}$$

$$\frac{\partial}{\partial t}(\rho c_p T) + \frac{\partial}{\partial x_j}\left(\rho c_p \overline{T' u'}_j\right) = -\frac{\partial p}{\partial t} + \frac{\partial}{\partial x_j}\left(k \frac{\partial T}{\partial x_j} - \rho c_p \overline{T' u'}_j\right) \tag{2.32}$$

For steady-state flow, the time derivative terms drop out. The Reynolds stresses $-\overline{u_i' u_j'}$, must be modeled in order to close Equation (2.31). All three turbulence models employ the Boussinesq hypothesis to relate the Reynolds stresses to the mean flow velocity gradients:

$$-\overline{u_i' u_j'} = \mu_t\left(\frac{\partial u_i}{\partial x_j} + \frac{\partial u_j}{\partial x_i}\right) - \frac{2}{3}\left(\rho k + \mu_t \frac{\partial u_k}{\partial x_k}\right)\delta_{ij} \tag{2.33}$$

The main issue in all the models that employ this hypothesis is how the turbulent viscosity μ_t is computed. The Boussinesq-type approximation is extended to evaluate the term $-\rho c_p \overline{T'u'_j}$ in the energy Equation (2.32):

$$-\rho c_p \overline{T'v'} = \frac{c_p \mu_t}{\mathrm{Pr}_t} \frac{\partial T}{\partial y}$$

where Pr_t is the turbulent Prandtl number. This number varies in the range from 0.6 to about 1.5, and in most engineering applications a value of 0.95 is used.

2.1.4.1 Standard k – ε Model

The standard k – ε model is based on Launder and Spalding (1974). In this model, the Boussinesq assumption is applied together with wall functions. The turbulent viscosity is expressed as:

$$\mu_t = \rho C_\mu \frac{k^2}{\varepsilon}$$

The transport equations for *k* and ε are:

$$\frac{\partial}{\partial t}(\rho k) + \frac{\partial}{\partial x_j}(\rho u_j k) = \rho P - \rho \varepsilon + \frac{\partial}{\partial x_j}\left[\left(\mu + \frac{\mu_t}{\sigma_k}\right)\frac{\partial k}{\partial x_j}\right] \qquad (2.34)$$

$$\frac{\partial}{\partial t}(\rho \varepsilon) + \frac{\partial}{\partial x_j}(\rho u_j \varepsilon) = C_{\varepsilon_1}\frac{\rho P \varepsilon}{k} - C_{\varepsilon_2}\frac{\rho \varepsilon^2}{k} + \frac{\partial}{\partial x_j}\left[\left(\mu + \frac{\mu_t}{\sigma_\varepsilon}\right)\frac{\partial \varepsilon}{\partial x_j}\right] \qquad (2.35)$$

with the production term *P* defined as:

$$P = \nu_t\left(\frac{\partial u_i}{\partial x_j} + \frac{\partial u_j}{\partial x_i} - \frac{2}{3}\frac{\partial u_m}{\partial x_m}\delta_{ij}\right)\frac{\partial u_i}{\partial x_j} - \frac{2}{3}k\frac{\partial u_m}{\partial x_m} \qquad (2.36)$$

The five constants used in this model are:

$$C_\mu = 0.09,\ C_{\varepsilon_1} = 1.44,\ C_{\varepsilon_2} = 1.92,\ \sigma_k = 1.0,\ \sigma_\varepsilon = 1.3$$

This model is a high Reynolds number model and is not intended to be used in the near-wall regions where viscous effects dominate the effects of turbulence. Instead, wall functions are used in cells adjacent to walls. Adjacent to the wall, the nondimensional wall parallel velocity is obtained from

$$u^+ = y^+; u^+ \le y_v^+ \tag{2.37}$$

$$u^+ = \frac{1}{k}\ln(Ey^+); y^+ > y_v^+$$

$$y^+ = y\frac{u_\tau}{V}; u^+ = \frac{u}{u_\tau}; u_\tau = C_\mu^{1/4}k^{1/2}; \kappa = 0.4 \tag{2.38}$$

$E = 9$ for smooth walls. Here, y_v^+ is the viscous sublayer thickness obtained from the intersection of Equations (2.37) and (2.38).

Similarly for heat transfer, a nondimensional temperature is defined:

$$T^+ = \frac{T - T_w \rho C_p u_\tau}{q_w} \tag{2.39}$$

and then the profiles of temperature near a wall are expressed as:

$$T^+ = Pr u^+ \ y^+ \le y_T^+ \tag{2.40}$$

$$T^+ = Pr_t(u^+ + P^+) \ y^+ > y_T^+ \tag{2.41}$$

where P^+ is a function of the laminar and turbulent Prandtl numbers (Pr and Pr_t) given by Launder and Spalding (1974) as:

$$P^+ = \frac{\pi/4}{\sin \pi/4}\left(\frac{A}{K}\right)^{1/2}\left(\frac{Pr}{Pr_t} - 1\right)\left(\frac{Pr_t}{Pr}\right)^{1/4}$$

Here, y_T^+ is the thermal sublayer thickness obtained from the intersection of Equations (2.40) and (2.41). Once T^+ has been obtained, its value can be used to compute the wall heat flux if the wall temperature is known, or to compute the wall temperature if the wall heat flux is known.

2.1.4.2 Standard k – ω Model

The standard $k-\omega$ model is a two-equation model that solves for the transport of ω, the specific dissipation rate of the turbulent kinetic energy, instead of ε. It is based on that of Wilcox (1998). The turbulent viscosity is expressed as:

$$\mu_t = \rho C_\mu \frac{k}{\omega},$$

where

$$\omega = \frac{\varepsilon}{k}$$

The transport equations for k and ω are:

$$\frac{\partial}{\partial t}(\rho k) + \frac{\partial}{\partial x_j}(\rho u_j k) = \rho P - \rho \varepsilon + \frac{\partial}{\partial x_j}\left[\left(\mu + \frac{\mu_t}{\sigma_k}\right)\frac{\partial k}{\partial x_j}\right] \qquad (2.42)$$

$$\frac{\partial}{\partial t}(\rho \omega) + \frac{\partial}{\partial x_j}(\rho u_j \omega) = C_{\omega_1}\frac{\rho P \omega}{k} - C_{\omega_2}\frac{\rho \omega^2}{k} + \frac{\partial}{\partial x_j}\left[\left(\mu + \frac{\mu_t}{\sigma_\omega}\right)\frac{\partial \omega}{\partial x_j}\right] \qquad (2.43)$$

with the production term P defined as:

$$P = \nu_t\left(\frac{\partial u_i}{\partial x_j} + \frac{\partial u_j}{\partial x_i} - \frac{2}{3}\frac{\partial u_m}{\partial x_m}\delta_{ij}\right)\frac{\partial u_i}{\partial x_j} - \frac{2}{3}k\frac{\partial u_m}{\partial x_m} \qquad (2.44)$$

The five constants used in this model are:

$$C_\mu = 0.09, C_{\omega_1} = 0.555, C_{\omega_2} = 0.833, \sigma_k = 2.0, \sigma_\omega = 2$$

The boundary conditions for k and ω at wall boundaries are:

$$K = 0 \text{ at } y = 0, \ \omega = 7.2\frac{v}{y^2} \text{ at } y = y_1$$

where y_1 is the normal distance from the cell center to the wall for the cell adjacent to the wall. The location of the cell center should be well within the laminar sublayer for best results ($y^+ \sim 1$). This model, therefore, requires very fine grids near solid boundaries.

2.1.4.3 $\overline{v^2} - f$ Model

According to Launder (1986), the normal stress $\overline{v^2}$, perpendicular to the wall plays the most important role to the eddy viscosity. Motivated by this idea, Durbin (1995) devised a "four-equation" model, known as the $k - \varepsilon - v^2$ model, or $\overline{v^2} - f$ model. It eliminates the need to patch models in order to predict wall phenomena like heat transfer or flow separation. It makes use of the standard $k - \varepsilon$ model but extends it by incorporating the anisotropy of near-wall turbulence and nonlocal pressure strain effects, while retaining a linear eddy viscosity assumption. The turbulent viscosity is expressed as:

$$\mu_t = \rho C_\mu \overline{v^2} T$$

The three transport equations are:

$$\frac{\partial}{\partial t}(\rho k) + \frac{\partial}{\partial x_j}(\rho u_j k) = \rho P - \rho \varepsilon + \frac{\partial}{\partial x_j}\left[\left(v + \frac{\mu_t}{\sigma_k}\right)\frac{\partial k}{\partial x_j}\right] \tag{2.45}$$

$$\frac{\partial}{\partial t}(\rho \varepsilon) + \frac{\partial}{\partial x_j}(\rho u_j \varepsilon) = C_{\varepsilon_1}\frac{\rho P \varepsilon}{k} - C_{\varepsilon_2}\frac{\rho \varepsilon^2}{k} + \frac{\partial}{\partial x_j}\left[\left(v + \frac{v_t}{\sigma_\varepsilon}\right)\frac{\partial \varepsilon}{\partial x_j}\right] \tag{2.46}$$

$$\frac{\partial}{\partial t}(\overline{v^2}) + (u_j \overline{v^2}) = k f_{22} - \overline{v^2}\frac{\varepsilon}{k} + \frac{\partial}{\partial x_j}\left[\left(v + \frac{v_t}{\sigma_k}\right)\frac{\partial \overline{v^2}}{\partial x_j}\right] \tag{2.47}$$

The elliptic equation for near-wall and nonlocal effects is given by:

$$L^2\ {}^2 f_{22} - f_{22} = (1 - C_1)\frac{\left[\frac{2}{3} - \frac{\overline{v^2}}{k}\right]}{T} - C_2\frac{P}{k} \tag{2.48}$$

$$L = C_L l$$

where

$$l^2 = \max\left[\frac{k^3}{\varepsilon^2}, C_\eta^2 \left(\frac{v^3}{\varepsilon}\right)^{\frac{1}{2}}\right]$$

$$T = \max\left[\frac{k}{\varepsilon}, 6\left(\frac{v}{\varepsilon}\right)^{\frac{1}{2}}\right]$$

The equation for production rate P is given by:

$$P = v_t \left(\frac{\partial u_i}{\partial x_j} + \frac{\partial u_j}{\partial x_i}\right)\frac{\partial u_i}{\partial x_j} \tag{2.49}$$

The constants used in this model are:

$$C_\mu = 0.19, C_{\varepsilon_2} = 1.9, C_1 = 1.4, C_2 = 0.3; \sigma_k = 1.0, \sigma_\varepsilon = 1.3, C_L = 0.3, C_\eta = 70.0$$

In order to model nonlocal characteristics of the near-wall turbulence, the $\overline{v^2} - f$ model uses an elliptic operator. This elliptic term is not confined to the wall; it becomes significant wherever there are significant inhomogeneities. As a result, $\overline{v^2} - f$ is valid throughout the whole flow domain, automatically becoming a near-wall model close to solid surfaces.

2.1.5 A Porous Media Model

The objective of this section is to describe an initial nonequilibrium porous-media model intended for use in multi-D Stirling codes (Dyson et al., 2005a, 2005b; Ibrahim et al., 2005; Tew et. al., 2006) for simulation of regenerators. More presentation will be made in Chapter 7 where porous media models are applied to specific cases. Experimental data from a regenerator research grant (Niu et al., 2003a, 2003b, 2003c, 2004, 2005a, 2005b, 2006) and experimentally based correlations derived from NASA/Sunpower Inc. (Athens, Ohio) oscillating-flow test-rig test data (Gedeon, 1999) were used in defining parameters needed in the nonequilibrium porous-media model. This effort was done under a NASA grant that was led by Cleveland State University with subcontractor assistance from the University of Minnesota, Gedeon Associates (Athens, Ohio), and Sunpower Inc. Determination of the particular porous-media model parameters presented were based on planned use in a computational fluid dynamics (CFD) model of Infinia Corporation's (Kennewick, Washington) Stirling Technology Demonstration Convertor (TDC). The nonequilibrium porous-media model presented is considered to be an initial, or "draft," model to be incorporated in commercial CFD codes (which now contain equilibrium porous-media models), with the expectation that it will likely need to be updated once resulting Stirling CFD model regenerator and engine results have been analyzed.

2.1.5.1 A Compressible-Flow Nonequilibrium Porous-Media Model for Computational Fluid Dynamics (CFD) Codes

Initial values of parameters needed for a macroscopic nonequilibrium porous-media model are defined below for use in modeling the TDC regenerators in CFD codes. Fluid continuity, momentum, and energy equations, and a solid-energy equation, are stated for reference in defining the parameters needed. Experimental and computational data generated under a NASA grant to Cleveland State University, and data from a NASA/Sunpower oscillating-flow test rig are used in defining this initial set of parameters (i.e., in defining closure models).

First define superficial (or Darcian), or approach, velocity as $U = \langle u \rangle = ua/a_{total} = u\beta$. This is also the volume-flow rate through a unit cross-sectional area of the solid-plus-fluid. It is determined by averaging the velocity over a region that is small with respect to the macroscopic flow dimension but large with respect to the matrix-flow-channel size. In the above equation for superficial velocity, a is the average fluid-flow area within the total solid-plus-fluid cross-sectional area, a_{total}, that the velocity is averaged over to determine the superficial velocity, U. β is the porosity. u, then, is the average-flow-channel fluid velocity (or local porous-media velocity) corresponding to the superficial velocity, $U = \langle u \rangle$, and $u = \langle u \rangle / \beta$.

2.2 Nonequilibrium Porous-Media Conservation Equations

The incompressible-flow, nonequilibrium, porous-media equations from Ibrahim et al. (2004c) were used as a starting point. Written for compressible flow, and including a hydrodynamic dispersion term from Ayyaswamy (2004), the fluid continuity, momentum, and energy equations, respectively, written in terms of the superficial, or approach, or Darcian velocity are as follows (where the brackets, < >, denote volume averaging):

$$\frac{\partial \langle \rho \rangle^f}{\partial t} + \frac{1}{\beta} \bullet [\langle \rho \rangle^f \langle \mathbf{u} \rangle] = 0 \tag{2.50}$$

$$\frac{1}{\beta} \frac{\partial (\langle \rho \rangle^f \langle \mathbf{u} \rangle)}{\partial t} + \frac{1}{\beta^2} \bullet [\langle \rho \rangle^f \langle \mathbf{u} \rangle \langle \mathbf{u} \rangle] = - \langle p \rangle^f + \bullet \left(\frac{\langle v_{eff} \rangle^f}{\beta} \langle \rho \rangle^f \langle \mathbf{u} \rangle - \frac{\langle \rho \rangle^f}{\beta} \langle \tilde{u}\tilde{u} \rangle \right)$$

$$- \frac{\langle \mu \rangle^f}{K} \langle \mathbf{u} \rangle - \langle \rho \rangle^f \frac{c_f}{\sqrt{K}} |\langle \mathbf{u} \rangle| \langle \mathbf{u} \rangle$$

$$\tag{2.51*}$$

$$\frac{\partial (\langle \rho \rangle^f \langle h \rangle^f)}{\partial t} + \frac{1}{\beta} \bullet [\langle \rho \rangle^f \langle \mathbf{u} \rangle \langle h \rangle^f] = \bullet \left[\overline{\overline{k}}_{fe} \bullet \langle T \rangle^f \right] + \left(\frac{\mu}{K} + \langle \rho \rangle^f \frac{c_f}{\sqrt{K}} |\mathbf{u}| \right) \mathbf{u} \cdot \mathbf{u}$$

$$+ H_{sf} \frac{dA_{sf}}{dV_f} (\langle T \rangle^s - \langle T \rangle^f) + \frac{d \langle p \rangle^f}{dt}$$

$$\tag{2.52†‡}$$

---

* The effective viscosity used in the "Brinkman term" of this momentum equation is different than the molecular viscosity, in general, as a result of the solid/fluid interfaces within the matrix, jetting of fluid from adjacent heat exchanger passages into the matrix, and so on. The Brinkman term computes momentum transport where there are velocity gradients within the flow.

† The pressure work term, $\frac{d\langle p \rangle^f}{dt}$, is the form for an ideal gas (helium, here). In general, it should be written $\beta_{cte} T \frac{d\langle p \rangle^f}{dt}$. For an ideal gas, the coefficient of thermal expansion $\beta_{cte} = \frac{1}{T}$.

‡ Burmeister (1993) suggests the form of the fluid energy viscous dissipation terms, $(\frac{\mu}{K} + \langle \rho \rangle^f \frac{c_f}{\sqrt{K}} |\mathbf{u}|) \mathbf{u} \cdot \mathbf{u}$, here—which is consistent with the Darcy-Forchheimer terms in the momentum equation.

where the wetted area per unit fluid volume, $\frac{dA_{sf}}{dV_f} = \frac{1}{r_h}$ (and r_h = hydraulic radius) and $\frac{d\langle p\rangle^f}{dt}$ is a substantial or material derivative. Finally, the solid energy equation is:

$$\frac{\partial(\rho_s C_s \langle T\rangle^s)}{\partial t} = \quad \bullet [\overline{\overline{k}}_{se} \bullet \ \langle T\rangle^s] - H_{sf} \frac{dA_{sf}}{dV_s} (\langle T\rangle^s - \langle T\rangle^f) \qquad (2.53)$$

where the wetted area per unit solid volume, $\frac{dA_{sf}}{dV_s} = \frac{1}{r_h} \frac{1}{1-\beta}$.

Kaviany (1995) has described a momentum equation similar to Equation (2.51) as a "semiheuristic" equation. That is, most of the terms are multidimensional terms based on derivation from fundamentals. However, the last two terms of Equation (2.51) were "heuristically" extended from a one-dimensional (1-D) momentum equation (i.e., this is a "speculative extension" of an experimentally based 1-D equation to serve as a guide in study of multidimensional problems, that has not been experimentally verified for multidimensional problems). Thus, in this initial or "draft" nonequilibrium porous-media model, we will assume that permeability is isotropic. (Even though we know that the wire-screen used in the University of Minnesota (UMN) experiments is not isotropic, and because of the method of fabrication of random fiber material, it is believed that most of the lengths of those fibers, also, lie in planes perpendicular to the flow direction.) Thus, as this porous-media model is improved upon, it may be desirable in the future to use anisotropic permeability. There are forms of Equation (2.51) (Ibrahim et al., 2005) with terms similar to the last two in Equation (2.51), but in which the permeability is a tensor quantity. But, in Equation (2.51), permeability is a scalar quantity assumed to be the same for all directions. Similar statements can be made for the inertial term multiplier of Equation (2.51)—that is, $C_f/\sqrt{K}$.

The expressions or parameters in the above equations needing definition, for use in modeling Stirling regenerators, are hydrodynamic dispersion, permeability and inertial coefficient, effective fluid and solid thermal conductivities, thermal dispersion conductivity, fluid-stagnant and solid-effective thermal conductivity, heat transfer coefficient between fluid and solid matrix elements. Note that the thermal conductivity expressions in the energy equations are tensors, though they are expected to have nonzero values only on the diagonals. The products of velocity vectors in the momentum equation are also tensors. Those parameters will be presented in more detail in the following sections.

2.2.1 Hydrodynamic Dispersion

In the fluid momentum equation, Equation (2.51) above, there is a hydrodynamic dispersion term:

$$\frac{1}{\beta}\langle\tilde{u}\tilde{u}\rangle \cong \tilde{u}\tilde{u} \qquad (2.54)$$

where β is porosity; $\mathbf{u}$ is the average-channel fluid, or local, velocity inside the matrix; and $\tilde{u}$ is the spatial-variation of the average-channel fluid velocity inside the matrix. $\tilde{u}\tilde{u}$ is a tensor quantity that can be expanded in the following form:

$$\tilde{u}\tilde{u} = \mathbf{ii}\,\tilde{u}\tilde{u} + \mathbf{ij}\,\tilde{u}\tilde{v} + \mathbf{ik}\,\tilde{u}\tilde{w} + \mathbf{ji}\,\tilde{v}\tilde{u} + \mathbf{jj}\,\tilde{v}\tilde{v} + \mathbf{jk}\,\tilde{v}\tilde{w} + \mathbf{ki}\,\tilde{w}\tilde{u} + \mathbf{kj}\,\tilde{w}\tilde{v} + \mathbf{kk}\,\tilde{w}\tilde{w}$$

(2.55)

For transport normal to the principal flow direction which is assumed to be the axial, $\mathbf{i}$, direction, the mean velocity gradient is in the radial, $\mathbf{j}$, direction and the term of interest (i.e., of significance) is $\tilde{u}\tilde{v}$, where v is the velocity component in that direction, $= |\mathbf{u}|\mathbf{j}$. Therefore, the term from the above expression for transport normal to the flow is $|\tilde{u}\tilde{v}|$ or $|\tilde{v}\tilde{u}|$. The dispersion represented by this term is of axial momentum, ρu, but this term is in the direction normal to u (in the direction of v).

In Niu et al. (2004), it was argued that this hydrodynamic dispersion could be equated to turbulent shear stress when the dispersion of interest is not in the direction of the mean flow, or $\langle \tilde{u}\tilde{v} \rangle = \langle \overline{u'v'} \rangle$ (where the prime refers to temporal variations about the temporal average, the overbar refers to a temporal average, and < > refers to spatial average), and that $\langle \overline{u'v'} \rangle$ could be modeled as $\varepsilon_M = \lambda d_h U$ where $\langle \overline{u'v'} \rangle = -\varepsilon_M \frac{\partial u}{\partial r}$ and $\lambda \cong 0.02$, and where $d_h =$ is the hydraulic diameter, u is the in-matrix average velocity, and $U = \langle u \rangle$ is the superficial, Darcian, or approach velocity. Therefore, for use in Equation (2.55) and in the momentum Equation (2.51):

$$\frac{\langle \tilde{u}\tilde{v} \rangle}{\beta^2} = \frac{\langle \tilde{v}\tilde{u} \rangle}{\beta^2} \approx \frac{\langle \overline{u'v'} \rangle}{\beta^2} = -\frac{1}{\beta^2}\varepsilon_M \frac{\partial U}{\partial r} = -\frac{1}{\beta^2} 0.02 d_h U \frac{\partial U}{\partial r} = -\frac{1}{\beta^2} \lambda d_h U \frac{\partial U}{\partial r} \quad (2.56)$$

2.2.2 Permeability and Inertial Coefficients

In the fluid momentum Equation (2.51), the permeability, K, and the inertial coefficient, c_f, need evaluation for each type of porous medium. These coefficients can be evaluated for particular types of Stirling engine regenerator porous media via use of the friction-factor data from the Sunpower/NASA oscillating-flow test rig (see Appendix A) and given in Gedeon (1999) under the assumption that the flow is quasi-steady (Ibrahim et al., 2005; Wilson et al. 2005). The Darcy-Forchheimer steady-flow form of the 1-D fluid momentum equation, and a similar pressure-drop equation, but in terms of the Darcy friction factor, can be written, respectively:

$$\frac{p}{L} = \frac{\mu}{K} u + \frac{C_f}{\sqrt{K}} \langle \rho \rangle^f u^2$$

(2.57)

$$\frac{p}{L} = \frac{f_D}{d_h} \frac{1}{2} \langle \rho \rangle^f u^2$$

(2.58)

From the Sunpower/NASA oscillating-flow test-rig data, as summarized by correlations given in the Sage manuals (Gedeon, 1999), it was determined that friction-factor correlations for random fiber and wire screen are of the following form:

$$f_D = \frac{\alpha}{Re} + \delta Re^\gamma \text{ where Reynolds number, Re} = \frac{\rho_f u d_h}{\mu} \quad (2.59)$$

and where, for random fiber, $\alpha = 192$, $\delta = 4.53$, $\gamma = -0.067$, and for woven screen, $\alpha = 129$, $\delta = 2.91$, $\gamma = -0.103$. Substituting the expression for friction-factor and the definition of Reynolds number from Equation (2.59), into Equation (2.58), and then equating the right-hand sides of Equations (2.57) and (2.58), it can be determined that:

$$\frac{K}{d_h^2} = \frac{2}{\alpha} \text{ and } C_f = \frac{\delta Re^\gamma}{\sqrt{2\alpha}} \quad (2.60)$$

A frequently used expression for hydraulic diameter of random fiber and wire screen in terms of porosity and wire diameter is:

$$d_h = \frac{\beta}{1-\beta} d_w \quad (2.61)$$

For the welded stacked screens used in the UMN regenerator test module, where $\beta = 0.9$ and $d_w = 0.81$ mm or 8.1E-4 m, then $d_h = 7.29$E-3 m. If it is also assumed that Reynolds numbers are in the range from 25 to 100, as expected in the TDC regenerators, then Equation (2.60) can be used to calculate values of permeability and inertial coefficient for TDC random fiber and the large screens used in the UMN test module. The resulting TDC random fiber values are given in Table 2.1 in italic font. The nonitalic TDC

TABLE 2.1

Comparison for Values of Permeability, K, and Inertial Coefficient, C_f, for the Large-Scale University of Minnesota (UMN) Wire Screen and Stirling Technology Demonstration Convertor (TDC) Random-Fiber Regenerator Materials

	UMN Large-Scale Screens ($d_w = 8.1$ E-4 m)				TDC Random Fiber	
Coefficient	UMN Old, Experimental	UMN New, Experimental	CSU Calculations	Sage Correction	Sage Correction	Unidirectional Flow Tests
K (m²)	1.07E-7	1.86E-7	8.9E-7	8.24E-7	*4.08E-10*	3.52E-10
K/d_w^2	0.163	0.283	1.36	*1.26*	—	—
C_f	0.049	0.052	0.14	*0.13–0.11*	*0.19–0.17*	0.154–0.095
				Re = 25–100	*Re = 25–100*	Re = 25–100

random fiber results given in Table 2.1, in the rightmost column, are based on unidirectional flow tests of the entire TDC heater head (including heater, regenerator, and cooler), taken from Wilson et al. (2005). The "UMN old" and "UMN new" values of permeability, in Table 2.1, are experimentally determined values of permeability and inertial coefficient, determined at an earlier and a later time, Simon (2003). The "CSU Calcs." are calculated values determined at Cleveland State University via microscopic CFD modeling of the UMN steady-flow test module.

2.2.3 Effective Fluid and Solid Thermal Conductivities

In the fluid energy Equation (2.52), $\overline{\overline{k}}_{fe} \equiv$, an effective fluid conductivity tensor, each element of which is, in general, a sum of components due to molecular conductivity, thermal tortuosity conductivity, and thermal dispersion conductivity. It can be broken down into these components as follows: from Equation (2.52),

$$\overline{\overline{k}}_{fe} \bullet \langle T \rangle^f = \overline{\overline{k}}_f \bullet \langle T \rangle^f + \frac{1}{V_f} \int_{Asf} \overline{\overline{k}}_f T \, dA - \rho_f C_p \langle \tilde{T} \tilde{u} \rangle \tag{2.62}$$

where

$$\frac{1}{V_f} \int_{Asf} \overline{\overline{k}}_f T \, dA \equiv \overline{\overline{k}}_{tor} \bullet \langle T \rangle^f \tag{2.63}$$

Equation (2.63) defines the thermal tortuosity thermal conductivity, $\overline{\overline{k}}_{tor}$.

$$-\rho_f C_p \langle \tilde{T} \tilde{u} \rangle \equiv \overline{\overline{k}}_{dis} \bullet \langle T \rangle^f \tag{2.64}$$

Equation (2.64) defines the thermal dispersion thermal conductivity, $\overline{\overline{k}}_{dis}$.

$$\therefore \quad \overline{\overline{k}}_{fe} \bullet \langle T \rangle^f = \overline{\overline{k}}_f \bullet \langle T \rangle^f + \overline{\overline{k}}_{f,tor} \bullet \langle T \rangle^f + \overline{\overline{k}}_{dis} \bullet \langle T \rangle^f \tag{2.65}$$

Equation (2.65) defines the effective fluid thermal conductivity, $\overline{\overline{k}}_{fe}$.

$$\overline{\overline{k}}_{fe} = \overline{\overline{k}}_f + \overline{\overline{k}}_{f,tor} + \overline{\overline{k}}_{dis} = \overline{\overline{k}}_{f,stag} + \overline{\overline{k}}_{dis} \tag{2.66}$$

where, in the above equation, the sum of the fluid molecular and thermal tortuosity conductivities are lumped together and called the fluid stagnant thermal conductivity.

It is assumed that only the diagonal elements of the effective fluid conductivity tensor are nonzero. Then in terms of 3-D cylindrical coordinates which are appropriate for Stirling engine simulation, it is further assumed that:

$$
\bar{\bar{k}}_{fe} \equiv
\begin{bmatrix}
k_{fe,rr} & 0 & 0 \\
0 & k_{fe,\vartheta} & 0 \\
0 & 0 & k_{fe,xx}
\end{bmatrix}
$$

$$
=
\begin{bmatrix}
k_f + k_{f,tor,rr} + k_{dis,rr} & 0 & 0 \\
0 & k_f + k_{f,tor,} + k_{dis,} & 0 \\
0 & 0 & k_f + k_{f,tor,xx} + k_{dis,xx}
\end{bmatrix}
$$

$$
=
\begin{bmatrix}
k_{f,stag,rr} + k_{dis,rr} & 0 & 0 \\
0 & k_{f,stag,} + k_{dis,} & 0 \\
0 & 0 & k_{f,stag,xx} + k_{dis,xx}
\end{bmatrix}
$$

$$(2.67)$$

In the above tensor equation for fluid effective thermal conductivity, (2.67), molecular conductivity is isotropic, and fluid tortuosity and dispersion conductivities will be assumed anisotropic. In general, the tortuosity conductivity is different in different directions when the alternating fluid/solid geometry is different when looking in different directions. Therefore, when fluid molecular and thermal tortuosity conductivities are lumped together to form the stagnant thermal conductivity, the stagnant thermal conductivity is also, in general, different in different directions—as dictated by the geometry of the matrix. The thermal dispersion conductivity has contributions from an advective term and an eddy term. In the flow direction, there would be both. Normal to the flow direction, there is only the eddy dispersion term.

In the solid energy Equation (2.53), the effective solid conductivity, $\bar{\bar{k}}_{se}$, is defined:

$$
\bar{\bar{k}}_{se} \equiv
\begin{bmatrix}
k_{se,rr} & 0 & 0 \\
0 & k_{se,} & 0 \\
0 & 0 & k_{se,xx}
\end{bmatrix}
=
\begin{bmatrix}
f(k_s, k_{s,tor,rr}) & 0 & 0 \\
0 & f(k_s, k_{s,tor,}) & 0 \\
0 & 0 & f(k_s, k_{s,tor,xx})
\end{bmatrix}
$$

$$(2.68)$$

The molecular thermal conductivities can be assumed to be well known. That leaves to be determined the (1) fluid thermal dispersion and (2) fluid stagnant conductivities and solid effective conductivities—which are each functions of the molecular conductivity and the thermal tortuosity conductivity, for the fluid or solid, and thus, the form of the last matrix of Equation (2.68).

2.2.4 Thermal Dispersion Conductivity

Niu et al. (2004) and other researchers measured eddy diffusivity (or transport) in, or very close to, porous media. If it is assumed that this is also equivalent to thermal dispersion due to eddies, as in Niu et al. (2004), based on the Reynolds Analogy, then values for thermal dispersion due to eddies are given in Table 2.2 in terms of porous media hydraulic diameter, d_h, and superficial, approach, or Darcian velocity, $U = <u>$. Although thermal dispersion by eddies is typically expected to be anisotropic, for this initial model, it may be adequate to use the same relationship to calculate eddy dispersion in all directions. It should be noted, however, that in the flow direction, total dispersion is by eddies and by advection, where the advection term is dominant. For a direction normal to the flow, dispersion is only by eddies. To explore further possible differences of eddy dispersion in different directions, refer to Niu et al. (2003a, 2003b, 2003c, 2004, 2005a, 2005b) and McFadden (2005).

TABLE 2.2

A Comparison of Thermal Dispersion Coefficients from Several Methods

	Estimated Thermal Dispersion	Porous Media
Direct measurements at University of Minnesota (UMN), Niu et al. (2004)	$\varepsilon_{M,eddy} = \dfrac{k_{dis,yy}}{\rho_f c_p} = 0.02 d_h U$ or $\dfrac{k_{dis,yy}}{k_f} = 0.02\,Pe$	Welded screen
Hunt and Tien (1988)	$\dfrac{k_{dis,yy}}{k_f} = 0.0011\,Pe$	Fibrous media
Metzger et al. (2004)	$\dfrac{k_{dis,yy}}{k_f} = (0.03 - 0.05)Pe$ and $\dfrac{k_{dis,xx}}{k_f} = 0.073\,pe^{1.59}$	Packed spheres
Gedeon (1999)	$\dfrac{k_{dis,xx}}{k_f} = 0.50\,Pe^{0.62}\beta^{-2.91}$ or $\dfrac{k_{dis,xx}}{k_f} \approx 0.06\,Pe$ for $\beta = 0.9$, $Pe = 560$	Woven screen

Source: Taken from Niu, Y., Simon, T., Gedeon, D., and Ibrahim, M., 2004, On Experimental Evaluation of Eddy Transport and Thermal Dispersion in Stirling Regenerators, *Proceedings of the 2nd International Energy Conversion Engineering Conference*, Paper No. AIAA-2004-5646, Providence, RI.

2.2.5 Fluid-Stagnant and Solid-Effective Thermal Conductivity

The fluid-stagnant and solid-effective thermal conductivities (each a function of the appropriate molecular and thermal tortuosity conductivities) are estimated based on the geometry of the matrix of interest. McFadden (2005) calculated a radial stagnant conductivity for the screens in the UMN test rig based on considerations of the geometry of the large-scale wire screens. Similar calculations are made for a fluid-saturated metal foam in Boomsma and Poulikakos (2001). In the case of the Stirling TDC that NASA modeled with CFD codes, a random fiber matrix is used in the regenerator. Because the details of the geometry for such a matrix are random, some assumptions must be made, as discussed below.

For the sake of comparison with the calculations in the UMN results mentioned above (McFadden, 2005), the molecular thermal conductivities will be used here (i.e., for air and stainless steel): So, assume that $k_s = 13.4$ W/m-K for stainless steel 316, $k_f = 0.026$ W/m-K for air at standard temperature, and $\beta = 0.90$ for porosity of the matrix, as in the UMN test rig.

In the random fiber matrix, most of the length of the fibers is believed to lie in planes perpendicular to the main flow axis. Therefore, initially make the assumption that for a three-dimensional (3-D) CFD model, the effective solid plus fluid conductivity in the radial and azimuthal directions (that would be appropriate for use in an equilibrium porous-media model) follows the parallel model defined below, and the effective solid plus fluid conductivity in the axial direction follows the series model, also defined below.

The parallel model for this lumped effective conductivity for fluid and solid, assuming all of the fibers run in the same direction (not including the fluid thermal dispersion) is:

$$k_{eff,s+f} = k_f\beta + k_s(1-\beta) \tag{2.69}$$

Therefore, this effective solid plus fluid conductivity for air and stainless steel combined, for a 90% porosity matrix, would be:

$$k_{eff,s+f} = (26x10^{-3}W/mK)(0.90) + (13.4W/mK)(0.1)$$

$$= 0.0234\ W/mK + 1.34\ W/mK = 1.36\ W/mk$$

As already mentioned, the above would be an appropriate effective solid plus fluid thermal conductivity for an equilibrium porous media model. However, for a nonequilibrium porous media model, assume that the two terms on the right of the above equation represent the fluid stagnant thermal conductivity and the solid effective conductivity, respectively. That is, assume:

$$k_{f,stag} = k_f\beta = (26x10^{-3}\ W/mK)(0.90) = 0.0234\ W/mK$$

$$k_{se} = k_s(1-\beta) = (13.4\ W/mK)(0.1) = 1.34\ W/mK$$

However, in accordance with McFadden (2005),

$$k_s/k_f = 13.4/26 \; x \; 10^{-3} = 515 \text{ and } k_{\textit{eff},s+f}/k_f = 1.36/26 \; x \; 10^{-3} = 52$$

52 is considerably larger than the 32.5 estimated by UMN for their wire screen in McFadden (2005), based on average geometrical considerations, rather than the parallel model. Because wire screen and random fiber are thought to have similar heat transfer properties, perhaps the solid part of the parallel model should be corrected to the UMN wire screen value by multiplication by the correction factor:

$$kcorrect = 32.5/52 = 0.625$$

Then the corrected parallel values for use for random fiber in the radial and azimuthal directions would be:

$$k_{\textit{eff},s+f} = 0.0234 + 1.36 \times 0.625 = 0.873 \text{ (for equil., model)}$$

$$k_{se} = k_s(1-\beta)0.625 = 1.34 \; W/mK \; x \; 0.625 = 0.838$$

(for nonequilibrium model, radial, and azimuthal directions)

For the axial direction, the lumped effective solid plus fluid effective conductivity, $k_{\textit{eff},s+f}$, (which also does not include thermal dispersion) should be substantially less than in the radial and azimuthal directions. Initially assume the series model mentioned in McFadden (2005). That is,

$$k_{\textit{eff},s+f} = \frac{1}{\left(\frac{\beta}{k_f} + \frac{1-\beta}{k_s}\right)} = \frac{1}{\left(\frac{0.90}{26x10^{-3}} + \frac{0.1}{13.4}\right)} = 0.0289 \text{ W/mK} \qquad (2.70)$$

This axial effective solid plus fluid thermal conductivity (not including thermal dispersion) is just slightly larger than the molecular fluid conductivity for air of 0.026 W/m-K and is probably too small because this series model assumes that the wires are not touching in the axial direction. 3-D CFD microscopic simulations of a representative elementary volume of the UMN regenerator by Rong (2005) suggest that the value shown in Equation (2.70) should be increased by a factor of 2.157. Therefore, assume that:

$$k_{\textit{eff},s+f} = \frac{2.157}{\left(\frac{\beta}{k_f} + \frac{1-\beta}{k_s}\right)} = \frac{2.157}{\left(\frac{0.90}{26x10^{-3}} + \frac{0.1}{13.4}\right)} = (0.0289 \text{ W/mK})(2.157) \qquad (2.71)$$

$$= 0.0623 \text{W/mK (axial direction)}$$

However it is obtained, this resulting effective solid plus fluid thermal conductivity in the axial direction would be appropriate for an equilibrium model. Seeing no obvious way to separate values for fluid and solid for a non-equilibrium macroscopic porous media model, based on the series model, we propose initially using this same value for both the fluid-stagnant and solid-effective conductivities in the axial direction in the nonequilibrium model, also—hoping that the overall axial effect might be reasonable.

Recall that to get the total effective fluid conductivities in different directions, the thermal dispersion conductivity should be added to the fluid stagnant conductivities in the radial, azimuthal, and axial directions. Also, for use in modeling the TDC regenerator, the above effective conductivities must be recalculated using the thermal conductivity of helium instead of air.

2.2.6 Heat Transfer Coefficient between Fluid and Solid Matrix Elements

A good source of heat transfer correlations for wire screen and random fiber Stirling regenerator materials is Gedeon (1999). These correlations are based on experimental data from the NASA/Sunpower oscillating-flow test rig; this rig was designed and fabricated specifically for the purpose of determining friction-factor and heat-transfer correlations for use in Stirling device design and modeling. Heat transfer correlations in terms of Nusselt number, Peclet number ($= Re \times Pr$), and porosity are as follows:

For wire screen

$$Nu = (1. + 0.99\ Pe^{0.66})\beta^{1.79} \tag{2.72}$$

For random fiber

$$Nu = (1. + 1.16\ Pe^{0.66})\beta^{2.61} \tag{2.73}$$

where

$$Nu = \frac{hd_h}{k}, Pe = Re\ Pr = \frac{\rho u d_h}{\mu} \frac{c_p \mu}{k}$$

Measurements by Niu et al. (2003b) indicated that the above correlation and its application in a quasi-steady fashion is suitable for the portion of an oscillatory cycle when the acceleration is sufficiently weak or the flow is decelerating, but during strong acceleration, the unsteady measurements indicate a violation of the quasi-steady-flow assumption characterized by a lag between the heat flux from one phase to the other and the temperature difference between phases. The measurements also showed an apparently poor mixing of the flow in the pore during this time in the cycle. The Valensi number of the flow was 2.1, which is a bit higher than values in the general operating range of an engine regenerator (~0.23 for one engine). Thus, this unsteady effect may have been overestimated. The data also indicate that over the full cycle, an estimate given by the correlation above and applied as if the flow were quasi-steady is a reasonable approach for this initial or "draft" model.

2.3 Summary

A set of transient, compressible-flow, conservation equations is summarized for reference in defining the parameters whose values are needed for a macroscopic, nonequilibrium porous media model. Such a porous media model is needed in existing commercial CFD codes (such as CFD-ACE and Fluent) in order to more accurately model the regenerator heat exchanger in Stirling engine devices (because only equilibrium porous-media models are now available in the CFD-ACE and Fluent codes). Available experimental information from large-scale wire screen testing is used to define a hydrodynamic dispersion term in the momentum equation. Experimental information is also used for definition of the permeability and inertial coefficients in the momentum equation and for the thermal dispersion conductivity for the regenerator fluid. Methods are also outlined for estimating the stagnant-fluid and effective-solid thermal conductivities. Thus, adequate information is presented for definition of an initial, or "draft," nonequilibrium porous-media model for use in CFD regenerator modeling of Stirling devices. It is anticipated that use of this initial model in CFD codes may demonstrate that further work on refinement of the nonequilibrium porous-media model and its parameters will be needed.

3

Correlations for Steady/Unsteady Fluid Flow and Heat Transfer

3.1 Introduction

The purpose of this chapter is to summarize all correlations available today for steady and unsteady (zero and nonzero mean) fluid flow and heat transfer. These are done under laminar, flow conditions. This theory also provides the designer with the equations to use in calculating heat transfer coefficient, wall heat flux, and so forth, under oscillatory (with zero mean) flow conditions. With this information, one can calculate the figure of merit, which permits balancing (or "trafe-off") heat transfer and flow losses, under different flow and heat transfer conditions.

This chapter will discuss correlations for basic geometries and make reference to similar correlations (with more details) that will be presented in Chapters 6, 7, 8, and 10.

3.2 Internal Fluid Flow and Heat Transfer

Two simple geometries are used as a reference for internal flow—namely, circular pipes and parallel plates. In the following section, fluid flow and heat transfer will be presented for fully developed flow as well as the entrance region. Also, presentation will be made for both steady flow and oscillatory flow (with zero mean).

3.2.1 Internal Fluid Flow

The friction factor, in internal flows, is basically a nondimensional viscous pressure drop. Under steady flow conditions, the friction factor is primarily dependent on the Reynolds number. As for oscillatory flow and from experimental observations (e.g., Taylor and Aghili, 1984), the friction factor depends on both Re_{max} (= $u_{max} Dh/\nu$) and Va (= $\omega Dh^2/4\nu$). For a fully developed laminar flow and at very low frequency of oscillation (low Va), the flow oscillates with a velocity profile similar to unidirectional flow (quasi-steady flow with a parabolic velocity profile), see Figure 3.1a (Simon and Seume, 1988). As the frequency gets higher (higher Va), the velocity profile becomes flatter and during some parts of the cycle the velocity direction near the wall is opposite to that of the mean flow direction (see Figure 3.1b). As the frequency gets even higher, the velocity profile becomes uniform in the core and the free shear layer near the wall becomes narrower (known as the Stokes layer) (see Figure 3.1c). The fluid mechanics differences between steady flow and oscillatory flow can be further elucidated by considering the control volumes shown in Figure 3.2. In Figure 3.2a, the pressure drop forces are balanced by the wall shear forces. Under oscillatory flow, however, additional forces are added (see Figure 3.2b) due to the fluid temporal acceleration (Simon and Seume, 1988). In the limit, at very high Va, the wall shear stress leads the mean velocity by 45 degrees, and the pressure drop leads the mean velocity by 90 degrees.

3.2.2 Friction Factor Correlations

For fully developed laminar pipe and parallel plate flows under unidirectional conditions, the friction factors are:

Pipe flow

$$f = 64/Re \tag{3.1}$$

Parallel plate flow

$$f = 96/Re \tag{3.2}$$

For fully developed laminar pipe and parallel plate flows under oscillatory flow conditions, the friction factors are as follows (Gedeon, 1986; Fried and Idelchik, 1989; Simon and Seume, 1988):

Pipe flow

$$f = \frac{64}{R_e} \quad if \ Va \leq 12.6 \tag{3.3a}$$

$$f = \frac{64}{R_e}(Va/12.6)^{0.45} \quad if \ Va > 12.6 \tag{3.3b}$$

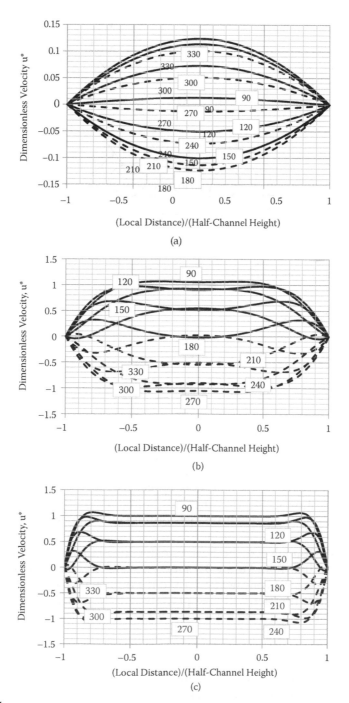

FIGURE 3.1
Velocity profiles in oscillating, laminar, fully developed channel flow: (a) $Va = 1$; (b) $Va = 100$; and (c) $Va = 1000$, at different crank angles, in degrees.

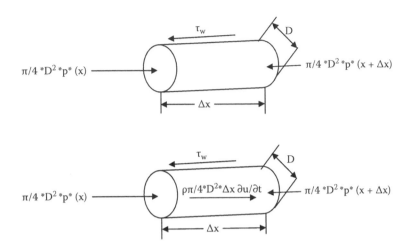

FIGURE 3.2
(a) Steady flow and (b) oscillatory flow.

Parallel plate flow

$$f = \frac{96}{R_e} \tag{3.4}$$

3.2.3 Internal Heat Transfer

The convective heat transfer in internal flows is basically dependent on the velocity profile. Therefore, for fully developed laminar flow, the same parameters that govern the fluid flow apply here for the heat transfer Re_{max} and Va in addition to Pr. Also, the wall boundary condition (e.g., uniform heat flux or uniform temperature distribution) will influence the heat transfer process.

Watson (1983) analyzed the heat transfer in oscillatory flow for a fully developed laminar flow. Figure 3.3 shows a sketch of an adiabatic pipe connecting two reservoirs: one hot (at T_h) and the other one cold at (T_c). If the flow is stagnant, heat transfer will take place only by conduction, and the axial heat flow per unit cross-sectional area will be:

$$q'' = k_f (T_h - T_c)/L \tag{3.5}$$

Watson showed that under oscillatory laminar incompressible flow, axial heat transfer can be expressed in a similar way by:

$$q'' = k_{eff} (T_h - T_c)/L \tag{3.6}$$

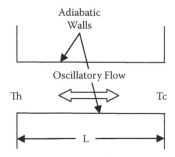

FIGURE 3.3
Oscillatory flow in a long duct connecting hot (T_h) and cold (T_c) reservoirs.

k_{eff} can be thousands of times larger than k_f depending upon Re_{max}, Va, Pr, and the ratios of fluid to wall thermal conductivity and thermal diffusivity (Watson, 1983; Kurzweg, 1985; Gedeon, 1986; Ibrahim et al., 1989). The physical explanation of this large augmentation in the heat transfer is due to the fact that steep velocity profiles near the wall (at higher Va) will enhance the conduction heat from the wall to fluid while the flow oscillation convects this heat from the hot end to the cold one. It should be noted that this physical observation holds only for laminar flow. In turbulent flow, convective eddy-transport in the cross-stream direction may reduce the oscillating cross-stream temperature gradients that are necessary for the axial heat transfer augmentation (Simon and Seume, 1988).

3.2.4 Nusselt Number Correlations

For fully developed laminar pipe and parallel plate flows under unidirectional conditions, the Nu number correlations are

Pipe flow

$$Nu_T = 3.66 \quad \text{(uniform wall temperature)} \tag{3.7a}$$

$$Nu_H = 4.35 \quad \text{(uniform wall heat flux)} \tag{3.7b}$$

Parallel plate flow

$$Nu_T = 7.54 \quad \text{(uniform wall temperature)} \tag{3.8a}$$

$$Nu_H = 8.23 \quad \text{(uniform wall heat flux)} \tag{3.8b}$$

For fully developed laminar pipe and parallel plate flows under oscillatory flow conditions, the Nu number correlations are given below (Gedeon, 1992; Swift, 1988); these correlations were taken for uniform wall heat flux (Nu is broken into separate parts, Nuo, Nuc, and Nua—real and complex Nu that apply, respectively, to the steady, compression-driven, and advection-driven components of the gas to wall temperature difference):

Pipe flow

$$Nuo = 6.0 \tag{3.9a}$$

$$\text{Real } (Nuc) = 6.0 \quad \text{if} \quad \sqrt{2Va\,Pr} < 6.0 \tag{3.9b}$$
$$= \sqrt{2Va\,Pr} \quad \text{otherwise}$$

$$\text{Imaginary } (Nuc) = \frac{1}{5}VaPr \quad \text{if} \quad Va\,Pr < \sqrt{2\,Va\,Pr} \tag{3.9c}$$
$$= \sqrt{2\,Va\,Pr} \quad \text{otherwise}$$

$$\text{Real } (Nua) = 4.2 \quad \text{if} \quad \sqrt{2\,Va\,Pr} < 8.4 \tag{3.9d}$$
$$= \frac{1}{2}\sqrt{Va\,Pr} \quad \text{otherwise}$$

$$\text{Imaginary } (Nua) = \frac{1}{10}VaPr \quad \text{if} \quad Va\,Pr < 5\sqrt{2\,Va\,Pr} \tag{3.9e}$$
$$= \frac{1}{2}\sqrt{Va\,Pr} \quad \text{otherwise}$$

Parallel plate flow

$$Nu_H = 8.23 \tag{3.10}$$

3.2.5 Entrance Effect

For the entrance length under oscillatory flow conditions, Peacock and Stairmand (1983) hypothesized that the entrance length is shorter than that in unidirectional flow. The main reason behind this is that the uniform velocity at the entrance will not need to change much because it will tend to be flatter under oscillatory flow. This hypothesis has not been supported, so far, experimentally. Therefore, in the following section, only steady flow data will be presented.

3.2.5.1 Circular Duct

The earlier discussion is confined to fully developed conditions (far away from the duct entrance). However, for a finite duct length, both entrance and exit effects should be accounted for.

Under steady flow conditions the hydrodynamic entrance length, L_{hy}, is defined, somewhat arbitrarily, as "the duct length required to reach a duct-section maximum velocity = 99% of the corresponding fully developed value when the entering flow is uniform" (Shah and London, 1978). Friedmann et al. (1968) solved the Navier-Stokes equations for a uniform entrance velocity profile in a circular pipe. Their solution is in agreement with the experimental data and is best approximated by Chen (1973) as:

$$\frac{L_{hy}}{D_h} = \frac{0.6}{0.035\,Re + 1} + 0.056\,Re \tag{3.11}$$

As for the friction factor in this entry region (Shah, 1978),

Hydrodynamically developing flow, the flow regime for the case studied:

$$f_F = \frac{1}{Re}\left(\frac{3.44}{(x^+)^{1/2}} + \frac{16 + \frac{1.25}{4x^+} - \frac{3.44}{(x^+)^{1/2}}}{1 + 0.00021(x^+)^{-2}}\right) \tag{3.12}$$

The thermal entrance for uniform temperature at the duct inlet depends on the thermal boundary conditions (e.g., uniform heat flux or uniform temperature). This thermal entry length is the distance from the start of application of the thermal boundary condition to the point where the flow becomes thermally fully developed. Typically, the thermal entry length is defined to be when the nondimensional temperature profile becomes independent of x (flow direction) to within a certain percent, typically to within 5% or smaller. This thermal entry length does not only depend on the wall boundary conditions but also on whether the flow is developed or developing. The latter case is defined as simultaneously developing flow. For fully developed flow, the following correlations exist:

Thermally developing flow, uniform wall temperature, Shah (1975):

$$\frac{L_{th,T}}{Pe^* D_h} = 0.0334654 \tag{3.13}$$

Thermally developing flow, uniform wall temperature, Churchill and Ozoe (1973b):

$$\frac{Nu_{x,T} + 1.7}{5.357} = \left[1 + \left(\frac{388}{\pi}x^*\right)^{-8/9}\right]^{3/8} \tag{3.14}$$

Thermally developing flow, uniform wall heat flux, Shah (1975):

$$\frac{L_{th,H}}{Pe^* D_h} = 0.0430527 \tag{3.15}$$

Thermally developing flow, uniform wall heat flux, Churchill and Ozoe (1973a):

$$\frac{Nu_{x,H} + 1}{5.364} = \left[1 + \left(\frac{220}{\pi}x^*\right)^{-10/9}\right]^{3/10}$$
(3.16)

Simultaneously developing flow, uniform wall temperature, $Pr = 0.7$, Shah and London (1978):

$$\frac{L_{th,T}}{Pe * D_h} = 0.037$$
(3.17)

Simultaneously developing flow, uniform wall temperature, $Pr = 0.7$, Churchill and Ozoe (1973b):

$$Nu_{x,T} = 1/2 Nu_{m,T} = \frac{0.6366[(4/\pi)x^*]^{-1/2}}{[1 + (Pr/0.0468)^{2/3}]^{1/4}}$$
(3.18)

Simultaneously developing flow, uniform wall heat flux, $Pr = 0.7$, Hornbeck (1965):

$$\frac{L_{th,H}}{Pe * D_h} = 0.053$$
(3.19)

Simultaneously developing flow, uniform wall heat flux, $Pr = 0.7$, Churchill and Ozoe (1973a):

$$Nu_{x,H} = \frac{[(4/\pi)x^*]^{-1/2}}{[1 + (Pr/0.0207)^{2/3}]^{1/4}}$$
(3.20)

3.2.5.2 Parallel Plates

Chen (1973) proposed a similar expression to Equation (3.11) for the hydrodynamic entrance length, parallel geometry:

$$\frac{L_{hy}}{D_h} = \frac{0.315}{0.0175\, Re + 1} + 0.011\, Re$$
(3.21)

And for the friction factor in this entry region, Shah (1978):

$$f_F = \frac{1}{Re}\left(\frac{3.44}{(x^+)^{1/2}} + \frac{24 + \frac{0.674}{4x^+} - \frac{3.44}{(x^+)^{1/2}}}{1 + 0.000029(x^+)^{-2}}\right)$$
(3.22)

Thermally developing flow, uniform wall temperature, Shah (1975):

$$\frac{L_{th,T}}{Pe * D_h} = 0.0079735 \tag{3.23}$$

Thermally developing flow, uniform wall temperature, Shah (1975):

$$Nu_{x,T} = 1.233(x^*)^{-1/3} + 0.4 \quad \textit{for } x^* \leq 0.001$$
$$= 7.541 + 6.874(10^3 x^*)^{-0488} e^{-245x^*} \quad \textit{for } x^* > .001 \tag{3.24}$$

Thermally developing flow, uniform wall heat flux, Shah (1975):

$$\frac{L_{th,H}}{Pe * D_h} = 0.0115439 \tag{3.25}$$

Thermally developing flow, uniform wall heat flux, Shah (1975):

$$Nu_{x,H} = 1.490(x^*)^{-1/3} \quad \textit{for } x^* \leq 0.0002 \tag{3.26a}$$

$$= 1.490(x^*)^{-1/3} - 0.4 \quad \textit{for} \quad 0.0002 < x^* \leq 0.0001 \tag{3.26b}$$

$$= 8.235 + 8.68(10^3 x^*)^{-0.506} e^{-164x^*} \quad \textit{for} \quad x^* > 0.0001 \tag{3.26c}$$

As for simultaneously developing flow for parallel plates under different boundary conditions, there are limited correlations available (Shah and London, 1978). Most of these results are for Nu_m (see Equation 8.26) rather than Nu_x and in tabulated form.

3.3 External Fluid Flow and Heat Transfer

The simplest geometry that can be considered in external flows is a single cylinder in cross-flow. In this case, the cylinder disturbs the flow not only in the vicinity, but also at large distances in all directions. At $Re = 100$, the vortices grow with time until a limit is reached, at which the wake becomes unstable. At this Reynolds number, which is above the critical Reynolds number, the viscous dissipation is too small to ensure the stability of the flow. The critical Reynolds number is defined as the Reynolds number after which the vortex starts shedding from the cylinder (it does not remain attached to the cylinder). The critical Reynolds number for a circular cylinder is $Re_{cr} \cong 46$ (Lange et al., 1998), for a square cylinder at $0°$ incidence $Re_{cr} \cong 51.2$ and for a square cylinder at 45° incidence $Re_{cr} \cong 42.2$ (Sohankar et al., 1998). The vortices are shed off by the main flow, forming a well-known Von Kármán vortex street. In the following section, different flow and heat transfer features will be presented for cylinders (different geometries) in cross-flow. See Section 6.5 for more detailed computational fluid dynamics (CFD) data for these cases.

3.3.1 Flow Structure

A CFD simulation, for steady flow and heat transfer, and for different cylinder geometries (circular, square with 0 and 45° incident angle, diamond and elliptic, with 90° incident angle) were conducted by Mudaliar (2003). The angle of incidence for the elliptic cylinder is measured between the flow direction and the major axis of the cylinder. For the diamond-shaped cylinder, the angle of incidence is measured between the flow direction and the long axis of the cylinder. The wake behind the cylinder becomes more violent as the characteristic length (the diameter in the case of a circular cylinder) increases. Alternate vortex shedding is also observed in temperature contours from the top and bottom of the cylinder. The vortex shedding frequency in temperature contours is out of phase with that of the velocity contours. The heat transfer in this fluctuation is limited, despite the violent wake observed in the flow field. This is attributed to the low flow field velocities encountered in the wake region and the domination of molecular conduction near the cylinder wall.

3.3.2 Drag and Lift Coefficient

One of the most important characteristic quantities of the flow around a cylinder is the drag coefficient. The resistance of a body as it moves through a fluid is of obvious technical importance in hydrodynamics and aerodynamics. The combined effect of pressure and shear stress (sometimes called skin friction) give rise to a resultant force on the cylinder. This result may be resolved into two components: a component in the direction of the flow F_D, called the drag force, and component normal to direction of flow, F_L, called the lift force. The components may be expressed in dimensionless terms by definition of drag and lift coefficients as follows:

$$C_D = \frac{F_D}{0.5\rho U^2 L} \tag{3.27}$$

$$C_L = \frac{F_L}{0.5\rho U^2 L} \tag{3.28}$$

where C_D and C_L are drag coefficient and lift coefficient, respectively. At $Re = 100$, when we can see flow separation, a recirculation zone develops behind the cylinder and moves gradually in the wake in the direction of the flow. The contribution of the pressure and viscous forces get out of

TABLE 3.1

Time-Averaged Drag Coefficient for Different
Geometries at $Re = 100$

Geometry	Coefficient of Drag
Circular cylinder	1.3667
Square cylinder at 0° incidence	1.4540
Square cylinder at 45° incidence	2.0028
Elliptic cylinder at 90° incidence	2.7219
Diamond cylinder at 90° incidence	3.8340

balance, and the pressure force tends to dominate the drag. The drag coefficient (as obtained from CFD computations by Mudaliar [2003], of the circular cylinder is in agreement with the results of Lange et al. [1998], Zhang and Dalton [1998], and Franke et al. [1995]; the drag coefficients for a square at 0° and 45° incidence are in agreement with the work of Sohankar et al. (1995, 1996, 1998). There is no literature for elliptic and diamond cylinders at 90° incidence. As expected, the drag coefficient for the diamond cylinder at 90° incidence is highest followed by the elliptic cylinder at 90° incidence, the square at 45° incidence, and the square at 0° incidence, and is minimum for the circular cylinder. Table 3.1 gives the values of the time-averaged drag coefficients for all the geometries.

3.3.3 Strouhal Number

The Strouhal number is proportional to the reciprocal of vortex spacing expressed as number of obstacle diameters and is used in momentum transfer in general von Karman vortex streets and unsteady-state flow calculations in particular. It is normally defined in the following form:

$$St = \frac{L}{U\tau} \tag{3.29}$$

where τ is time period of vortex shedding in seconds, L is characteristic length in m, and U is the mean flow velocity in m/s. In the present study, the Strouhal number is calculated based on the periodic fluctuation of lift forces, as they are a consequence of the vortex shedding. The time period for one cycle is calculated from the graph of lift forces plotted against the time. The reciprocal of this time will give the required frequency. In the transient flow regime, downstream of a cylinder, the Strouhal number constitutes an additional characteristic quantity. Experimental values of Williamson (1996), for circular cylinder, are represented by the following:

$$St = -\frac{3.3265}{Re} + 0.1816 + 1.6 \times 10^{-4} \, Re \tag{3.30}$$

TABLE 3.2

Strouhal Number for Different Geometries at $Re = 100$

Geometry	Strouhal Number
Circular cylinder	0.1656
Square cylinder at 0° incidence	0.1311
Square cylinder at 45° incidence	0.1656
Elliptic cylinder at 90° incidence	0.2266
Diamond cylinder at 90° incidence	0.2098

Table 3.2 shows CFD results from Mudaliar (2003) for different geometries at $Re = 100$. These results are in agreement with Equation (3.30) for a circular cylinder.

3.3.4 Nusselt Number

The Nusselt number is defined as:

$$Nu = \frac{h L}{k} \tag{3.31}$$

where h is the heat transfer coefficient calculated from:

$$\dot{q}_w = h\,(T_w - T_\infty) \tag{3.32}$$

Usually, just the mean value of the Nusselt number is needed—namely, when no local effect on the cylinder surface is of particular interest. In this case, the value of Nu is averaged over the whole cylinder perimeter. In the remaining section, we consider only the surface averaged value of Nu.

Table 3.3 shows the average Nu over circumference of different cylinder geometries.

TABLE 3.3

Nusselt Number for Different Geometries at $Re = 100$

Geometry	Nusselt Number
Circular cylinder	5.1904
Square cylinder at 0° incident	3.5133
Square cylinder at 45° incident	5.5949
Elliptic cylinder at 90° incident	6.9539
Diamond cylinder at 90° incident	6.8692

The CFD results of Mudaliar's (2003) study for a circular cylinder were in agreement with the computational study of Lange et al. (1998). The study was conducted for a circular cylinder for $10^{-4} \leq Re \leq 200$, representing Nu in terms of Reynolds number.

$$Nu = 0.082\,Re^{0.5} + 0.734\,Re^x$$

$$\text{where} \quad x = 0.05 + 0.226\,Re^{0.085}$$

(3.33)

Table 3.3 indicates that average Nu over the cylinder is nearly the same for elliptic and diamond cylinders at 90° incidence, probably because the Strouhal number is also nearly the same which means that Nu is function of vortex shedding frequency (Strouhal number). Similarly, the average values Nu are close for a circular and the square cylinder at 45° incidence, both of them having the same Strouhal number at $Re = 100$. Average Nu is lowest for the square cylinder at 0° incidence, and the square cylinder also has the lowest Strouhal number. This implies that the vortex shedding frequency (Strouhal number) affects the heat transfer from the cylinder to the fluid.

3.3.5 Coefficient of Pressure

The following is the definition of the pressure coefficient:

$$C_p = \frac{P}{0.5\rho U^2}$$

(3.34)

There are no relevant data, in the open literature, to compare with the CFD results of Mudaliar (2003). The time-averaged pressure coefficient values of Mudaliar's (2003) study for all the geometries are given in Table 3.4. Table 3.4 indicates that the time-averaged pressure coefficient over the cylinder is minimum for elliptic cylinder at 90° incidence with value of −0.58, and for all the other geometries the values vary from −0.26 to −0.29.

TABLE 3.4

Pressure Coefficient for Different Geometries at $Re = 100$

Geometry	Coefficient of Pressure
Circular cylinder	−0.29569
Square cylinder at 0° incidence	−0.26671
Square cylinder at 45° incidence	−0.2885
Elliptic cylinder at 90° incidence	−0.58644
Diamond cylinder at 90° incidence	−0.26587

3.4 Fluid Flow and Heat Transfer in Regenerators

The regenerator geometry is complex; randomly stacked metal fibers, stacked, woven wire screens, folded sheet metal, metal sponge, and sintered metals are some of the matrices used in Stirling machines. These types of matrices are used in other applications such as gas turbines, combustion processes, catalytic reactors, packed bed heat exchangers, electronics cooling, heat pipes, thermal insulation engineering, nuclear waste repository, and miniature refrigerators, to name a few. Also, these geometries can be modeled as porous media.

More information about these matrices (and others) will be discussed in Chapter 5. Furthermore, a newly developed matrix, "Segmented-Involute-Foil Regenerator" will be discussed in detail in Chapter 8. Eleven different Stirling engines were examined by Simon and Seume (1988). The operating conditions for these engines were presented in terms of similarity parameters (Re_{max}, Va, and A_R) for the heaters, coolers, and regenerators separately. The dimensionless amplitude ratio, A_R, is used to describe the fluid displacement in a heat exchanger. It is defined as the fluid displacement in half a cycle divided by the tube length. This is based on the assumption that the fluid moves as a plug at the mean velocity. $A_R \ll 1$ indicates that the fluid oscillates inside the tube (of a heat exchanger) without exiting. On the other hand, $A_R \gg 1$ indicates that the fluid oscillates quickly inside the tube while residing upstream and downstream most of the cycle.

Figure 3.4a shows where the different engine operating conditions will exist in the Re_{max}, and Va plane. Figure 3.4b shows where the different engine operating conditions will exist in the A_R, and Va plane. The plots show that most of the engines run in the laminar flow regime with $A_R < 1$ (where axial augmentation would be important). $A_R < 1$ indicates that some fluid does not leave the regenerator.

3.4.1 Steady Flow

3.4.1.1 Friction Factor Correlations

Correlations for flow through stacked screens are provided by Kays and London (1984). Additional experimental data for the stacked screens are given by Miyabe et al. (1982). Data for foamed-metal matrices are presented by Takahachi et al. (1984). A comprehensive review of regenerator pressure drop and heat transfer correlations for steady flow was presented by Finegold and Sterrett (1978).

3.4.1.2 Heat Transfer Correlations

Kays and London (1984), Walker and Vasishta (1971), Finegold and Sterrett (1978) and Miyabe et al. (1982) provided heat transfer correlations for

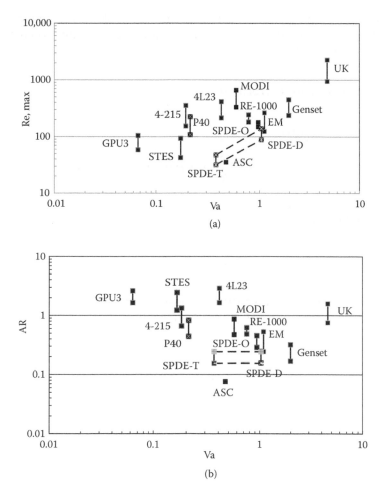

FIGURE 3.4

(a) Re_{max} versus Va for regenerators and (b) A_R versus Va for regenerators.

stacked screens. These data are from experiments and utilized the single-blow technique described by Kays and London (1984). On the other hand, Takahachi et al. (1984) calculated the heat transfer correlations from oscillatory flow data.

3.4.2 Oscillatory Flow and Heat Transfer

In this section, different correlations for friction factor and Nusselt number are presented (Gedeon, 1999) for different geometries: woven screen, random fiber, packed sphere, and rectangular channels. Additional descriptions of these matrices are given in Chapter 5. More recent experimental correlations data for the random fiber matrix are presented in Table 6.15, Chapter 6. Also, more correlations for the newly developed segmented-involute-foil regenerator are given in Chapter 8.

3.4.2.1 Friction Factor Correlations

Woven screen matrix

$$f = \frac{129}{R_e} + 2.91 R_e^{-0.103} \tag{3.35}$$

Random fiber matrix

$$f = \frac{192}{R_e} + 4.53 R_e^{-0.067} \tag{3.36}$$

More attention has been given to the random fiber matrix because it is widely used in Stirling machines. In Chapter 6, more recent test data are provided.

Packed sphere matrix

$$f = \left(\frac{79}{R_e} + 1.1\right)\beta^{-0.6} \tag{3.37}$$

Rectangular channels

$$f = \frac{64b}{R_e} \qquad if\ Va \le 69.4 \tag{3.38a}$$

$$f = \frac{64b}{R_e}(Va/69.4)^{0.5} \qquad if\ Va > 69.4 \tag{3.38b}$$

Where $b = 1.47 - 1.48a + 0.92a^2$ and $a \equiv$ the smaller ratio of the two sides of the rectangle.

3.4.2.2 Nusselt Number Correlations

Woven screen matrix

$$N_u = (1 + 0.99 P_e^{0.66})\beta^{1.79} \quad and \quad N_k = 0.73 + 0.50 P_e^{0.62} \beta^{-2.91} \tag{3.39}$$

Random fiber matrix

$$N_u = (1 + 1.16 P_e^{0.66}) \beta^{2.61} \tag{3.40}$$

Packed sphere matrix

$$N_u = 0.1 + c R_e^m \qquad \begin{aligned} m &= 0.85 - 0.43\beta \\ c &= 0.537\,\beta \end{aligned} \tag{3.41}$$

Rectangular channels

See Section 3.24 (Equation 3.9) for an explanation of *Nuo*, *Nuc*, and *Nua*.

$$Nuo = 10.0c \tag{3.42a}$$

(*c* is defined below.)

$$\text{Real}\,(Nuc) = 10.0c \qquad \text{if } \sqrt{2\,Va\,Pr} < 10.0c \tag{3.42b}$$

$$= \sqrt{2\,Va\,Pr} \qquad \text{otherwise,}$$

$$\text{Imaginary}\,(Nuc) = \frac{1}{5} Va\,Pr \qquad \text{if} \qquad Va\,Pr < 5\sqrt{2\,Va\,Pr} \tag{3.42c}$$

$$= \sqrt{2\,Va\,Pr} \qquad \text{otherwise,}$$

$$\text{Real}\,(Nua) = 8.1c \qquad \text{if} \qquad \sqrt{2\,Va\,Pr} < 16.2c \tag{3.42d}$$

$$= \frac{1}{2} \sqrt{Va\,Pr} \qquad \text{otherwise,}$$

$$\text{Imaginary}\,(Nua) = \frac{1}{10} Va\,Pr \qquad \text{if} \qquad Va\,Pr < 5\sqrt{2\,Va\,Pr} \tag{3.42e}$$

$$= \frac{1}{2} \sqrt{Va\,Pr} \qquad \text{otherwise,}$$

where *a* is the aspect ratio (the smaller of the ratio between the two sides of the rectangular geometry), and $c = 0.438 + 0.562 \times (1 - a)^3$.

3.5 Summary

See Tables 3.5 and 3.6.

TABLE 3.5

Summary of Correlations for Steady/Oscillatory *f* and *Nu*, Internal Flow (Pipe and Parallel Plate)

Geometry	Correlation	Steady/ Oscillatory	Entrance/Fully Developed Region	Equation(s)
Circular pipe	Friction factor	Unidirectional flow	Entrance length	3.11
			f, fully developed	3.1
			f, entrance region	3.12
		Oscillatory flow	Entrance length	NA
			f, fully developed	3.3
			f, entrance region	NA
	Nusselt number	Unidirectional flow	Entrance length	3.13, 15, 17, and 19
			Nu, fully developed	3.7
			Nu, entrance region	3.14, 16, 18, and 20
		Oscillatory flow	Entrance length	NA
			Nu, fully developed	3.9
			Nu, entrance region	NA
Parallel plate	Friction factor	Unidirectional flow	Entrance length	3.21
			f, fully developed	3.2
			f, entrance region	3.22
		Oscillatory flow	Entrance length	NA
			f, fully developed	3.4
			f, entrance region	NA
	Nusselt number	Unidirectional flow	Entrance length	3.23, 25
			Nu, fully developed	3.8
			Nu, entrance region	3.24, 26
		Oscillatory flow	Entrance length	NA
			Nu, fully developed	3.10
			Nu, entrance region	NA

Note: NA, not available.

TABLE 3.6

Summary of Correlations for Oscillatory *f* and *Nu*, Different Matrices

Geometry	Correlation	Equation
Woven screen	Friction factor	3.35
Random fiber		3.36
Packed sphere		3.37
Rectangular channels		3.38
Woven screen	Nusselt number	3.39
Random fiber		3.40
Packed sphere		3.41
Rectangular channels		3.42

4

Fundamentals of Operation and Types of Stirling Devices, with Descriptions of Some Sample Devices (Including Power and Cooling Levels)

4.1 Introduction

Brothers Robert and James Stirling worked together on the development of the original engine now known as the Stirling engine. The Reverend Robert Stirling patented the engine in 1816. Development continued off and on over the years; Stirling engines and coolers are presently being used or developed for space and terrestrial applications.

This chapter provides a very brief description of the fundamentals of operation of Stirling engines, coolers, and heat pumps—with references to more complete sources of information. Several different structural configurations of Stirling engines are also discussed. Examples of a few Stirling engines that have been designed, fabricated, and tested are briefly discussed—to provide some information about the practical range of Stirling engines. Finally, a Stirling cooler discussion is included; this includes a general schematic of a Stirling cooler, plus a description of a Stirling cryocooler that is being integrated into an experiment for use in the International Space Station. Very brief descriptions of the types of regenerator used in these Stirling devices are included, when we were able to identify these characteristics.

4.2 Fundamentals of Operation of Stirling Engines, Coolers, and Heat Pumps

The pressure-volume (P-V) diagrams and the temperature-entropy (T-S) diagrams of Figure 4.1, and the Stirling device schematics of Figure 4.2 with piston and displacers in position at cycle states 1, 2, 3, and 4 can be used to

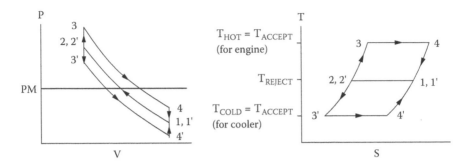

FIGURE 4.1
Pressure-volume (P-V) diagrams (left) and temperature-entropy (T-S) diagrams (right) for ideal
Stirling engine and refrigeration/cooling cycles.

describe ideal Stirling engine and refrigeration/heating cycles. (A general
discussion of the operation of Stirling engines and coolers can be found in
Bowman, 1993). The Stirling device schematics of Figure 4.2 imply three different heat exchange regions in the following locations: The cylinder wall at
the top of the displacer acts as an acceptor heat exchanger at temperature,
T_{ACCEPT}, where heat is accepted into the cycle from the outside. The cylinder

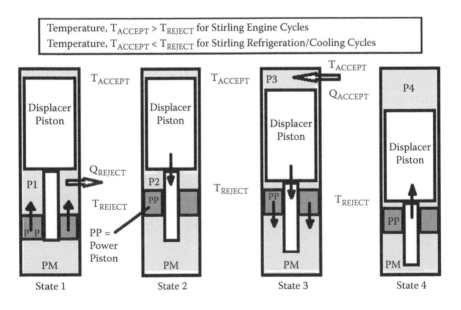

FIGURE 4.2
Stirling engine-refrigeration/cooling at four different times during ideal cycles (P1 indicates
correspondence to state points 1 and 1′, P2 to state points 2 and 2′, and so forth. The bounce
space, underneath the power piston, is at the mean pressure, PM.

wall between the displacer and the power piston acts as a rejector heat exchanger at temperature, T_{REJECT}, which rejects heat from the cycle. The gap between the displacer and the cylinder wall and the surfaces of the gap act as a regenerator heat exchanger; the regenerator solid surfaces store some of the cycle's heat as relatively warm gas passes through it in one direction, and gives up heat to the cycle when relatively cool gas passes through it in the other direction. Note that in these ideal cycles, portions of the cylinder wall act as the heat exchangers, or part of the heat exchanger in the case of the regenerator. (Most practical Stirling cycles require a separate heat exchanger gas circuit containing acceptor, regenerator, and rejector heat exchangers in series—through which the gas oscillates—in order to achieve adequate heat transfer surface areas between gas and solid.)

Operation of an ideal Stirling engine is as follows, with $T_{ACCEPT} > T_{REJECT}$ for the engine case. At cycle state 1 of Figure 4.2, all of the working space gas is in the relatively cold compression space between the power piston and the displacer. This corresponds to the maximum-volume, minimum-pressure thermodynamic state point 1 on the P-V diagram of Figure 4.1, and also to point 1 on the T-S diagram. In going from cycle state 1 to 2 of Figure 4.2, with the displacer remaining stationary, the power piston moves up and compresses the gas; work is done on the gas transitioning the thermodynamic states from 1 to 2 in the P-V and T-S diagrams. In passing from cycle state 2 to 3, with the power piston stationary, the displacer moves down and pushes the gas from the cold compression space, through the displacer-cylinder gap, to the hot expansion space above the displacer; this produces thermodynamic state changes from points 2 to points 3 on the P-V and T-S diagrams. In passing from cycle state 3 to 4, both power piston and displacer move down together, expanding the gas in the hot expansion space above the displacer; as the expansion of the gas tends to cool it, heat is absorbed from the outside via the acceptor wall; the state points change from 3 to 4 on the P-V and T-S diagrams. The cycle is completed via the change from cycle state 4 back to 1; here, with the power piston stationary, the displacer moves from its bottom to its top position, pushing the gas from the hot space through the displacer-cylinder gap to the cold space between the piston and displacer; this change also corresponds to changes in thermodynamic state points from 4 to 1 on the P-V and T-S diagrams. The area inside the completed P-V diagram represents the net work done on the power piston over one cycle. The area inside the completed T-S diagram represents the net heat added to the engine from the outside over one complete engine cycle. The waste or rejected cycle heat must be dissipated to the engine surroundings via the wall of the rejector.

Operation of an ideal Stirling refrigerator, or cooler, is as follows, with $T_{ACCEPT} < T_{REJECT}$ in this case: The general description of the piston motions between the various cycle states (1 through 4, and back to 1) is the same for the cooler as for the engine (described in the above paragraph). But, the overall process is different for the cooler than for the engine due to the relative differences in acceptor and rejector temperatures; in the cooler case, instead of raising the temperature of the acceptor end above that of the surroundings by application of an external heat source as in the engine case, this end is now insulated from the surroundings; thus, the working gas's absorption of heat from the acceptor end during the expansion process (between cycle states 3 and 4 in the engine case) gradually reduces the temperature of the acceptor end below that of the surroundings. Thus, in the steady ideal-cyclic process, the thermodynamic state points between cycle states 2 and 3 for the engine, for example, change from 2′ to 3′ in the P-V and T-S diagrams, for a cooling process; note that this is consistent with reductions instead of increases in working-gas pressure and temperature relative to state 2.

Operation of a Stirling heat pump is similar to that of the Stirling cooler; for both cooler and heat pump, the acceptor temperature is less than the rejector temperature. The major differences are that in the case of the cooler, the useful quantity is the heat absorbed at the acceptor which accomplishes the cooling of an insulated region below the temperature of the surroundings. In the case of the heat pump, the useful quantity is the heat rejected at the rejector end, which is usually used to heat a space above that of the outside temperature. Thus, a useful figure of merit for the ideal Stirling cooler is the cooling coefficent of performance (COP) defined as:

$$\text{Cooling COP} = \frac{\text{Heat Accepted}}{\text{Work Input}} = \frac{T_{\text{ACCEPT}}}{T_{\text{ACCEPT}} - T_{\text{REJECT}}}$$

whereas the appropriate figure of merit for the Stirling heat pump is:

$$\text{Heating COP} = \frac{\text{Heat Rejected}}{\text{Work Input}} = \frac{T_{\text{REJECT}}}{T_{\text{ACCEPT}} - T_{\text{REJECT}}}$$

In contrast, the useful quantity for the ideal Stirling engine is the work produced per cycle. Thus, the useful figure of merit for the ideal Stirling engine is:

$$\text{Work Efficiency} = \frac{\text{Work Output}}{\text{Heat Input}} = \frac{T_{\text{ACCEPT}} - T_{\text{REJECT}}}{T_{\text{ACCEPT}}}$$

Also, see Bowman (1993) for the above definitions and discussion of Stirling engine fundamentals. Some useful books about Stirling engines and Stirling engine analysis are those by Walker (1980), Urieli and Berchowitz (1984), West (1986), Hargreaves (1991), and Organ (1997).

The Stirling configuration shown in Figure 4.2 is a beta-type configuration, one of several configurations discussed in Section 4.3.

4.3 General Structural Configurations of Stirling Engines

Four general types of structural configurations of Stirling engines are shown in Figure 4.3. In the alpha-type configuration, the working space gas volume (including expansion-space, heater, regenerator, cooler, and compression-space) is located between two power pistons; the two power pistons are located in separate cylinders. The beta-type configuration has a displacer and a power piston that are located in the same cylinder; the function of the displacer is to displace the gas between the hot expansion space and the cold compression space; the difference in areas on the two sides of the displacer is generally designed to overcome the overall pressure drop through the three heat exchangers (particularly for free-piston engines); the power piston is the primary means of extracting mechanical work for this Stirling configuration. The gamma-type configuration has a displacer and a power piston, with displacer and power piston located in separate cylinders; note in the beta-type schematic of Figure 4.1b, if the power piston cylinder, located directly under the displacer cylinder, should be of different diameter from the displacer cylinder—then this would be a gamma-type rather than a beta-type configuration. The double-acting type of configuration, shown in Figure 4.1d, can be thought of as a c

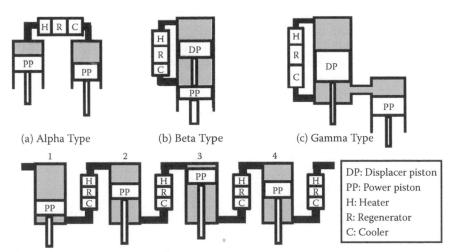

(a) Alpha Type   (b) Beta Type   (c) Gamma Type

(d) Double-Acting, or rinia, configuration.
    Cylinder 4 connects via the heat exchanger circuit on the right to Cylinder 1.
    Each heater, regenerator, cooler unit, plus the adjacent variable volumes, acts like
    an alpha configuration.
    Each power piston is 90 degrees out of phase with each of the pistons in the
    two adjacent variable volumes.

FIGURE 4.3
Four types of Stirling engine configurations.

ompound alpha-type configuration, with four double-acting pistons, each of which interacts with two of the four separate working spaces. From the point of view of each of the separate working spaces, there is an alpha-type configuration process ongoing. One goal of using double-acting pistons is to reduce the weight per unit power of the Stirling engine. This general type of four-piston, four-working-space, double-acting configuration (as shown in Figure 4.1d) was used in the Stirling automotive engines (Nightingale, 1986).

4.4 Methods of Getting Power Out of Stirling Engines

Mechanical power was extracted from early Stirling engines, and still is from some modern Stirling engines, via piston-rod linkages with a crankshaft; an illustration of one such power take-off arrangement is shown in Figure 4.4. These engines are now frequently called kinematic Stirling engines. Generally, kinematic engines require a lubricant, mechanical seals, and mechanical bearings, which can cause undesirable limits in performance and life, particularly for some types of applications (e.g., long-duration space applications).

For space-power applications of Stirling engines, which may require maintenance-free operation for a decade or two, free-piston/linear alternator designs are most appropriate for production of electrical power (Chan et al., 2007). These engines are hermetically sealed so that gas cannot leak from the working space through seals. There are piston-clearance seals inside the working space to minimize leakage by the piston and displacer, but these "leakages" are between different portions of the internal working space and

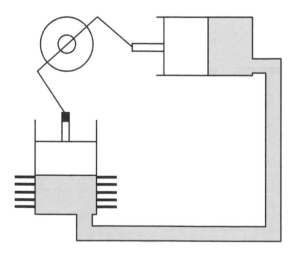

FIGURE 4.4
V-crank-arrangement of an alpha-configuration kinematic, or crank-drive, Stirling engine.

so do not deplete the gas inventory of the engine. Unique gas bearings have been developed which prevent any rubbing between piston/displacer and the cylinder, under normal operation. Thus, these free-piston designs are not subject to the life-limiting characteristics of mechanical seals and bearings.

4.5 Power Outputs of Some Stirling Engines That Have Been Fabricated and Tested

Powers of some free-piston Stirling engines intended for practical applications which have been fabricated and tested range from the ~88 We per cylinder advanced Stirling convertor (engine/alternator) (Chan et al., 2007) to the 12.5 kWe per cylinder component test power convertor (CTPC) (Dhar, 1999). Some kinematic, or crank-drive, Stirling engines that have been fabricated and tested are (1) the ~7.5 kW or 10 hp (mechanical power) single-cylinder rhombic-drive GPU-3 (Ground Power Unit 3) (Cairelli et al., 1978; Tew et al., 1979; Thieme, 1979, 1981; Thieme and Tew, 1978); (2) the ~54 kW (mechanical) four-cylinder, double-acting, MOD II automotive Stirling engine (Nightingale, 1986); and (3) the 25 kWe Stirling engine used in the SES (Stirling Engine Systems) dish-Stirling, or solar-Stirling, system. These engines will be discussed below. Many model Stirlings of fractional *W* power levels, both kinematic and free piston, have been fabricated and operated; information about many of these can be obtained by doing a search via the Internet. Also, ~75 kW Stirling (kinematic, double-acting) Air Independent Propulsion systems, manufactured by the Swedish company, Kockums, help power several Swedish submarines (see: www.stirlingengines.org.uk/manufact/manf/misc/subm.html).

4.5.1 Free-Piston Advanced Stirling Convertor (Engine/Alternator), or ASC, 88 We

A schematic of the free-piston advanced Stirling convertor (ASC) for space applications is shown in Figure 4.5 (Chan et al., 2007). These engines currently used random-fiber regenerators of ~90% porosity.

Figure 4.6 shows a photograph of the ASC with overall dimensions. The engine was developed by Sunpower Inc. (Athens, Ohio). Wong et al. (2008) report that an updated version of the ASC, the ASC-E2, will have the following operating conditions: indicated power of 84.5 W_{AC} with heat input from the radioisotope general purpose heat source (GPHS) of 224 $W_{thermal}$ for an efficiency of 37.7%; maximum heater head temperature will be 850°C, reject temperature will be 90°C, convertor mass is 1.32 kg, mean pressure of the helium working space gas is 3.65 MPa absolute, and operating frequency is 102 Hz. The engine is being used in a radioisotope power system for future space missions, under the joint sponsorship of the U.S. Department of Energy and the

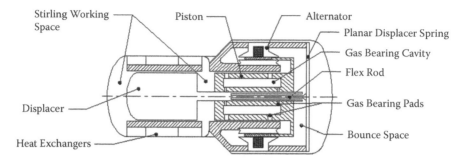

FIGURE 4.5
Internal features of Sunpower 88 We free-piston advanced Stirling convertor (engine/alternator).

National Aeronautics and Space Administration (NASA). Lockheed Martin is the system integrator. Figure 4.7 shows an engineering version of this system, the advanced Stirling radioisotope generator (ASRG), from Chan et al. (2007).

4.5.2 Free-Piston Component Test Power Convertor (Engine/ Alternator), or CTPC, 12.5 kWe per Cylinder (Dhar, 1999)

The first generation of hardware in the large free-piston Stirling engine program was the space power demonstrator engine (SPDE). The SPDE was a free-piston Stirling engine coupled to a linear alternator. It was a double-cylinder, opposed-piston convertor designed to produce 25 kWe at 25% overall efficiency. After a demonstration, the SPDE was modified to form two separate, single-cylinder power convertors called the space power research engines (SPREs).

The SPRE had a design operating point of 15 MPa using helium as the working fluid. The 650°K hot-end temperature and 325°K cold-end temperature provided an overall temperature ratio of 2. Piston stroke was 20 mm, and operating frequency was about 100 Hz. The SPRE incorporated gas springs,

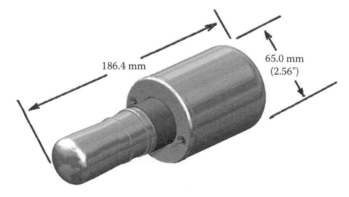

186.4 mm

65.0 mm
(2.56")

FIGURE 4.6
Photograph of 88 We advanced Stirling converter (ASC) with overall dimensions.

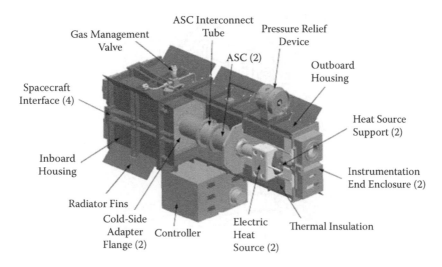

FIGURE 4.7
Engineering unit of the advanced Stirling radioisotope generator (ASRG).

hydrostatic gas bearings, centering ports, and close clearance noncontacting seals.

The second generation of hardware, the component test power convertor (CTPC), was to be a 25 kWe modular design consisting of two 12.5 kWe/cylinder opposed piston convertors. Only one-half of the CTPC was fabricated and tested. Details of the design, fabrication, and early testing are reported in Dhar (1999). The CTPC used a wire-screen regenerator of ~73% porosity. During the first fully functional test of the CTPC, the design goals of 12.5 kWe output power and 20% overall efficiency were easily surpassed.

The mean helium working gas pressure of the CTPC was also 15 MPa. Heater temperature was 1050°K, and cooler temperature was 525°K. The pistons oscillated at 70 Hz. The novel "Starfish" heat-pipe heater head design greatly reduced the number of braze joints, relative to the SPDE tubular heater design. Figures 4.8 and 4.9 show an overall CTPC engine layout and the hot-end assembly; many more detailed diagrams are shown in Dhar (1999). Mechanical Technology, Inc. (Albany, New York) designed and built both the SPDE and the CTPC, under NASA contracts.

4.5.3 General Motors Ground Power Unit (GPU)-3 Rhombic-Drive Stirling Engine, ~7.46 kW Mechanical Power

A ~10 horsepower (7.46 kW) single-cylinder rhombic-drive Stirling engine was converted to a research configuration to obtain data for validation of Stirling computer simulations. Test results were reported in several DOE and NASA reports (Cairelli et al., 1978; Tew et al., 1979; Thieme, 1979, 1981; Thieme and Tew, 1978). The engine was originally built by General Motors

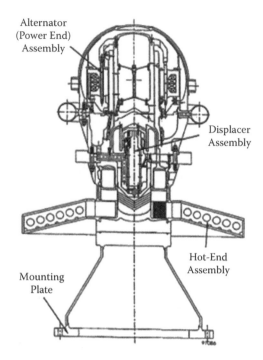

FIGURE 4.8
Component test power convertor (CTPC) engine layout.

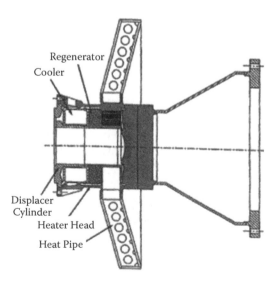

FIGURE 4.9
CTPC hot-end assembly.

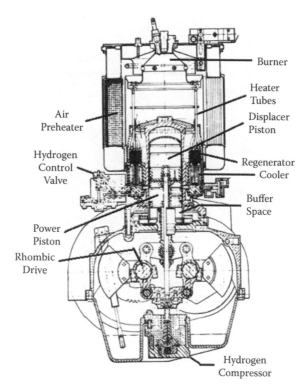

FIGURE 4.10
GPU-3 Stirling engine schematic.

Research Laboratories for the U.S. Army in 1965 as part of a 3 kWe engine-generator set, designated the GPU-3.

An engine schematic is shown in Figure 4.10.

Hydrogen was the working fluid. Design speed was 3000 RPM (50 Hz). Mean pressure was ~6.9 MPa. Hot-end temperature was 677°C, and cold-end temperature was 37°C. There were eight separate regenerator/cooler modules located around the piston/displacer cylinder. Eighty heater tubes interfaced with the regenerators; thus, there were 10 heater tubes connecting with each regenerator unit. The regenerators consisted of wire cloth, or screen, of ~71% porosity. Very detailed information about the engine design can be found in the NASA reports.

4.5.4 Mod II Automotive Stirling Engine

The Mod II automotive Stirling (Nightingale, 1986) engine utilized a four-cylinder V-block design with a single crankshaft and an annular heater head. It developed a maximum power of 62.3 kW (83.5 hp) and had a maximum speed of 4000 rpm (66.67 Hz). It was a four-cylinder double-acting configuration. There were three basic engine systems. (See Figure 4.11.) First, the

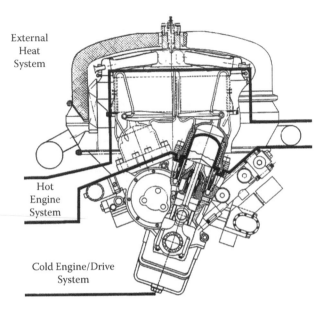

External
Heat
System

Hot
Engine
System

Cold Engine/Drive
System

FIGURE 4.11
MOD II automotive Stirling engine cross section.

external heat system converted energy in the fuel to heat flux. Next, the hot engine system contained the hot hydrogen in a closed volume to convert this heat flux to a pressure wave that acted on the pistons. Finally, the cold engine/ drive system transferred piston motion to connecting rods, and the recipro-cating rod motion was converted to rotary motion through a crankshaft. The engine was also equipped with all the controls and auxiliaries necessary for automotive operation. A relatively complicated hydrogen-gas inventory control system was used to shuttle gas into and out of the engine to meet the varying power demands of the automotive driving cycle. Figure 4.11 shows a MOD II automotive Stirling engine cross section. The automotive Stirlings were installed in and tested in several vehicles.

4.5.5 SES 25 kWe Stirling Engine System for Dish Solar Stirling Application

A photograph of four SunCatcher™ solar-dish Stirling systems is shown in Figure 4.12; the picture is from a Web site of Sandia National Laboratory (U.S. Department of Energy). (A statement on the Web site indicates the fig-ure and information in the accompanying article can be freely downloaded and published.) The article accompanying the picture on the Sandia Web site indicates the following:

> Stirling Energy Systems (SES) and Tessera Solar recently (in 2009, apparently) unveiled the four newly designed solar power collection

FIGURE 4.12
Four Suncatcher™ (25 kWe) solar-dish Stirling systems. (From the U.S. Department of Energy's Sandia National Laboratory Web site: https://share.sandia.gov/news/resources/news_releases/new-suncatcher-power-system-unveiled-at-national-solar-thermal-test-facility-july-7-2009.)

dishes at Sandia National Laboratories' National Solar Thermal Test Facility (NSTTF) [i.e., those shown in Figure 4.12]. [Six older SunCatchers had been producing 150 kWe during the day (i.e., 25 kWe per Suncatcher).]

"The modular CSP [Concentrating Solar-thermal Power] SunCatcher uses precision mirrors attached to a parabolic dish to focus the sun's rays onto a receiver, which transmits the heat to a Stirling engine. The engine is a sealed system filled with hydrogen. As the gas heats and cools, its pressure rises and falls. The change in pressure drives the piston inside the engine, producing mechanical power, which in turn drives a generator and makes electricity."

According to the information on Sandia's Web site:

Tessera Solar, the developer and operator of large-scale solar projects using the SunCatcher technology and sister company of SES, is building a 60-unit plant generating 1.5 MW (megawatts) by the end of the year either in Arizona or California [seemed to imply at end of 2010]. One megawatt powers about 800 homes. The proprietary solar dish technology will then be deployed to develop two of the world's largest solar generating plants in Southern California with San Diego Gas & Electric in the Imperial Valley and Southern California Edison in the Mojave Desert, in addition to the recently announced project with CPS Energy in West Texas. The projects are expected to produce 1,000 MW by the end of 2012.

Last year one of the original SunCatchers set a new solar-to-grid system conversion efficiency record by achieving a 31.25 percent net efficiency rate, toppling the old 1984 record of 29.4.

A good general discussion of dish Stirling technology is found in Stine and Diver (1994). Fraser (2008) appears to be the most recent publication about solar-dish Stirling technology.

4.6 Stirling Coolers

In a Stirling engine, the acceptor heat exchanger temperature is higher than that of the rejector heat exchanger, and the heat input to the engine is converted to power (typically mechanical or electrical). In a Stirling cooling device, or a heat pump, the acceptor temperature is lower than that of the rejector temperature, and input power to the device is used to drive the cooling process in the case of a cooler, or the heating process in the case of a heat pump. In a Stirling cooler, the cooled region is cooled below ambient, and heat is rejected to ambient. In a Stirling heat pump, heat is absorbed from the outside environment, is increased by conversion of the driving, or input, power to heat—and the sum total of this heat is used for interior heating.

Figure 4.13 shows the general arrangement of a free-piston Stirling cooler, or heat pump. Also, see Berchowitz (1993) for a discussion of Stirling coolers.

Note that the power piston is driven by a linear motor. The object of a Stirling cooling device is to produce cooling in the region of the expansion space, via absorption by the acceptor heat exchanger. The object of a Stirling heat pump is to combine the heat absorbed in the expansion space region

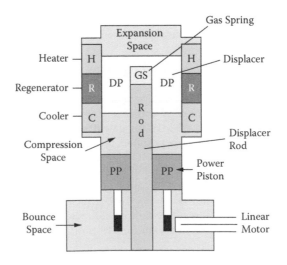

FIGURE 4.13
Schematic of a free-piston Stirling cooler driven by a linear motor.

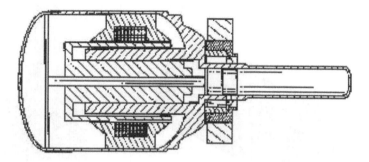

FIGURE 4.14
Schematic of Sunpower M87 linear free-piston integral cooler. (Reprinted from *Cryogenics*, 46, K. Shirey et al., Design and Qualification of the AMS-02 Flight Cryocoolers, 142–148. Copyright 2006, with permission from Elsevier.)

(via the acceptor), with heat generated by conversion of the input power of the motor, and to reject the sum total of this heat via the rejector—for heating purposes.

4.6.1 Sunpower M87 Cryocooler

The M87, shown in Figure 4.14, is a single, free-piston, integral Stirling-cycle cryocooler designed for high volume manufacturing (Shirey et al., 2006). The M87 was designed to provide 7.5 watts of cooling at 77°K with 150 watts input power while operating at a reject temperature of 35°C. This cryocooler has a design lifetime of >40,000 hours. The M87's original intended use was as an oxygen liquefier at a patient's home.

The Sunpower M87N cryocooler is a modified M87 with enhancements targeted to better suit the NASA Alpha Magnetic Spectrometer-02 (AMS-02). The AMS-02 is a state-of-the-art particle physics detector containing a large superfluid helium-cooled superconducting magnet. Highly sensitive detector plates inside the magnet measure a particle's speed, mass, charge, and direction. The AMS-02 experiment, which will be flown as an attached payload on the International Space Station, will study the properties and origin of cosmic particles and nuclei including antimatter and dark matter. Four commercial Sunpower M87N Stirling-cycle cryocoolers will be used to extend the lifetime of the AMS-02 experiment. The baseline performance requirement for the four cryocoolers is a total of 9.4 W of heat lift at 60°K with 400 W of input power.

Shirey et al. (2006) describe operation of the cryocooler as follows: A pressure oscillation generated in the compression space drives the displacer through its rod, which extends through the piston and into the bounce space. While the piston relies on the gas spring in the compression space for its resonance, the displacer is attached to a planar spring through a compliant member. The displacer, containing the random fiber regenerator, shuttles

FIGURE 4.15

Sunpower Inc. , M87 cryocooler, three-dimensional cutaway view (left) and photograph (right). (With kind permission from Springer Science+Business Media: *Proceedings of the International Cryocooler Conference,* The Advent of Low Cost Cryocoolers, 11, 2002, 79–86, R.Z. Unger, R.B. Wiseman, and M.R. Hummon, Figure 5. A. M87 Cryocooler, 3-D cutaway view [left], B. photograph [right]. Copyright Kluwer Academic/Plenum.)

the gas in between the cold-end and the warm-end heat exchanger. The gas bearing systems of the commercial cooler were redesigned for the M87N to provide better performance regardless of orientation. The gas bearings systems are used to radially center the piston and displacer and thus prevent contact of the moving parts. Cooler vibrations generated by the piston and displacer are countered by a passive (tuned spring-mass) balancer system.

A three-dimensional (3-D) cutaway view and a photograph of an M87 cryocooler are shown in Figure 4.15 (Unger et al., 2002).

5

Types of Stirling Engine Regenerators

5.1 Introduction

The regenerator is one of three types of heat exchangers that occur in Stirling devices. The various Stirling engine structural configurations of Figure 4.3 (in Chapter 4) include the typical series configurations of a heater or acceptor (H), regenerator (R), and cooler or rejector (C). Regenerators are crucial in achieving good performance. In an engine, the regenerator solid surfaces store heat from the gas as relatively hot gas passes through the regenerator from the hot heater/expansion space to the cold cooler/compression space; the regenerator solid surfaces then restore heat to the gas as relatively cold gas returns from the cold cooler/compression space to the hot heater/expansion space. The amount of heat that the regenerator saves from the gas, and then restores to the gas, during one cycle is typically on the order of four times the amount of heat that enters through the heater/acceptor during one cycle. Thus, if a regenerator were removed from such an engine, the heater/acceptor would need to absorb five times as much heat during a cycle to maintain the same power output as it did with a regenerator; because engine efficiency is power out divided by heat in, removal of the regenerator with consequent increase of heat into the heater/acceptor by a factor of 5 would result in a five times decrease in engine efficiency. Of course, the original heater could not be expected to achieve such an increase in heat transfer at the same operating conditions, so in a practical engine design—removal of such a regenerator would be very destructive to the performance of an engine—so destructive that it would very likely not operate. The regenerator types discussed in this chapter have been tested, and the data have been modeled via friction-factor and heat-transfer correlations. Correlations for the more conventional random-fiber and wire-screen matrices are given in Chapter 3. Correlations for the new segmented-involute-foil regenerators are given in Chapters 8 and 9.

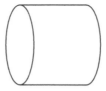

FIGURE 5.1
Cylindrical canister. (From Gedeon, D., 2010, Sage User's Guide, Electronic Edition for Acrobat Reader, Sage v7 Edition, available on Web site: http://sageofathens.com/Documents/SageStlxHyperlinked.pdf, Gedeon Associates, Athens, Ohio.)

5.2 Regenerator Envelope (Canister or Volume) Configurations

Regenerator envelopes, or overall containment volumes, are typically of cylindrical or annular shape, as shown in the schematics of Figures 5.1 and 5.2 (Gedeon, 2010). The Gedeon (2010) reference is a manual for use of the Sage one-dimensional (1-D) Stirling computer code that is used by Sunpower Inc. (Athens, Ohio), Infinia Corporation (Kennewick, Washington), and others, to aid in the design of Stirling engines; it is also the code that the National Aeronautics and Space Administration (NASA) Glenn Research Center uses to support the NASA/U.S. Department of Energy (DOE) Stirling development contracts. Engines that use cylindrical regenerator volumes sometimes have multiple regenerator canisters. Engines that use annular regenerator canisters typically have the regenerator annulus surrounding the pistons (e.g., double-acting automotive engines), or displacers. It is possible there might be some benefit in using variable diameter regenerator envelopes, or volumes, such as illustrated in Figures 5.3 and 5.4 (Gedeon, 2010).

Gedeon (2010) calls these "tubular-cone canisters" and "annular-cone canisters," respectively. Sage allows for the modeling of variable diameter (cone-shaped) cylindrical and annular canisters; however, these would be more difficult to fabricate, and the authors are not aware of Stirling devices that have used such canisters. It would be interesting to use Sage, or some

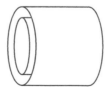

FIGURE 5.2
Annular canister. (From Gedeon, D., 2010, Sage User's Guide, Electronic Edition for Acrobat Reader, Sage v7 Edition, available on Web site: http://sageofathens.com/Documents/Sage StlxHyperlinked.pdf, Gedeon Associates, Athens, Ohio.)

FIGURE 5.3
Variable diameter cylindrical canister. (From Gedeon, D., 2010, Sage User's Guide, Electronic Edition for Acrobat Reader, Sage v7 Edition, available on Web site: http://sageofathens.com/ Documents/SageStlxHyperlinked.pdf, Gedeon Associates, Athens, Ohio.)

other computer modeling software, to determine if there may be performance benefits in using variable diameter (along the flow axis) regenerators. The authors are not aware if anyone has investigated the potential benefits of such variable diameter regenerators.

5.3 Regenerator Porous Material Structures

5.3.1 No Porous Matrix: Walls Act as Heat Storage Medium

The simplest type of regenerator is just a gaseous "gap," as shown between the displacer and the cylinder walls of Figure 4.2. Here the displacer and cylinder wall surfaces act as the heat-storage solid that exchanges heat with the gas as it oscillates through a clearance gap. Small annular-regenerator engines, on the order of 10 W of output or less, may perform satisfactorily with such a simple regenerator. According to Infinia Corporation's Web site (www.infiniacorp.com/accomplishments.html), during the period of 1991 to 1993, "Infinia developed a 10W Stirling generator prototype for the Pacific Northwest National Laboratory." The authors believe that this engine had a simple "gap regenerator."

FIGURE 5.4
Variable diameter annular canister. (From Gedeon, D., 2010, Sage User's Guide, Electronic Edition for Acrobat Reader, Sage v7 Edition, available on Web site: http://sageofathens.com/ Documents/SageStlxHyperlinked.pdf, Gedeon Associates, Athens, Ohio.)

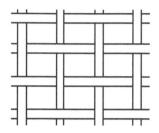

FIGURE 5.5
Woven-wire screen. (From Gedeon, D., 2010, Sage User's Guide, Electronic Edition for Acrobat Reader, Sage v7 Edition, available on Web site: http://sageofathens.com/Documents/ SageStlxHyperlinked.pdf, Gedeon Associates, Athens, Ohio.)

5.3.2 Stacks of Woven-Wire Screens

As Stirling machines become larger, they need more than a simple gap regenerator to provide enough solid surface and heat capacity for heat storage. Many Philips and General Motors cylindrical-canister regenerator Stirling engines used wire screens as a regenerator "matrix," or porous material. For example, the General Motors GPU-3 (Ground Power Unit-3) rhombic-drive Stirling (~7.45 kW or 10 hp, mechanical power) used eight small stacked wire-screen Stirling regenerators in cylindrical canisters located symmetrically around the central Stirling engine cylinder (Cairelli, 1978; Thieme, 1979, 1981; Thieme and Tew, 1978); the schematic of Figure 4.10 shows two of these canister regenerators. Figure 5.5 (Gedeon, 2010) shows conceptually how wires are woven to form individual screens.

The four-cylinder double-acting automotive Stirling engines also used wire screen (stainless-steel) in their regenerators (Nightingale, 1986). Individual wire-mesh screens were stacked, pressed, and vacuum sintered into a single annular "biscuit"—each of which fit around a piston cylinder.

Some engines (e.g., the GPU-3) used stacks of wire screens inserted into canisters, which are then inserted into the engine. Simpler engines with regenerators occupying annular volumes around the displacer may just use press-fitting of the matrix material into the annular volume. During the automotive Stirling engine program (Nightingale, 1986), it was found by trial and error that it was best to braze the porous regenerator material to the walls to avoid gaps opening up between the porous material and the wall and allowing regenerator bypass flow—which can have a serious impact on engine performance; sintering of the stacks of wire screen was also found advisable, to avoid disintegration of the stacks of screen due to vibrations of the screens during engine operation.

5.3.2.1 Sample Wire-Screen Dimensions

Table 5.1, taken from Gedeon (2009), documents the geometric characteristics of stacks of woven-wire screen that were tested in the NASA/Sunpower oscillating-flow test rig (Appendix A).

TABLE 5.1

Woven-Wire Screen Geometric
Characteristics (Tested in NASA/
Sunpower Inc. Oscillating-Flow Rig)

Mesh/Inch	Wire Diameter, Microns (in)	Porosity
200	53.3 (0.0021)	0.6232
100	55.9 (0.0022)	0.7810
80	94.0 (0.0037)	0.7102

5.3.3 Random-Fiber Porous Material

Random fiber regenerators are almost as good as wire screens and are much cheaper to fabricate. Figure 5.6 (Gedeon, 2010) shows a conceptual schematic of a layer of random fiber (see also, Figures 1.2 and 1.3 in Chapter 1). Because of the way random-fiber materials are fabricated, most of the fibers are oriented in a plane perpendicular to the primary flow direction (thus, fiber orientation relative to the primary flow axis is similar to that of stacked wire screens).

Most of the modern free-piston Stirling devices—such as those manufactured by Sunpower Inc., Infinia Corporation, and Global Cooling Manufacturing Co. (Global Cooling, Athens, Ohio) use annular regenerators that fit around the displacer cylinder. The ~80 We advanced Stirling convertor (ASC), a convertor being an engine/alternator, that is used in the advanced Stirling radioisotope generator (Chan et al., 2007), uses a sintered random-fiber annular regenerator.

Chapter 6 reports on experimental and computational investigations of actual-size random-fiber regenerators supported by the U.S. Department of Energy and NASA. Chapter 7 reports on that part of the effort devoted to

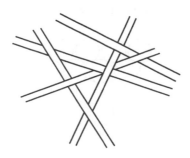

FIGURE 5.6
Conceptual random-fiber schematic. (From Gedeon, D., 2010, Sage User's Guide, Electronic Edition for Acrobat Reader, Sage v7 Edition, available on Web site: http://sageofathens.com/Documents/SageStlxHyperlinked.pdf, Gedeon Associates, Athens, Ohio.)

TABLE 5.2

Random-Fiber Geometric Characteristics (Tested in NASA/Sunpower Inc. Oscillating-Flow Rig)

Nominal Wire Diameter	Manufacturer	Material	Measured Average Wire Diameter, Microns	Porosity
2 mil	Brunswick	Inconel	52.5	0.688
1 mil	Brunswick	Stainless steel	27.4	0.820
12 micron	Bekaert	Stainless steel	13.4	0.897
30 micron	Bekaert	Stainless steel	31.0	0.85
30 micron	Bekaert	Stainless steel	31.0	0.90
30 micron	Bekaert	Stainless steel	31.0	0.93
30 micron	Bekaert	Stainless steel	31.0	0.96
24 micron	Bekaert	Oxidation-resistant material	24.3	0.909

testing of a large-scale representation of a random-fiber regenerator, and supporting computational efforts. The purpose of these investigations was to try to understand how to improve Stirling regenerator performance. Knowledge acquired during these efforts led to a microfabricated regenerator development program, which is reported on in detail in Chapters 8 and 9 (actual-scale and large-scale investigations, respectively).

5.3.3.1 Sample Random-Fiber Dimensions

Table 5.2, taken from Gedeon (2009), documents the geometric characteristics of random-fiber regenerator materials that were tested in the NASA/Sunpower oscillating-flow test rig (Appendix A).

5.3.4 Packed-Sphere Regenerator Matrix

Packed spheres have been used in some Stirling cooling devices; the conceptual packed-sphere diagram of Figure 5.7 is taken from Gedeon (2010). The

FIGURE 5.7
Packed-sphere conceptual diagram. (From Gedeon, D., 2010, Sage User's Guide, Electronic Edition for Acrobat Reader, Sage v7 Edition, available on Web site: http://sageofathens.com/Documents/SageStlxHyperlinked.pdf, Gedeon Associates, Athens, Ohio.)

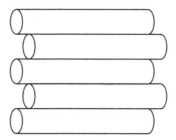

FIGURE 5.8
Conceptual tube-bundle regenerator. (From Gedeon, D., 2010, Sage User's Guide, Electronic Edition for Acrobat Reader, Sage v7 Edition, available on Web site: http://sageofathens.com/ Documents/SageStlxHyperlinked.pdf, Gedeon Associates, Athens, Ohio.)

authors are not aware of any testing of packed-sphere regenerators in the NASA/Sunpower oscillating-flow rig.

5.3.5 Tube-Bundle Regenerator Concept

A bundle of tubes could serve as a regenerator matrix; the conceptual diagram of Figure 5.8 is taken from Gedeon (2010). Sage allows modeling of tube-bundle regenerators, providing parallel gas flow paths—having some circular and some noncircular cross sections. Possible advantages of such a geometry include avoidance of the flow separations that occur with flow over wire screens and random fibers (which are typically fabricated so that most fibers are oriented perpendicular to the flow axis); these wire-screen and random-fiber flow separations, due to flow across the cylindrical fibers, tend to increase pressure drop losses. (See Chapter 1, "Introduction," for an additional discussion of these flow separations.) Disadvantages of continuous tube bundles that run the entire flow axis of the regenerator are (1) relatively large axial conduction losses and (2) inability of flow to redistribute itself in the radial direction.

The authors are not aware of any data for testing of a tube-bundle regenerator.

5.3.6 Wrapped-Foil Regenerators

Figure 5.9 shows a conceptual diagram of a wrapped-foil matrix. "Wrapping" refers to the fabrication procedure of wrapping, or winding, the foil around a central structural element. Some method must be used to try to ensure uniform foil separations. One approach is to use uniform-sized "protrusions" fabricated on one surface, or dimples stamped into a metal foil.

Global Cooling's Stirling-cooler annular regenerators are made from wrapped-plastic "foils," with uniform part-of-a-sphere protrusions on one surface. (Global Cooling develops Stirling coolers for food-storage-type

FIGURE 5.9
Conceptual diagram of a wrapped-foil matrix. (From Gedeon, D., 2010, Sage User's Guide, Electronic Edition for Acrobat Reader, Sage v7 Edition, available on Web site: http://sageofathens. com/Documents/SageStlxHyperlinked.pdf, Gedeon Associates, Athens, Ohio.)

temperatures, not cryocoolers.) Gedeon (2005) indicates that the common wisdom in the cryocooler [Stirling] industry is that foil regenerators (parallel-plates) do not work. Theoretically speaking they should work just fine. In fact they should work better than any other know regenerator type. Also, the authors have heard that wrapped-metal foils have been tried in Stirling engines, without success. Apparently, due to large spatial temperature gradients across the wrapped metal foils, both during steady operation and during startup and shutdown, the wrapped foils become distorted, apparently causing nonuniform gaps and flow and serious degradation of engine performance.

Gedeon (2005) (also see Appendix D) modeled the effects of nonuniform foil spacing on cryocooler performance. He found that the problem worsens with decreasing temperature and increasing foil spacing. In the temperature range of 30–100 K it is easily possible for a foil regenerator to degrade overall cooling efficiency (heat lift/compressor PV power) by 15% or more for a foil spacing variation of only ±10%, between different parts of the regenerator.

5.3.7 Parallel-Plate Regenerators

Parallel-plate regenerators can be represented by the conceptual schematic of Figure 5.10 (Gedeon, 2010).

If continuous along the entire axial length, parallel-plate regenerators would also suffer large axial conduction losses, as do continuous tube bundles. They would also not allow redistribution of flow from one channel to the next, if needed—due to a nonuniform flow entering the channels. Also, parallel-plate channels fitted into cylindrical or annular canisters provide different wall geometries at the canister for each channel, and different

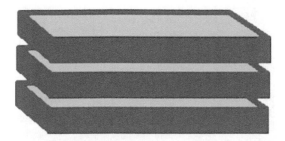

FIGURE 5.10
Conceptual parallel-plate regenerator.

channel perimeters, which would tend to make it difficult to get uniform axial flow over a cross section of the regenerator.

Backhaus and Swift (2001) reported on use of a parallel-plate regenerator in a thermoacoustic-Stirling heat engine at Los Alamos National Laboratory (New Mexico). Their results indicated significant enhancement in engine performance relative to earlier tests with a screen regenerator. They also explained how variations in plate spacing can have a significant impact on cryogenic refrigerator performance. Gedeon (2007) later reported to the authors of this book a conversation with Backhaus about the parallel-plate regenerator after the Los Alamos test program was completed; Backhaus reportedly offered the parallel-plate regenerator for possible testing in the NASA/Sunpower oscillating-flow test rig but informed Gedeon that examination of the regenerator indicated that distortion of the plates had occurred.

5.3.8 Segmented-Involute-Foil Regenerators

Much of this book is focused on the development and testing of segmented-involute-foil regenerators (Ibrahim et al., 2007, 2009a, 2009b; Tew et al., 2007). Figure 5.11 illustrates the segmented-involute-foil concept. There were two types of disks that alternated in the stack of foils. The best disk thickness of the two thicknesses tested in the NASA/Sunpower oscillating-flow test rig was ~250 microns.

Chapter 8 focuses on the design, development, and testing of segmented-involute-foil regenerators; other detailed dimensions of the disks fabricated and tested are given there. Chapter 9 focuses on design, development, and testing of a large-scale (~30× actual design size) segmented-involute-foil regenerator.

The segmented-involute-foils were conceived and designed to minimize the effects of cylinder cross-flow in increasing pressure drop losses (as occurs in wire-screen and random-fiber regenerators). They were also designed to avoid the large axial conduction losses of continuous (in the flow-axis direction) parallel plates and tube bundles, and to allow for flow distribution in the radial direction, to keep flow as uniform as possible throughout the

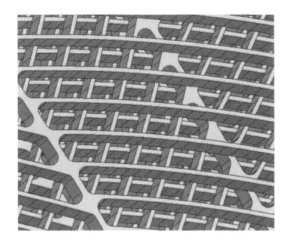

FIGURE 5.11
Three-dimensional (3-D) view of four layers of microfabricated segmented-involute-foil channels.

cross sections perpendicular to the flow axis. They are also thought to be less subject to matrix element breakage than random fibers that might let debris escape into the engine working space. The test results suggest that the segmented-involute-foils approach the superior (e.g., to wire screens and random fibers) desirable features of theoretical parallel plate regenerators, while avoiding some of the practical disadvantages of continuous parallel-plate configurations (discussed in Section 5.3.7, above). The biggest disappointment of the research effort, discussed briefly in Chapter 8 and at length in the references (Ibrahim et al., 2007, 2009a, 2009b)—is that the LiGA technique chosen for development did not prove to be adequately cost effective. So, more work needs to be done to identify a cost-effective development process for segmented-involute-foil regenerators.

5.3.9 "Mesh Sheet" and Other Chemically Etched Regenerators

Professor Noboru Kagawa of the National Defense Academy of Japan has led the development of mesh-sheet regenerators over a period of several years (Furutani et al., 2006; Kitahama et al., 2003; Matsuguchi et al., 2005; Takeuchi et al., 2004; Takizawa et al., 2002). Chemical etching has been used to fabricate mesh sheets whose detailed dimensions have evolved with the purpose of improving performance in a particular Stirling engine. Figure 5.12 shows a detail of the mesh sheet concept as presented in Takeuchi et al. (2004).

A mesh sheet is a chemically etched geometry that appears to be similar in concept to a wire screen but without the overlapping wires (i.e., the flow openings constitute a field of ~ square openings, located in the plane of a

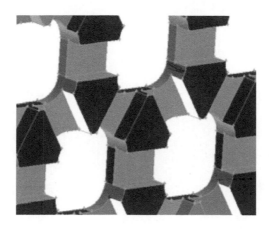

FIGURE 5.12
The mesh sheet concept from Takeuchi et al. (2004). (From Takeuchi, T. et al., 2004, Performance of New Mesh Sheet for Stirling Engine Regenerator, Paper AIAA 2004-5648, A Collection of Technical Papers, 2nd International Energy Conversion Engineering Conference, Providence, RI, August 16–19. Reprinted with permission of the American Institute of Aeronautics and Astronautics.)

circular disk). The evolution of the mesh sheets has appeared to be toward trying to optimize the detailed dimensions of the mesh sheets in order to maximize the performance of the test engine. The mesh sheet regenerator is treated in more detail in Chapter 10.

Mitchell et al. (2005, 2007) has also used chemical etching to develop a unique regenerator geometry. His concept is discussed in more detail in Chapter 10, where photographs of three different etched foils are shown in Figure 10.26.

6

Random-Fiber Regenerators—Actual Scale

6.1 Introduction

Much of the actual scale Stirling regenerator research work reported here was done under contract (Ibrahim et al., 2004a, 2004b, 2004c) with the U.S. Department of Energy (DOE). That joint effort was carried out via principal investigators at Cleveland State University (the lead), the University of Minnesota, and Gedeon Associates (Athens, Ohio); National Aeronautics and Space Administration Glenn Research Center (NASA GRC) participated via a Space Act Agreement with DOE. The principle investigators worked with three Stirling-engine development companies over the course of the research: Stirling Technology Co. (STC) (now Infinia Corporation, Kennewick, Washington), Sunpower Inc. (Athens, Ohio), and Global Cooling Manufacturing, Inc. (Global Cooling, Athens, Ohio). These companies helped identify the important issues they faced concerning the regenerator and provided specific engine parameters required to design the experiments. They also assisted with fabrication and testing of actual-size regenerator test samples. Once the DOE regenerator research effort was completed, this work was continued for several more years under NASA grant funding. The latest NASA/Sunpower oscillating-flow test-rig results were reported in Ibrahim et al. (2009a, 2009b). Some of those results are also included in this chapter.

For a general discussion of the function of the regenerator, its location in a Stirling engine, some common types of regenerators (random fibers, wire screens, and foil-type regenerators), and some problems with and desirable features of regenerators—see the introductory discussion in Chapter 1. Figure 1.1 shows the location of a regenerator in a small radioisotope-type Stirling engine. Figures 1.2 and 1.3 show a random-fiber–type regenerator and a micrograph that shows the microfeatures of a particular random-fiber regenerator, respectively.

Early in the DOE program, a decision was made to focus attention on regenerators of the random-fiber type, rather than woven screens or wrapped foils. Random-fiber felts are much cheaper to manufacture than woven screens and perform about as well. And they generally outperform wrapped foils. According to idealized one-dimensional theories, it should be the other way around, with foil regenerators providing the best performance (measured by heat transfer per unit flow resistance). But, in practice, wrapped-foil regenerators are difficult to manufacture in such a way that the theoretically good performance can be achieved. In particular, it is difficult to maintain the uniform spacing between the wrapped-foil layers, required to achieve uniform flow. However, in spite of the lack of success applying wrapped-foil regenerators to Stirling engines, "plastic" wrapped-foil regenerators have performed well enough to achieve standard usage in some Stirling coolers developed by Global Cooling.

Random-fiber matrices offered many opportunities for improvement. They are amenable to changes in porosity, packing structure, fiber shape, and fiber orientation, all of which are likely to affect performance. One of the goals of this DOE project was to discover exactly how. Late in the DOE program, consideration was again given to foil-type regenerators, but never with the same intensity as directed toward random-fiber regenerators.

After the DOE program was completed, a NASA-funded project started to focus on advanced techniques for manufacturing the next generation Stirling regenerator. This new microfabricated foil-regenerator will be discussed in detail in Chapters 8 and 9.

6.1.1 Metallic Random-Fiber Regenerators

In this section, different random-fiber matrices that were tested will be presented. Table 6.1 shows the main dimensions of these matrices, and more details about each are provided in the following sections.

6.1.1.1 Bekaert Stainless Steel

6.1.1.1.1 90%-Porosity Stainless Steel (Ibrahim et al., 2004a, 2004b, 2004c)

Sunpower donated material cut from a sheet of Bekipor (Bekaert trade name) 316 stainless steel random-fiber metal felt, 12 micron nominal fiber diameter,

TABLE 6.1

Dimensions of Different Random-Fiber Regenerator Matrices Tested

Item	Bekaert Fiber, ~90% Porosity	Bekaert Fiber, ~96% Porosity	Bekaert Fiber, ~93% Porosity	Oxidation-Resistant Fiber, ~90% Porosity
Matrix length (mm)	15.2	33.0	37.8/18.8	18.83
Matrix diameter (mm)	19.05	19.05	19.05	19.05
Fiber material	316 stainless	316 stainless	316 stainless	Oxidation resistant
Measured wire diameter (mm)	12–15 micron	31	31	22
Calculated porosity	0.897	0.96	0.93	~0.90

FIGURE 6.1
Completed Bekaert fiber matrix on left. Close-up on the right showing edge compression at the canister boundary. Small subdivisions on the reticule scale is equal to about 33 microns.

already pressed and sintered to 90% porosity. The sheet had been prepared at a fiber mass density of 600 g/m², which resulted in a thickness of about 0.74 mm after sintering. From this material, Gedeon Associates punched a number of disks using a custom-made punch and die. In order to ensure a good edge seal, the die was oversized by a factor of about 1.004 compared to the canister ID (about 0.08 mm [0.003"] oversize). The disks were then press fit and stacked into the canister. The final matrix porosity turned out to be 89.7%, very close to the target of 90%. Porosity was calculated from the measured mass and volume of the matrix and the looked up density of 316 stainless of 8027 kg/m³. Dimensions of the matrix are shown in Table 6.1. Figure 6.1 shows the completed matrix and a close-up micrograph. The micrograph is for a rough visual impression only. The electron microscope pictures shown later give a more detailed look at the fibrous microstructure.

6.1.1.1.2 96%-Porosity Stainless Steel (Ibrahim et al., 2009a, 2009b)

In 2006, three 96%-porosity regenerator samples were fabricated, one "full-length" sample and two "half-length" samples, named "A" and "B." They were all made from the same batch of random-fiber material. Their purpose was to learn something about repeatability of test results at high porosity.

In 2006, the full-length sample and half-length sample "A" were tested. It was discovered there was a length dependence in the heat-transfer testing that was ultimately traced to the inability of the test rig to resolve the low thermal losses of the long regenerator sample at low Reynolds numbers (low piston amplitudes in the 10-bar helium tests). In 2008, half-length sample "B" was tested to compare with the 2006 tests of nominally identical sample "A." Sample "A" was also retested to see if the previous results could be duplicated. The two half-length, 96% porosity, random-fiber regenerator samples, named "A" and "B" showed nearly identical friction factors but differing thermal performance, as measured by overall figures of merit. The thermal discrepancy was most pronounced at high Reynolds numbers, amounting to about 19% at $R_e = 400$. But the data are clouded somewhat by small modifications made to the rig since testing of sample "A" in 2006. Nominal specifications for both half-length samples are shown in Table 6.1.

6.1.1.1.3 93%-Porosity Stainless Steel (Ibrahim et al., 2009a, 2009b)

In 2006, a length dependence in the test data for 96%-porosity regenerator samples was discovered. It was concluded that it was caused by the inability of the test rig to resolve the low thermal losses associated with the 10-bar helium tests. At that time, the experimenters were curious if the same sort of length dependence would show up for 93%-porosity samples, where thermal losses are generally higher. However, in 2006, there was only funding to test one 93%-porosity sample, the "full-length" sample. Then, in 2008, funding became available for fabricating and testing a half-length, 93%-porosity sample made from the original lot of material.

Compared to the full-length, 93%-porosity, random-fiber regenerator sample tested in 2006, the half-length sample tested in 2008 showed a nearly identical friction factor but lower thermal performance, as measured by an overall figure of merit (see Section 8.2.3 and Appendix D). The thermal discrepancy was only a few percent when comparing 10-bar helium test data but about 40% for 50-bar helium data and about 20% for 50-bar nitrogen data, leading to suspicions that the full-length test data for 50-bar helium and nitrogen might be bad.

Retesting the full-length, 93%-porosity, random-fiber regenerator sample in 2008 gave different results than the original 2006 tests and came much closer to matching the figure of merit for the recently tested half-length sample. The two tests done in 2008 demonstrate good length independence for the test procedure and good repeatability for two different random-fiber regenerators compressed to the same porosity. Thus, the full-length 2006 results for 93% porosity will be not be presented below (for additional details see Ibrahim et al., 2009a, 2009b). The definitive test for 93%-porosity random fibers is probably the half-length sample test of August 2008.

Nominal specifications for the full- and half-length 93%-porosity samples are shown in Table 6.1.

6.1.1.2 Oxidation-Resistant Alloy Random Fiber (Ibrahim et al., 2009a, 2009b)

Sunpower Inc., and NASA GRC provided the ~22-micron oxidation-resistant fiber diameter and the material density, respectively. The sample characteristics are shown in Table 6.1.

TABLE 6.2

Polyester-Fiber Regenerator

Matrix length	20.3 mm
Matrix diameter	18.5 mm
Fiber material	Polyester
Measured fiber diameter	19–30 micron
Calculated porosity	0.85

6.1.2 Polyester-Fiber Regenerator

Global Cooling fabricated a regenerator similar to the above random-fiber regenerator except using polyester fibers and a different packing technique.

The matrix was assembled by stacking a number of disks into a container (not the final canister), pressing the entire stack to the desired length, then heating the assembly to stress-relieve the fibers so that they would retain their compressed length and diameter. This is similar to the sintering process used to prepare the sheet of Bekaert material, except that there is no evidence that the temperature was high enough to actually bond the fibers to each other, as is the case in the Bekaert sintering process.

As a result of sintering, the matrix shrank in diameter so it was a loose fit in the final canister (about 0.5 mm smaller diameter). To prevent "blow by" (see Section 6.4.3), the edge was sealed by a thin coating of silicone room-temperature-vulcanizing (RTV) adhesive applied to the canister inner diameter (ID) and matrix outer diameter (OD) prior to assembly. Another alternative would have been to oversize the sintering container by 0.5 mm so that the sintered matrix would be a tight fit in the canister. We did not try that approach. Dimensions of the matrix are as given in Table 6.2.

The fiber diameters are not uniformly the same. This is especially evident in the electron micrographs shown later. Figure 6.2 shows the assembled canister and a close-up view of the matrix where it joins to the canister.

6.1.3 Electron Microscope Fiber-Diameter Measurements

6.1.3.1 Stainless Steel and Polyester Fiber

NASA GRC supported the DOE regenerator research effort by photographing several regenerator matrices in their SEM (scanning electron microscope) facility. For this purpose, several random-fiber regenerator test samples were selected, including the above stainless-steel and polyester samples, and some early random-fiber samples that had been previously tested in the NASA/ Sunpower test rig. The goal was to more accurately measure fiber diameters than had previously been possible and to see details about fiber shape and

FIGURE 6.2
Completed polyester-fiber regenerator and close-up view showing edge sealing. Small subdivisions on the reticle scale is equal to about 33 microns.

matrix structures at high resolution. The procedure for determining fiber diameters from the SEM photos was as follows:

1. Print out full-page size SEM photos of appropriately magnified wires.
2. Select at random about 60 measurement locations for fibers that stand out in good resolution.
3. Number these locations by writing directly on the printouts.
4. Use a dial caliper to measure fiber diameters at each selected location and log them in an Excel spreadsheet file.
5. Within Excel, convert the measurements to microns (based on the scale superimposed at the bottom of the photos) and perform the calculations for mean effective wire diameter.

The formula for calculating the *mean effective* fiber diameter is:

$$d_e = \frac{\Sigma_i d_i^2}{\Sigma_i d_i} \tag{6.1}$$

where the d_i values are the sampled diameter measurements. The mean effective diameter is the diameter that makes the hydraulic diameter calculation come out right. In other words, a matrix of uniform wires of diameter d_e has the same hydraulic diameter (void volume/wetted surface) as the actual matrix, provided that the porosity is the same. Mean effective diameters so computed were different than the previously assumed wire diameters. The typical discrepancy was on the order of 10%, as can be seen in Table 6.3 (by comparing nominal diameters with measured mean effective diameters).

Figures 6.3 through 6.6 show representative micrographs for different random-fiber matrices.

TABLE 6.3

Matrix Mean Effective Diameter

Nominal	Number of Samples	Mean Diameter (μm)	Standard Deviation	Mean Effective Diameter (μm)
Bekaert 12 micron	60	13.3	0.96 (7.2%)	13.4
Brunswick 1 mil	60	27.0	3.1 (11.3%)	27.4
Brunswick 2 mil	60	52.4	2.0 (3.9%)	52.5
Polyester fibers (variable, 19–30 μm)	60	21.6	2.9 (13.6%)	22.0

6.1.3.2 Oxidation-Resistant Material Fiber

To double-check the oxidation-resistant fiber diameter, the SEM micrograph of Figure 6.7 was imported into Adobe Acrobat, and the diameter was measured at about 30 locations using the measurement tool. This gave a mean effective fiber diameter of 22.48 microns, confirming the above 22-micron value reasonably well. For data reduction purposes, it was assumed that the 22-micron diameter was correct. There were not enough Acrobat measurements to justify changing it.

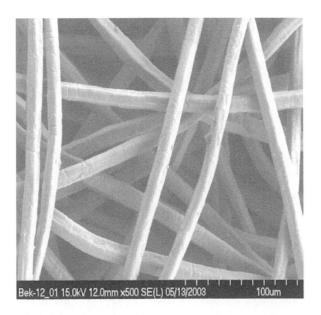

FIGURE 6.3
Bekaert nominal 12-micron round fibers, 316 stainless steel. The measured mean effective diameter was 13.4 microns. This was from a 90%-porosity regenerator matrix made and tested under the U.S. Department of Energy regenerator research program. (The micrograph is courtesy of NASA GRC.)

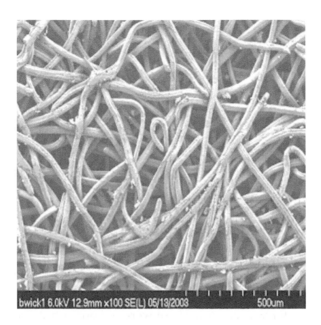

FIGURE 6.4
Brunswick nominal 1-mil stainless-steel fibers. The measured mean effective diameter was 27.4 microns. This was from a historical 82%-porosity regenerator matrix tested in the oscillating-flow regenerator test rig in 1992, under NASA funding. (The micrograph is courtesy of NASA GRC.)

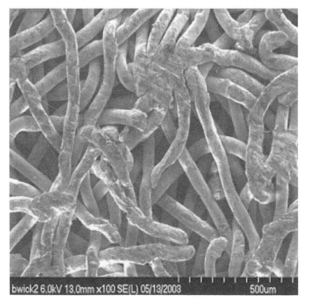

FIGURE 6.5
Brunswick nominal 2-mil stainless-steel fibers. The measured mean effective diameter was 52.5 microns. This was from a historical 69%-porosity regenerator matrix tested in the oscillating-flow regenerator test rig in 1992–1993, under NASA funding. (The micrograph is courtesy of NASA GRC.)

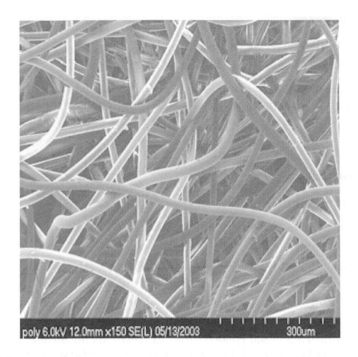

FIGURE 6.6
Polyester fibers. The measured mean effective diameter was 22 microns. It is from an 85%-porosity regenerator matrix made and tested under the U.S. Department of Energy regenerator research program. Matrix separation at the canister wall prevented completion of testing. (The micrograph is courtesy of NASA GRC.)

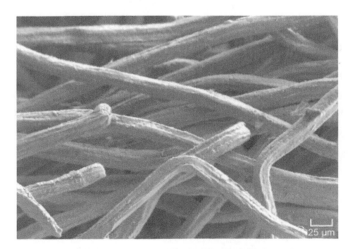

FIGURE 6.7
Scanning electron microscope (SEM) photo of oxidation-resistant fiber at 400x magnification. (Photo by Randy Bowman of NASA GRC. The micrograph is courtesy of NASA GRC.)

6.2 NASA/Sunpower Oscillating-Flow Test-Rig and Test-Rig Modifications

Testing of the regenerator samples was in the NASA/Sunpower oscillating-flow test facility at Sunpower Inc., Athens, Ohio. The test rig and modifications made to the test rig are described in Appendix A.

6.3 Random-Fiber Test Results

6.3.1 Bekaert 90%-Porosity Stainless-Steel Fiber Regenerator Test Results

The goal was to derive more accurate correlations for the 90% Bekaert regenerator. The derived correlations for the "12 micron" Bekaert material are reported below.

6.3.1.1 Friction-Factor and Heat-Transfer Correlations for 90%-Porosity Bekaert Stainless Steel (Ibrahim et al., 2004a, 2004b, 2004c)

6.3.1.1.1 Friction-Factor Correlation

The final recommended correlation for the "12 micron" 90%-porosity Bekaert regenerator was:

$$f = \frac{263}{R_e} + 5.83 R_e^{-0.151} \tag{6.2}$$

The range of key dimensionless groups for the tests is given in Table 6.4.

6.3.1.1.2 Heat Transfer Correlation

The final recommended correlations for simultaneous Nusselt number, N_u, and enhanced conductivity (thermal dispersion) ratio, N_k, derived for the

TABLE 6.4

Range of Dimensionless Groups Used for 90%-Porosity Bekaert Friction-Factor Correlation, Equation (6.2)

Peak R_e range	0.95–760
V_a range (Valensi number)	0.05–1.9
δ/L range (tidal amplitude ratio)	0.04–0.8

TABLE 6.5

Range of Dimensionless Groups for
90%-Porosity Bekaert Matrix Heat-
Transfer Correlation, Equation (6.3)

Peak R_e range	7.9–640
V_a range (Valensi number)	0.031–1.23
δ/L range (tidal amplitude ratio)	0.098–1

"12 micron" 90%-porosity Bekaert regenerator were as follows (see Section 8.4.1.4.5, "Heat-Transfer Correlations: Simultaneous N_u and N_k"):

$$N_u = 1 + 1.29 P_e^{0.60} \quad \text{and} \quad N_k = N_{k0} + 1.03 P_e^{0.60} \tag{6.3}$$

The range of key dimensionless groups for the tests is given in Table 6.5.

The final correlation for effective Nusselt number, assuming zero enhanced conductivity, derived for the "12-micron" 90%-porosity Bekaert regenerator was:

$$N_{ue} = 1 + 0.594 \, P_e^{0.73} \tag{6.4}$$

This correlation did not fit the data very well.

6.3.2 Bekaert 96%-Porosity Stainless-Steel Fiber Regenerator Test Results *(Ibrahim et al., 2009a, 2009b)*

The results presented below are only for the half-length samples (one tested in 2006, one tested in 2008—after some modifications had been made to the test rig)—because, as explained earlier in this section, the test rig could not resolve the low thermal losses of the long 96%-porosity regenerator sample. The 2006 half-length sample was retested in 2008, but some damage to the sample, and the results, led to the decision to eliminate those results from consideration. See Ibrahim et al. (2009a, 2009b) for more details.

Regarding the results below, the friction factors are quite close to each other. Regarding the differences in the two heat-transfer correlations, one problem is that the simultaneous reduction of N_u and N_k is sensitive to small errors in the data, and the data reduction process may be shifting the effects of thermal loss between the two mechanisms. So even though the b coefficients, in the heat transfer correlations to be discussed, vary quite a bit, the bottom-line implications for regenerator thermal losses do not appear to be very divergent. (See Ibrahim et al., 2009a, 2009b, for more information about these results.)

TABLE 6.6

Equation (6.5): Friction-Factor Coefficients for 96%-
Porosity Bekaert Stainless-Steel Half-Length Samples

Friction-Factor Coefficients	a_1	a_2	a_3
2008 half-length sample "B"	704.1	7.245	–0.131
2006 half-length sample "A"	633.1	7.506	–0.136

6.3.2.1 Friction-Factor and Heat-Transfer Correlations for 96%-Porosity Bekaert Stainless Steel

6.3.2.1.1 Friction-Factor Correlations

The form of the correlations for the 96%-porosity Bekaert stainless-steel half-length regenerator samples is:

$$f = \frac{a_1}{R_e} + a_2 R_e^{a3} \tag{6.5}$$

The coefficients for the friction-factor correlation of Equation (6.5) are given in Table 6.6.

6.3.2.1.2 Heat-Transfer Correlations

The forms of the correlation equations for simultaneous Nusselt number and enhanced conductivity (thermal dispersion) ratio derived for the 96%-porosity Bekaert stainless-steel half-length regenerator samples are:

$$N_u = 1 + b_1 P_e^{b2} \quad \text{and} \quad N_k = 1 + b_3 P_e^{b2} \tag{6.6}$$

The coefficients for the heat-transfer correlations of Equations (6.6) are given in Table 6.7.

TABLE 6.7

Equation (6.6): Heat-Transfer Correlation
Coefficients for 96%-Porosity Bekaert
Stainless-Steel Half-Length Samples

Heat-Transfer Coefficients	b_1	b_2	b_3
2008 sample "B"	4.22	0.545	0.866
2006 sample "A"	8.60	0.461	2.498

TABLE 6.8

Equation (6.5): Friction-Factor
Coefficients for 93%-Porosity Bekaert
Stainless-Steel Half-Length Samples

Friction-Factor Coefficients	a_1	a_2	a_3
2008 half length	466.3	5.710	−0.104
2008 full length	437.8	5.482	−0.103

6.3.3 Bekaert 93%-Porosity Stainless-Steel Fiber Regenerator Test Results (Ibrahim et al., 2009)

The definitive test for 93%-porosity random fibers is probably the half-length sample test of August 2008. (For further explanation of this, see the earlier 93% porosity discussion in this section.)

6.3.3.1 Friction-Factor and Heat-Transfer Correlations for 93%-Porosity Bekaert Stainless Steel

6.3.3.1.1 Friction-Factor Correlations

The form of the correlation equation is the same as for the previous samples (see Equation 6.5). The 93%-porosity sample coefficients for the friction-factor correlation of Equation (6.5) are given in Table 6.8.

6.3.3.1.2 Heat-Transfer Correlations

The form of the correlation equations for simultaneous Nusselt number and enhanced conductivity (thermal dispersion) ratio are the same as for the previous samples (see Equation 6.6).

The 93%-porosity sample coefficients for the heat-transfer correlations of Equation (6.6) are given in Table 6.9.

TABLE 6.9

Equation (6.6): Heat-Transfer-Correlation
Coefficients for 93%-Porosity Bekaert
Stainless-Steel Half-Length Samples

Heat-Transfer Coefficients	b_1	b_2	b_3
2008 half length	2.428	0.542	1.100
2008 full length	3.289	0.508	2.561

TABLE 6.10

Equation (6.5): Friction-Factor
Correlation Coefficients for
90%-Porosity Oxidation-Resistant
Sample

a_1	a_2	a_3
283	4.920	−0.109

6.3.4 Oxidation-Resistant Alloy 90%-Porosity Regenerator Test Results *(Ibrahim et al., 2009a, 2009b)*

The friction-factor and heat-transfer oscillating-flow rig correlation results below are for the 90%-porosity, 22-micron width, oxidation-resistant alloy regenerator sample.

6.3.4.1 Friction-Factor and Heat-Transfer Correlations for 90% Oxidation-Resistant Alloy Regenerator

6.3.4.1.1 Friction-Factor Correlation

The form of the correlation equation is the same as for the previous samples. See Equation (6.5). The 90%-porosity oxidation-resistant sample coefficients for the friction-factor correlation of Equation (6.5) are given in Table 6.10.

The ranges of key dimensionless groups for the flow-resistance oxidation-resistant sample tests are given in Table 6.11.

6.3.4.1.2 Heat-Transfer Correlation

The form of the correlation equations for simultaneous Nusselt number and enhanced conductivity (thermal dispersion) ratio are the same as for the previous samples (see Equation 6.6).

The 90%-porosity oxidation-resistant sample coefficients for the heat-transfer correlations of Equation (6.6) are given in Table 6.12.

The ranges of key dimensionless groups for the heat-transfer oxidation-resistant sample tests are given in Table 6.13.

TABLE 6.11

Range of Dimensionless Groups Used
for 90%-Porosity Oxidation-Resistant
Friction-Factor Correlation, Equation
(6.5) with Table 6.10 Coefficients

Peak R_e range	3.85–1460
V_a range (Valensi number)	0.19–6.9
δ/L range (tidal amplitude ratio)	0.06–0.63

TABLE 6.12

Equation (6.6): Heat-Transfer
Correlation Coefficients for
90%-Porosity Oxidation-
Resistant Sample

b_1	b_2	b_3
1.822	0.538	1.661

6.3.5 General, Porosity-Dependent, Friction-Factor, and Heat-Transfer Correlations (*Ibrahim et al., 2004a, 2004b, 2004c, 2009a, 2009b*)

The general (i.e., porosity dependent) correlations defined below are based on data from the following tests:

- 2008 tests of 30-micron, stainless-steel, half-length samples at 96% and 93% porosities, and a 22-micron ~90%-porosity sample made from an oxidation-resistant material
- 2006 tests of 30-micron Bekaert stainless-steel fibers at porosities ranging from 85% to 96%
- 2003 tests of 12-micron Bekaert stainless-steel fibers at 90% porosity
- 1992–1993 tests for Brunswick stainless-steel fibers at 69% and 82% porosities

6.3.5.1 Porosity-Dependent Friction-Factor Correlations

The general, porosity-dependent, friction-factor correlation is of the same form as for the above specific samples (i.e., the form of Equation 6.5). However, the coefficients in this equation are now variables of the form:

$$a_1 = 22.7x + 92.3$$

$$a_2 = 0.168x + 4.05$$

$$a_3 = -0.00406x - 0.0759$$

(6.7)

TABLE 6.13

Range of Dimensionless Groups Used for
90%-Porosity Oxidation-Resistant
Heat-Transfer Correlation, Equation (6.6)
with Table 6.12 Coefficients

Peak R_e range	3.14–1165
V_a range (Valensi number)	0.12–4.3
δ/L range (tidal amplitude ratio)	0.07–0.83

where the variable x is a function of porosity, β, as defined by:

$$x = \beta/(1 - \beta) \tag{6.8}$$

6.3.5.2 Porosity-Dependent Heat-Transfer Correlations

The general, porosity-dependent, heat-transfer correlations are of the same form as for the above specific samples (i.e., the form of Equations 6.6). However, the coefficients in these equations are now variables of the form:

$$
\begin{aligned}
b_1 &= (0.00288x + 0.310)x \\
b_2 &= -0.00875x + 0.631 \\
b_3 &= 1.9
\end{aligned}
\tag{6.9}
$$

where the variable x is the same function of porosity, β, as defined above in Equation (6.8), for the general porosity-dependent friction-factor correlation.

6.3.6 Polyester-Fiber Regenerator Test Results

Our polyester-fiber regenerator developed problems. The rig operator noticed a high-frequency pressure oscillation during heat-transfer testing. It seems some gaps opened up at the matrix wall, between the RTV sealant and the canister ID. Even though these gaps did not extend the full length of the matrix, they seem to have short-circuited the flow to some degree. The high-frequency pressure oscillations were probably caused by vibration of the flexible RTV surface at the matrix/wall interface.

Another indication that something was amiss, was that data reduction suggested large thermal loss at low Reynolds numbers compared to thermal losses in stainless-steel felt regenerators. Some of this could have been due to low signal (cooler heat rejection) compared to noise due to the lower than normal ΔT across the matrix (about 30°C compared to our usual 170°C). But it also could have been due to nonuniform flow caused by edge blow-by (see Section 6.4.3).

Although the results are preliminary, the friction factor came out somewhat higher than expected. We measured $f\,Re = 335 \pm 2.96$ compared to $f\,Re = 204 \pm 2.55$ for combined metal felt at 85% porosity, according to the above correlation. The discrepancy may have been due to the wide variation of fiber diameters in the sample which may be an indication of our failure to measure the mean diameter correctly. At this point, we abandoned further testing on the polyester-fiber regenerator.

6.4 Theoretical Investigations

During the 3-year effort of the DOE project, we completed several theoretical investigations concerning random-fiber regenerators. These were a study of the effects of porosity on regenerators, calculation of viscous and thermal eddy transports, investigation of edge blow-by, and documentation of regenerator instability. Below are summaries of these investigations:

1. The investigation into the effects of porosity for random-wire and woven-screen regenerators indicates that as the porosity increases, we obtain higher Nusselt numbers and lower drag coefficients. These results support the hypothesis that high porosity is beneficial for regenerator performance.

2. As for the effective thermal conductivity in the regenerator matrix, the ratio N_k was obtained. It may be defined as the ratio of the magnitude of total molecular plus eddy thermal conduction divided by the magnitude of molecular thermal conduction, or

$$N_k = 1 + \frac{c_p}{k} \frac{\overline{|(\rho v)'T'|}}{|\overline{T}|} \tag{6.10}$$

In the numerator of the above equation are some spatial averages, which can be computed experimentally and in our computational fluid dynamics (CFD) models. In our CFD modeling, we utilized this expression to estimate the effective axial thermal conduction (see Section 6.5).

3. The presence of the wall could produce an edge blow-by (see Section 6.4.3). From our preliminary observations, it appears that the flow resistance through a porous matrix may decrease near the wall boundary even though there may be no actual gap there. Even though decreased flow resistance is overall a good thing, it is bad if it occurs in only part of the regenerator matrix, because it tends to produce nonuniform flow.

4. Experimental evidence has suggested the existence of regenerator flow instabilities under certain conditions. Possible reasons for such instabilities were investigated (see Section 6.4.4).

6.4.1 Effects of Porosity on Regenerators

Discussions of the effects on enhanced thermal diffusion of changing matrix porosity led us to take a closer look at the available literature. What we found is interesting. The plots in Figures 6.8 and 6.9 compare two important dimensionless groups, Nusselt number and enhanced conductivity ratio, for packed beds, screens, and random fibers, as a function of porosity. Reynolds number, based on hydraulic diameter, is fixed at 100, which is in the range of typical Stirling engine and cooler practice and high enough that some interesting flow eddies should be present in the flow. Prandtl number is assumed to be 0.7.

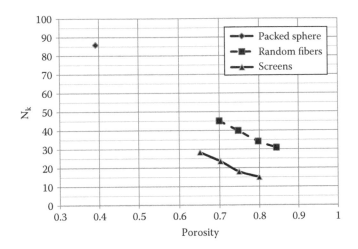

FIGURE 6.8
Enhanced conductivity for $Re_{dh} = 100$ versus porosity.

The evidence of Figures 6.8 and 6.9 seems to suggest a correlation of the Nusselt number and enhanced conductivity ratio dimensionless groups with porosity.

In the case of Nusselt number (a measure of the inverse of gas-to-matrix heat transfer resistance), most striking is the steep rise in Nusselt number with porosity for the screen and random-fiber data. In the case of enhanced

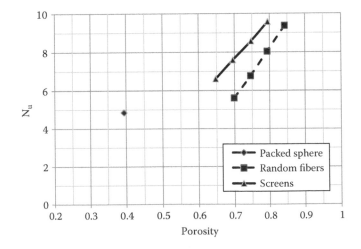

FIGURE 6.9
The Nusselt number for $Re_{dh} = 100$ versus porosity. Note that Re_{dh} in these two figures is the Reynolds number based on the hydraulic diameter.

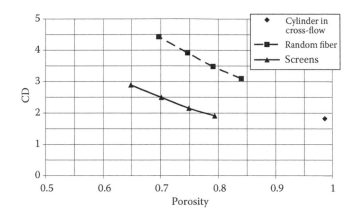

FIGURE 6.10
Drag coefficient for $Re_d = 100$ versus porosity. Re_d is the Reynolds number based on the wire diameter.

conductivity (a direct measure of potential for axial thermal conduction loss) the plot shows a definite inverse correlation with porosity, with the data for packed spheres and random fibers seeming to lie along the same smooth curve. That curve tends to approach zero at a porosity of one, as it should (nothing to produce flow eddies). Because high Nusselt number and low axial conduction are good, both curves tend to support the hypothesis that high porosity is beneficial for regenerator performance.

The plots referred to above are for matrices with porosity below 0.84. The plots of Figures 6.10 and 6.11 compare drag coefficient (measure of potential for flow resistance) and Nusselt number for random wire and woven screen matrices against the limiting case for a single wire in cross-flow, where porosity is 1.

Figure 6.10 gives more evidence that high porosity is good, because matrix flow resistance subtracts directly from engine power output. It is interesting that the drag coefficient for a single cylinder in cross-flow appears to lie at the precise point where the extrapolated drag coefficients for random-fiber matrices are heading. It would be surprising if this were not the case. After all, random fibers in these sorts of matrices are generally oriented transverse to the flow to a high degree, just like small cylinders in cross-flow. And as fiber spacing increases, they can be expected to behave more and more like isolated cylinders. The drag coefficients for screen matrices, however, are below those of random fibers. Perhaps this is due to the wavy nature of the wires in the weave pattern. They are not always transverse to the flow but generally lie more obliquely. Or maybe it has something to do with the flow channels through the weave pattern.

In the Nusselt-number plot of Figure 6.11, the value for a single cylinder in cross-flow lies well below the point where either random-fiber or screen

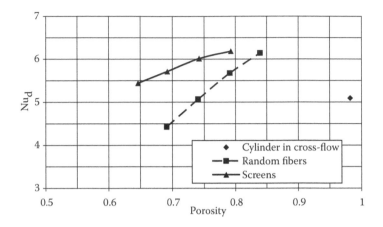

FIGURE 6.11
The Nusselt number for $Re_d = 100$ versus porosity. Re_d is the Reynolds number based on the wire diameter.

values are headed. This suggests that there is something interesting to learn about the range of porosities between 0.84 and 1.0, which is where we are focusing our efforts. Based on the drag-coefficient plot (Figure 6.10), one might guess that the random-fiber Nusselt-number data would approach the single-cylinder data at a porosity of unity. Note that the Nusselt number of Figure 6.11 is based on wire diameter, and the previous Nusselt number plot of Figure 6.8 is based upon hydraulic diameter.

6.4.2 Calculating Viscous and Thermal Eddy Transports

We spent considerable time discussing how to quantify enhanced thermal diffusion in porous materials—both in our experiments and in our CFD investigations. We were guided by a paper by Hsu and Cheng (1990), which discusses the relevant formulation for porous materials in incompressible flows. We extended this formulation to the case of compressible flow and included a formulation for measuring heat transfer and friction factor. We also developed recipes for computing the required spatial integrals of various quantities in curvilinear computational coordinate systems.

Physically, what we are talking about are microscopic flow structures within a porous regenerator that are impossible to resolve in a one-dimensional (1-D) computer simulation like Sage (Gedeon, 1999, 2009, 2010). Microscopic eddies transport momentum and thermal energy beyond what would be expected in laminar flows. We are trying to relate our microscopic measurements to the macroscopic concepts a program like Sage uses to model regenerator flows. These concepts are the friction factor, f, Nusselt number, N_u, and enhanced conduction ratio, N_k.

For example, the subgrid scale eddy transport of thermal energy (eddy thermal conductive heat-flux) in a compressible ideal gas may be written as:

$$c_p \overline{(\rho v)' T'} \tag{6.11}$$

where the overbar denotes spatial average (not time average) over a representative macroscopic volume element and the primed quantities denote local deviations from the spatial average. The product of density and velocity are grouped together, as in $(\rho v)'$, to represent the mass flux vector, which is considered to be a fundamental quantity.

It is convenient to lump the mean molecular conduction $\overline{q} = -k \ \overline{\nabla T}$ and eddy conduction together into an effective overall heat-flux vector defined by:

$$q_e = \overline{q} + c_p \overline{(\rho v)' T'} = k_e \ \overline{\nabla T} \tag{6.12}$$

where k_e is an overall effective conductivity. Under these provisions, the effective thermal conductivity ratio, N_k, may be defined as the ratio of the magnitude of total molecular plus eddy thermal conduction divided by the magnitude of molecular thermal conduction—as shown in Equation (6.10).

In the numerator of Equation (6.10) are some spatial averages, which can be computed experimentally and in our CFD models.

Experimentally, our plan was to evaluate the numerator in the second term of the right-hand side of Equation (6.10) by averaging over the exit plane of the matrix (farthest from the tube entrance). We can evaluate the denominator by differencing the average temperature at the exit plane with that of a plane some distance within the matrix.

In our CFD modeling, we can, in principle, average over any plane in the regenerator, except that we are unable to model more than a few wires in cross-flow at the present time. For the cases we can model, we must sort out over which regions to average and which temperature gradients correspond to the mean axial temperature gradient in a complete regenerator. Depending on the scale of the volumes over which we average, there may also be significant (even dominant) mean temperature gradients corresponding to the local variations around the wires or their wakes. In any event, of interest will be whether the eddy thermal transport will always be proportional to the local mean temperature gradient. Or, if not, what the tensor equivalent of the effective conductivity, k_e, might be.

We developed similar formulations for calculating Nusselt number and friction factor, the two other quantities of interest in macroscopic computational models. We will not go into the mathematical details because they are probably not of general interest.

6.4.3 Edge Blow-By

We also became aware of the possibility of matrix edge blow-by as a significant loss mechanism in regenerators. Based on preliminary observations, it appears that the flow resistance through a porous matrix may decrease near the wall boundary—even though there may be no actual gap there. For example, in the Minnesota test matrix (see Figure 7.1) the wires extend all the way to the wall with any gap much smaller than the spacing between wires. Yet preliminary evidence showed increased flow near the wall. Even though decreased flow resistance is overall a good thing, it is bad if it occurs in only part of the regenerator matrix because it tends to produce nonuniform flow.

Possibly, the local suppression near the wall of the flow eddies present in the matrix interior is the cause. This would tend to decrease the local viscous and thermal eddy transports near the wall, reducing apparent friction factor and thermal diffusion. If further evidence supports this view, then it could be an important discovery. This would be especially important if we can develop ways to eliminate the problem. Roughening the wall surface so that near-wall eddies are more similar in size to flow eddies and gradually reducing matrix porosity near the wall are two possibilities.

Edge blow-by was later made the subject of an experimental study at the University of Minnesota. See a discussion of this study, and the results, in Section 7.2.6.

6.4.4 Regenerator Instability Investigation

Large-scale Stirling engines and coolers often do not perform as predicted by one-dimensional models. Typically, the length of components, like the regenerator, are similar to smaller machines, but diameters and cross-sectional areas are much larger. The flow between the expansion and compression spaces tends to take the path of least resistance making it difficult to distribute flow to the outermost reaches of the regenerator or heat exchangers. But there may be other problems as well. Reports from the field sometimes suggest unexplained flow phenomena within the regenerator. These flows resemble convection cells. Often the evidence is based on temperature measurements on the outside surface of the regenerator pressure wall. The temperature distribution is not uniform and may go unstable with time. There are several possible reasons for these observations, ranging from poor flow distribution at the two ends of the regenerator to nonuniformities of the matrix structure, including actual movement of the porous material leading to progressive deterioration.

But there may be other, more subtle, reasons for flow instability rooted in the fact that zero-mean oscillating mass flows do not always produce zero-mean pressure drops. In viscous flows, pressure drops are dependent on velocity shear rates, whereas mass flows are dependent on the product of density and velocity. So variations in fluid density over the course of an oscillating-flow cycle may cause a net mass imbalance even though pressure drops are equal in the two directions. Or if that is prevented by the lack of a return path for the surplus mass flow, there may arise a time-average pressure bias from one end of the flow path to the other. This subject is the topic of a Gedeon paper (1997) describing the potential of DC flow loops in certain types of cryocoolers employing closed-loop flow paths.

To make some headway in understanding the problem, consider a regenerator consisting of parallel plates, divided into two parts of equal flow area. Subject the two parts to equal but opposite perturbations in flow gap and ask what happens in the context of an operating Stirling cycle. Unpublished memoranda (Gedeon, 2004a, 2004b) derive the equation governing DC flow for such a two-part, parallel-plate regenerator in detail (also see Appendix D). The procedure is briefly summarized here. Begin with the equation governing Darcy flow for a parallel-plate regenerator:

$$u = -\frac{g^2}{12\mu}\frac{\partial P}{\partial x} \tag{6.13}$$

where u is the section-mean velocity, g is the flow gap (see Figure 8.27), μ is the viscosity, and $\partial P/\partial x$ is the axial pressure gradient. Then assume that both velocity and density are the sum of a constant plus phasor part (sinusoidally varying), $u = u_m + \vec{u}$, $\rho = \rho_m + \vec{\rho}$. It was concluded after very little math that the DC flow per unit flow area may be written as follows:

$$(\rho u)_{dc} = \rho_m u_m + \tfrac{1}{2}\vec{\rho}\cdot\vec{u} \tag{6.14}$$

where $\vec{\rho}\cdot\vec{u}$ denotes the usual dot product for vectors. Use a linear superposition principle to apply the Darcy flow Equation (6.13) to both the mean and phasor velocity components, giving:

$$(\rho u)_{dc} = -\frac{g^2}{12\mu}\left(\rho_m \frac{\partial P_m}{\partial x} + \tfrac{1}{2}\vec{\rho}\cdot\frac{\partial \vec{P}}{\partial x}\right) \tag{6.15}$$

Then make a few assumptions in order to conclude that there is a point, roughly the regenerator midpoint, where the local pressure gradients may be replaced by regenerator averages—that is, $\partial P/\partial x \approx P/L$, where L is regenerator length. Denote that point by subscript c, and write the previous equation as:

$$(\rho u)_{dc} \approx \frac{g^2}{12\mu_c L}\left(\rho_c \ P_m + \tfrac{1}{2}\vec{\rho}_c \cdot \ \vec{P}\right) \tag{6.16}$$

Apply this equation to the two regenerator parts, denoting one part by subscript A and the other by subscript B; then add the equations. In order to conserve mass, the left-hand side must sum to zero (assuming there is no DC flow beyond the regenerators). Solve the resulting equation for DC pressure drop ΔP_m and then substitute the result back into the equation. Replace phasor density variation $\tilde{\rho}$ with an approximation based on the gas energy equation. Also assume that the temperature distribution in the regenerator is approximately linear, that an ideal-gas equation of state applies, and that gas viscosity and conductivity vary as $T^{0.7}$ (reasonably accurate for helium in the range of 10° to 1000°K). At this point, the equation gets messy, but it can be simplified considerably if one is willing to assume a few things. First, assume that the regenerator pressure-drop phasor is 90° out of phase with the pressure phasor (pressure and velocity in phase). Second, assume that the regenerator compliance (volume) is small. The first assumption is reasonable because optimal regenerator performance requires that pressure and velocity fluctuations are roughly in phase (due to the low Valensi number of the flow oscillations). The second assumption is equivalent to saying that velocity phase does not vary much through the regenerator; this is not a very good approximation but it will have to do. The final oversimplified but instructive equation governing DC flow in regenerator A is:

$$\frac{(\rho u)_{dc}}{(\rho u)_1} \approx \frac{\gamma P_m}{48(\gamma-1)N_u k_r \mu_r L} \frac{P_1 T_r^{1.4}}{\partial x} \frac{\partial T}{\partial x}\left[\frac{g^4}{T_c^{3.4}}\right]_B^A \qquad (6.17)$$

where the $[\cdot]_B^A$ notation means evaluate the difference $(\cdot)_A - (\cdot)_B$. The quantity on the left is the ratio of the DC mass flux to the amplitude of the oscillatory mass flux for the baseline unperturbed regenerator. Quantities on the right are defined in the Nomenclature.

A similar equation governs the equal and opposite DC flow in regenerator B. Note that the DC flow depends strongly on the gap difference between regenerator parts (g^4 dependence) and that the direction of DC flow reverses, depending on the sign of the temperature gradient. For a cryocooler, where $\partial T/\partial x$ is negative, a positive gap perturbation in regenerator part A results in a negative DC flow. For an engine, where $\partial T/\partial x$ is positive, a positive gap perturbation results in a positive DC flow. In either case, a positive gap perturbation produces DC flow from the colder end toward the warmer end of the regenerator. This DC flow combined with the regenerator temperature gradient tends to remove heat from the interior of the regenerator, perturbing the temperature distribution so that it is colder in the part with the larger gap.

Also to note in the last equation is that DC flow depends roughly on the central regenerator temperature raised to the −3.4 power. This means that the drop in central regenerator temperature noted in the previous paragraph tends to amplify the DC flow. The exact gain factor for the amplification is difficult to evaluate analytically. The heat removal produced by DC flow is counteracted by other thermal-energy transport mechanisms in the regenerator that include gas conduction, solid conduction within the regenerator foil, time-averaged enthalpy flux produced by the regenerator AC flow, and any transverse thermal conduction between the two regenerator parts. All of these effects tend to restore the temperature distribution to its baseline roughly linear state.

6.5 Computational Fluid Dynamics (CFD) Simulation for Cylinders in Cross-Flow

6.5.1 Flow and Thermal Field

In this section, CFD analyses were conducted for a single cylinder in cross-flow at $Re = 100$ and for different cylinder shapes (circular, square at $0°$ and $45°$ incidence, elliptic at $90°$ incidence, and diamond at $90°$ incidence). The goal is to get a relative magnitude of the axial thermal dispersion (see Equation 6.10) for a simplified geometry, single cylinder in cross-flow.

In cross-flow over a cylinder, the cylinder disturbs the flow not only in the vicinity, but also at large distances in all directions. At $Re = 100$, the vortices grow until a limit at which the wake becomes unstable. At this Reynolds number, which is above the critical Reynolds number, the viscous dissipation is too small to ensure the stability of the flow.

The critical Reynolds number is defined as the Reynolds number after which the vortex starts shedding from the cylinder. (It does not remain attached to the cylinder.) The critical Reynolds number for a circular cylinder is $Re_{cr} \cong 46$ (Lange et al., 1998), for square cylinder at $0°$ incidence $Re_{cr} \cong 51.2$ and for a square cylinder at $45°$ incidence $Re_{cr} \cong 42.2$ (Sohankar et al., 1998). The vortices are shed off by the main flow, forming a well-known von Kármán vortex street.

Figures 6.12 and 6.13 are taken from animation files generated in the CFD-ACE+ Code (for more details, see Mudaliar, 2003). Figures 6.12 and 6.13 show the velocity vectors and temperature contours over a cycle for the circular geometry only. The cycle time is derived from the time period of the fluctuating lateral forces (lift forces), which are responsible for vortex shedding. Figure 6.12 shows the velocity vectors and stream function over a cycle that is divided in five parts. From the figure, it is clear that there is vortex shedding from the rear end of the cylinder. Alternate vortex shedding has been observed from the top and bottom end of the cylinder for all the geometries. The size of the wake changes with change in geometry, and it is maximum for elliptic and diamond cylinders at $90°$ incident angle. Figure 6.13 shows the temperature contours for the circular cylinder geometry; it clearly shows that there is temperature difference in only the wake behind the cylinder. Alternate vortex shedding is also observed in temperature contours from the top and bottom of the cylinder. The vortex shedding frequency in temperature contours is out of phase with velocity contours. The heat transfer in this fluctuation is limited, despite the violent wake observed in the flow field.

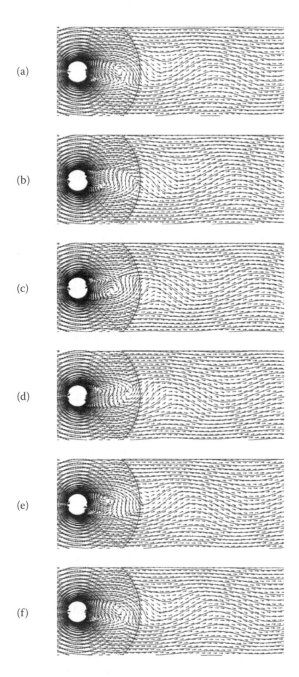

FIGURE 6.12
(a) Velocity vector and stream-function contours at 0-cycle time. (b) Velocity vector and stream-function contours at 1/5-cycle time. (c) Velocity vector and stream-function contours at 2/5-cycle time. (d) Velocity vector and stream-function contours at 3/5-cycle time. (e) Velocity vector and stream-function contours at 4/5-cycle time. (f) Velocity vector and stream-function contours at 1-cycle time.

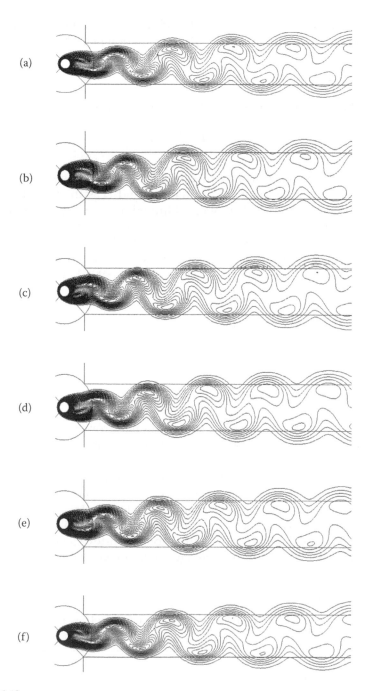

FIGURE 6.13
(a) Temperature contours at 0-cycle time. (b) Temperature contours at 1/5-cycle time. (c) Temperature contours at 2/5-cycle time. (d) Temperature contours at 3/5-cycle time. (e) Temperature contours at 4/5-cycle time. (f) Temperature contours at 1-cycle time.

This is attributed to the low flow field velocities encountered in the wake region and the domination of molecular conduction near the cylinder wall.

6.5.2 Strouhal Number Calculations

The Strouhal number is defined in the following form:

$$St = \frac{L}{U\tau} \qquad (6.18)$$

where τ is time period of vortex shedding in seconds, L is the characteristic length (e.g., cylinder diameter) in meters, and U is the velocity in m/s. Table 6.14 shows the Strouhal numbers obtained from this study for different geometries. The CFD values compare very well with data available for circular cylinders (Franke et al., 1995; Lange et al., 1998; Norberg, 1993; Zhang and Dalton, 1998). As for the square cylinder cases (0 and 45° incidence), the CFD data compare very well with that from Knisely (1990), Franke et al. (1995), Okajima et al. (1990), and Sohankar et al. (1995, 1998).

6.5.3 Nusselt Number Calculations

The Nusselt number is defined as:

$$N_u = \frac{hL}{k} \qquad (6.19)$$

where h is the heat transfer coefficient as obtained from $\dot{q}_w = h(T_w - T_\infty)$. Table 6.14 shows the average Nu over the circumference of different cylinder geometries.

The result of the present study for a circular cylinder was in agreement with the computational study of Lange et al. (1998). Table 6.14 indicates that the average Nu over the cylinder is nearly the same for elliptic and diamond cylinders at 90° incidence, probably because the Strouhal number is also nearly the same. This means that Nu is a function of vortex shedding frequency (Strouhal number). Similarly, the average values of Nu are close for a circular

TABLE 6.14

Strouhal and Nusselt Numbers for Different Geometries at $Re = 100$

Geometry	Strouhal Number	Nusselt Number
Circular	0.1656	5.1904
Square at 0° incidence	0.1311	3.5133
Square at 45° incidence	0.1656	5.5949
Elliptic at 90° incidence	0.2266	6.9539
Diamond at 90° incidence	0.2098	6.8692

and a square cylinder at 45° incidence, both of them having the same Strouhal number at $Re = 100$. The average Nu is at its lowest for a square cylinder at 0° incidence. This square cylinder also has the lowest Strouhal number.

6.5.4 Axial Thermal Dispersion

In this study, the computational domain was chosen to be 40 L (L is the characteristic length, the cross-section width) in the stream-wise direction and 20 L in the cross-stream direction. This domain was chosen to be large enough to enable obtaining fluctuating velocities and temperatures in the flow stream. Of particular interest in this investigation is the effect of porosity on thermal dispersion (enhanced thermal conduction). Therefore, we decided to use a simple approach (to at least show the trend of the results) by focusing on three boxes (see Figure 6.14) surrounding the cylinder. The dimensions of these boxes were chosen to provide porosities of 0.85, 0.90, and 0.95, respectively.

For calculating the fluctuating velocities and temperatures, the following method was adopted: Taking, for example, porosity 0.9 (the middle box in Figure 6.14), the side of the box is calculated to be equal to 2.8 L. We then take a node on the east face of the box (i.e., downstream), compute the spatial average velocities, U and V, and average temperature in a Y-domain 1.4 L above and 1.4 L below this node. This provides an average quantity in space with the

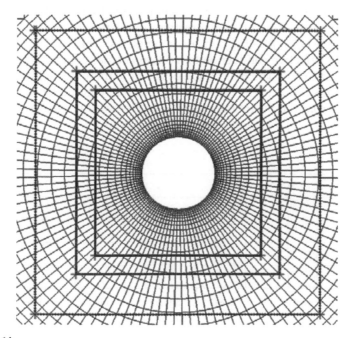

FIGURE 6.14
The computation domain near the circular cylinder. The three boxes show three different domains chosen to represent porosities: 0.85 (inner one), 0.9, and 0.95.

same dimension as 2.8 L. We take the spatial average, for the other nodes that are located beyond the dimension 2.8 L (1.4 L above and 1.4 L below). Their values are mirror reflections of the values at the node locations within the original 2.8 L dimension. The idea behind taking the average this way is to give a more realistic picture, as if another cylinder is located above and below this cylinder. The fluctuating quantities were then calculated from the difference between the instantaneous value and the corresponding average value just obtained. The time average values for fluctuating components were taken by averaging the value over the vortex shedding cycle. Similar computations were made based at various nodes along the east face of the 0.9 porosity box. This was repeated for the left edge of the box (upstream) where, as expected, the fluctuating values are virtually zero. Now we can calculate the product of $\rho C_p U' T'$ which is the enhanced axial thermal conduction (where ρ is density, C_p specific heat of the fluid, U' is the fluctuating component of the axial velocity, and T' is the fluctuating component of temperature). To repeat, the fluctuating components for a particular location are obtained by subtracting the spatial average velocity component from the corresponding local velocity component (e.g., $U' = U - \langle U \rangle$), and the spatial average velocity component is obtained by taking an average from 1.4 L above to 1.4 L below the particular location along the Y direction on the downstream edge of the porosity box. A similar procedure is applied for calculating $\langle T \rangle$ and T'. From the local mean temperature, we can calculate the molecular thermal conduction and then obtain the ratio of the effective thermal conduction to the molecular one.

We evaluated the ratio of axial effective conduction to molecular conduction $(\rho C_p U' T' + (-k \frac{\partial T}{\partial x}))/(-k \frac{\partial T}{\partial x})$ in all geometries. The values of averaged ratios of axial effective conduction to axial molecular conduction are shown in Figures 6.15 to 6.17. The magnitude of the ratio increases with an increase in porosity; this is due to the decrease in molecular conduction with increase in porosity. From the Stirling engine point of view, this quantity should be

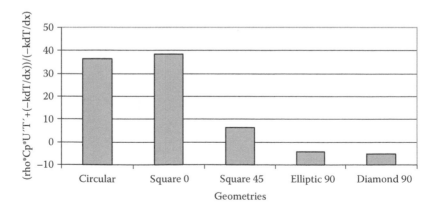

FIGURE 6.15
Averaged ratio of axial effective conduction to molecular conduction for porosity is 0.85.

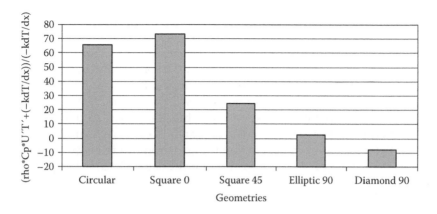

FIGURE 6.16
Averaged ratio of axial effective conduction to molecular conduction for porosity is 0.90.

as small as possible in order to improve the efficiency by way of minimizing the axial conduction losses. The diamond and the elliptic cylinder at 90° incidences have negative values at porosity of 0.85, and the diamond and 90° incidence also has negative values for porosity at 0.90 and 0.95. The negative values of the ratio might be advantageous.

6.6 Concluding Remarks and Summary of Experimental Correlations

Chapter 6 summarized test results of actual-scale random-fiber regenerators and results of CFD simulations of cylinders in cross-flow. The random-fiber regenerators were tested in the NASA/Sunpower oscillating-flow test rig on

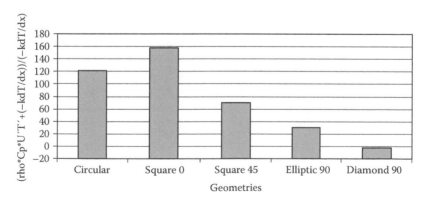

FIGURE 6.17
Averaged ratio of axial effective conduction to molecular conduction for porosity is 0.95.

loan to Sunpower in Athens, Ohio. Friction-factor and heat-transfer correlations developed from this oscillating-flow-rig test data are used in the Sage computer design code, for example, in designing the Stirling devices that are used by Sunpower, NASA, and others. The latest stainless-steel porosity-dependent random-fiber correlations, developed from oscillating-flow rig data taken through 2008, are summarized by Equations (6.20) through (6.22). The development of these equations, and the data used to develop them, were discussed earlier in this chapter, and are also discussed in Appendix I; the data that were used in developing the equations were based on oscillating-flow rig testing of stainless-steel random-fiber matrices with porosities ranging from 0.69 to 0.96.

Porosity-dependent friction-factor correlation (same form as Equation 6.5):

$$f = \frac{(22.7x + 92.3)}{Re} + (0.168x + 4.05) \cdot Re^{(-0.00406x - 0.0759)} \tag{6.20}$$

Porosity-dependent heat-transfer correlations (same form as Equation 6.6):

$$N_u = 1 + (0.00288x^2 + 0.310x) \cdot Pe^{(-0.00875x + 0.631)}$$
$$N_k = 1 + (1.9) \cdot Pe^{(-0.00875x + 0.631)} \tag{6.21}$$

where the parameter, x, in the above equations is a function of porosity, β, as in the following equation:

$$x = \frac{\beta}{1 - \beta} \tag{6.22}$$

For a brief discussion of the relationship between the dimensionless heat-transfer parameters, Nusselt number, N_u, and enhanced conductivity ratio, N_k, see Section 8.4.1.4. Coefficients for friction-factor and heat-transfer correlations for a single-porosity (90%) oxidation-resistant random-fiber were given in Section 6.3.4.1 of this chapter.

The purpose of this DOE- and NASA-funded regenerator research was to determine how to improve the design and performance of Stirling regenerators. This random-fiber/wire-screen–based research led to the development of the segmented-involute-foil regenerators discussed in Chapters 8 and 9.

7

Random-Fiber Regenerator—Large Scale

7.1 Introduction

Much of the material reported in this chapter is based on editing of information from a final report to the U.S. Department of Energy (DOE) (Ibrahim et al., 2003). That DOE effort was aimed at improving performance of Stirling devices (engines in particular) via better understanding of regenerator losses, so that Stirling regenerator performance could be improved. The effort was aimed, primarily, at the widely used random-fiber regenerators, even though the large-scale test specimens were stacked wire screens. Because it was decided that insights into regenerator operation could be enhanced by measurements within the regenerator matrices, it was necessary to scale up the geometries sufficiently to enable measurements within large-scale matrices. Scaling of the geometries also required scaling of the experimental parameters, in order to maintain the magnitude of the important dimensionless parameters close to those expected in the Stirling engines of interest.

The goals of the DOE large-scale regenerator test effort were to reveal fundamentals of fluid mechanics and thermal behavior of oscillatory flow within a regenerator matrix by experimental measurements and to improve the development of advanced computational fluid dynamics (CFD) models and design rules for the next generation of Stirling regenerators and engines. The regenerator fundamentals learned should also apply to the regenerators of Stirling coolers. However, the regenerator testing was done for porosities of ~90%, which are characteristic of modern small Stirling engines; Stirling cooler regenerator porosities are usually much lower than this.

7.2 Major Aspects and Accomplishments of the Large-Scale Regenerator Test Program

7.2.1 Construction of the Test Section

In order to simulate the fluid mechanics and heat-transfer characteristics, a representative Stirling engine was chosen for selecting the test section geometry and operating parameters. Application of dynamic similitude to develop a mock-up of this representative engine for test purposes is documented by Niu et al. (2002). The mock-up regenerator is shown in Figure 7.1, the drive for the oscillatory flow is shown in Figure 7.2, and the other components of the test facility are shown in Figure 7.3. Table 7.1 shows dimensions of the University of Minnesota (UMN) test facility of importance for CFD calculations, which are discussed later in Chapter 7. A discussion of dynamic similitude in general and with reference to Stirling engines, including the introduction of a number of dimensionless parameters used for general dynamic simulation of systems and discussions of their physical importance, is presented in Appendix B.

7.2.2 Jet Spreading into the Regenerator Matrix

Design rules suggest that the thickness of each plenum between the regenerator and the heat exchangers, cooler, or heater, within a Stirling engine influences the engine performance. This is, probably, by virtue of the effects it has

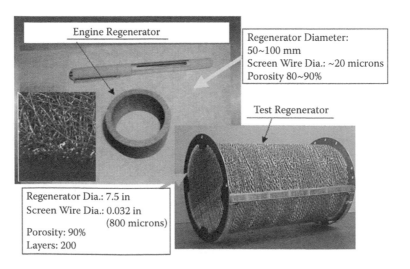

FIGURE 7.1
An engine regenerator and the large-scale mock-up; the test regenerator at the University of Minnesota.

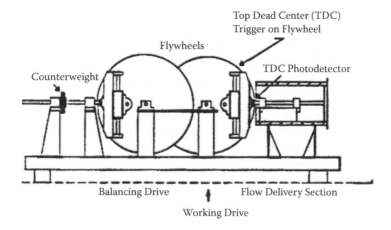

FIGURE 7.2
The oscillatory-flow generator. (From Seume, J.R., Friedman, G., and Simon, T.W., 1992, Fluid Mechanics Experiments in Oscillating Flow. Volume I—Report, NASA Contractor Report 189127.)

on depth of penetration of the jets that emerge from the heat exchanger and penetrate into the regenerator matrix. This penetration depth is the depth to which the jets of higher-speed fluid from the discrete heat exchanger tubes enter the matrix before they have merged to give a uniform flow. Also of interest is the fraction of the matrix material that is not participating fully in the heat transfer process within that depth of penetration. There is a finite length required for the jets to spread and fill the entire flow area of the matrix. The study was conducted to show the effects of the plenum thickness on jet

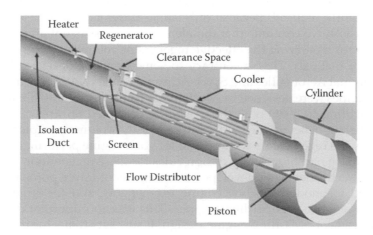

FIGURE 7.3
The cylinder (right), distributor, cooler, regenerator, heater, and a duct that isolates the test fluid from the ambient.

TABLE 7.1

Dimensions of the University of Minnesota
(UMN) Test Rig (Used in Computational Fluid
Dynamics Calculations Discussed in Section 7.2.7)

Piston diameter	14 inch (355.6 mm)
Piston stroke	14 inch (355.6 mm)
Regenerator	
Diameter	7.5 inch (190 mm)
Length	12.8 inch (325 mm)
Matrix porosity	0.9
Cooler	
Tube diameter	0.75 inch (19 mm)
Tube length	35 inch (888 mm)
Tube spacing	2.2 inch (55.8 mm)
Tube number	9
Heater	
Duct length	48 inch (1218 mm) (changed to 2020 mm later)
Duct diameter	7.5 inch (190 mm)
Isolation Duct	
Length	12 inch (300 mm)
Diameter	7.5 inch (190 mm)

penetration using three different plenum sizes. It appears that for all three cases, the jets merge and lose their individual identities at an axial distance equal to 3.33 times the hydraulic diameter of the regenerator matrix. As the plenum thickness increases, the fraction of the material that is ineffective is only slightly reduced. Because only small differences are observed for the three cases, we conclude that the plenum thickness is not an important parameter for modifying the effective regenerator volume at the ends of the regenerator matrices. This is contrary to some previous design rules. The DOE/National Aeronautics and Space Administration (NASA) Stirling automotive engines were designed by Mechanical Technology, Inc. (Albany, New York) (Nightingale, 1986) with the assistance of United Stirling of Sweden. United Stirling believed the configurations of the volumes, or plenums, at the ends of the regenerators were critical for good engine performance. And, United Stirling had proprietary design rules for these regenerator plenums. However, such rules were developed for matrices of lower porosities than those used for modern Stirling engine regenerators, and for large, crank-driven, automotive-type Stirling engines. This was an important discovery for improvement of Stirling engine design; it suggests that—for these relatively small, high-regenerator porosity, free-piston Stirling engines, at least—plenum thickness should be small to minimize the negative impact of dead volume on engine performance.

The effects of thermal dispersion are estimated based upon thermal measurements of the jet spreading within the regenerator matrix. A model for thermal dispersion was developed. It was found that the effective spreading of these jets was captured with a very large dispersion coefficient, larger than published values (Schlichting, 1979) that were developed from experience with porous matrices of lower porosity and different geometries than those presently in use for modern Stirling engines. This also is an important finding. Our model was formulated for general use but is recommended specifically for flow conditions and geometries representative of the interface between the Stirling engine heat exchanger and high-porosity regenerator matrix. It has been verified only for high-porosity matrices (i.e., ~90%), representative of modern small engines. More details are presented in Niu et al. (2003a, 2003c).

7.2.3 Evaluations of Permeabilities and Inertial Coefficients of the Regenerator Matrix

In support of computational work, the permeabilities (used in the porous medium flow equation, Darcy term) and the inertial coefficients (used in the porous medium flow inertial term) of the regenerator matrix were obtained by fitting measured pressure drops at various velocities through the matrix into the Forchheimer-extended Darcy's equation. The results were consistent with published values. Useful to the analysis and engine design is the transition Reynolds number range, from the Reynolds number at which the inertial term begins to become significant to the Reynolds number at which the inertial term dominates the Darcy term. See the discussion in Chapter 9, under Section 9.4 (Section 9.4.4, in particular).

7.2.4 Unsteady Heat-Transfer Measurements

Ensemble-averaged and time-resolved flow and matrix solid component temperatures and solid-to-fluid temperature differences were taken at the radial centerline of the matrix at the 60th, 80th and 100th layers of screen within the 200-layer deep matrix under oscillatory flow conditions. Instantaneous, unsteady heat-transfer coefficients were calculated from the temperature history of the solid component and the energy equation applied to a single screen wire of the matrix. Unsteady Nusselt numbers were obtained and compared to values computed from correlations for woven-screen matrices using the assumption of quasi-static flow and heat transfer. Comparisons show agreement during the deceleration part of the flow oscillation cycle, from 90° to 180° and from 270° to 360°. During acceleration, a significant disagreement between the unsteady and quasi-steady flow results was observed. See Niu et al. (2003b) for more details on the unsteady heat-transfer measurements.

It is hypothesized that because the flow is not hydrodynamically and thermally in equilibrium during the acceleration portion of the cycle, the measured flow temperature is not representative of the temperature of the convective film immediately surrounding the screen wire. In the deceleration portion of the cycle, the effective eddy transport mixes the flow within a pore of the matrix.

7.2.4.1 Effects of Flow Transition (Laminar-to-Turbulent) on Unsteady Heat-Transfer Measurements within the Regenerator Matrix

7.2.4.1.1 Background

Our heat-transfer coefficient measurements have shown that at the beginning of each half-cycle, from 0° to about 40°, the Nusselt number, Nu, is negative and approaches negative infinity (see Figure 7.4). This is due merely to a phase shift between the time of heat flow reversal and time of temperature difference reversal, as sensed by two thermocouples within the matrix, one measuring the regenerator-solid-material temperature and the other measuring the fluid temperature within a pore that is adjacent to the solid material thermocouple. At a crank angle of about 40°, the Nu number jumps to positive infinity as a result of the nonzero heat-transfer rate (see Figure 7.5) and the zero temperature difference (see Figure 7.6). Afterward, the Nu number falls until it reaches zero at a crank angle of about 180°. During the second half of the cycle, from 180° to 360°, the Nu number repeats the same trend.

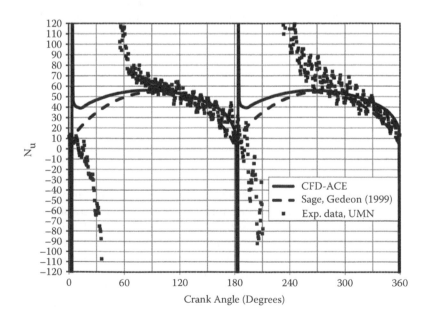

FIGURE 7.4
Instantaneous Nusselt numbers, Nu; measured values and values computed assuming unidirectional flow at the instantaneous velocity using two different correlations.

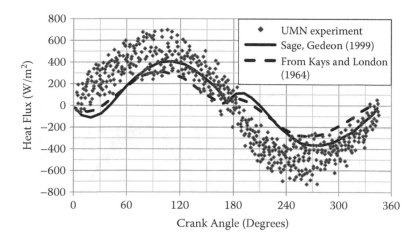

FIGURE 7.5
Instantaneous heat-transfer rates; measured values and values computed assuming unidirectional flow at the instantaneous velocity using two different correlations.

These trends give rise to the following question: What is happening from the fluid mechanics point of view during the period when the crank angle range is from 0° to 40°?

7.2.4.1.2 Theory

To answer this question, we take a close look at the instantaneous velocity within the plenum between the cooler and the regenerator. We can gain access to this region with a hot-wire probe.

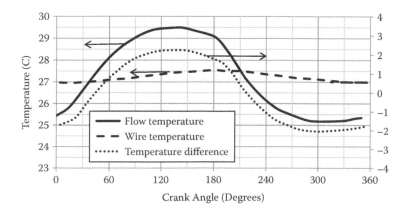

FIGURE 7.6
Instantaneous temperatures and temperature difference at the axial center of the matrix and at the 80th layer of screen (200 layers, total).

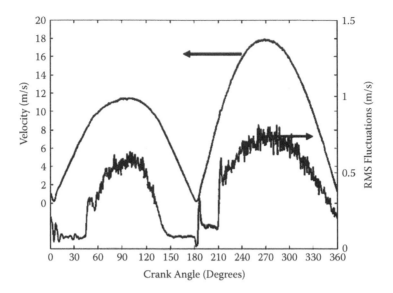

FIGURE 7.7
Instantaneous velocity and rms of the velocity ($\sqrt{u'^2}$) at the centerline of a cooler tube within the cooler-to-regenerator plenum. Taken at $Va = 2.1$.

Figure 7.7 shows the instantaneous velocity and rms fluctuation at the centerline of one cooler tube within the plenum between the cooler and the regenerator matrix under the same flow conditions as those for the above heat-transfer measurements. During the first half of the cycle, the flow approaching the probe is traversing the plenum space and is accelerating toward the cooler channel hole, having just come through the regenerator matrix. For this discussion, we are interested in only the first half of the cycle.

Figure 7.7 shows that the rms fluctuation suddenly jumps to a high value (>0.3 m/s) at the crank angle of about 45°. This is when the temperature difference shown in Figure 7.6 changes sign. Prior to 45°, the rms fluctuation, $\sqrt{\overline{u'^{-2}}}$ is low and remains about 0.1 m/s. This indicates that, at the beginning of the half cycle when the velocity is low, the eddy transport is relatively weak compared to its strength during the part of the cycle from 45° to 150°. Therefore, the momentum and thermal transport are hypothesized to be dominated by molecular transport, which is a less effective transport mechanism than if the transport were by turbulent eddies. As a result, the fluid surrounding the screen wire is not mixed well within the pore (is not in thermal equilibrium on the pore scale). One may conclude that the measured temperature (taken at a certain distance away from the screen wire) does not represent the heat-transfer sink temperature of the convective film, the temperature that describes heat transfer with the screen wire. This might be the reason for the negative heat-transfer coefficients during the beginning of acceleration and the lag in the measured temperature difference behind the heat flux computed from the measured solid matrix temperature history. If we were capable of locating a thermocouple close enough to the wire to take the sink temperature of the convective film, we should not get this anomalous behavior.

At about 45°, transition to a turbulent-like flow takes place. Afterward, eddy transport is dominant. At this point, it is believed that the flow is hydrodynamically and thermally in equilibrium at the pore scale, and the measured temperatures and heat fluxes can be used in the standard expression for Newton's law of cooling to derive the traditional heat-transfer coefficients and Nusselt numbers. During this time, we can expect the flow and heat transfer to behave as though they were quasi-steady.

It was observed by Ward (1964) that in a porous medium, transition from the Darcy regime (viscous effects dominate) to the Forchheimer regime (inertial effects dominate) takes place at the Reynolds number, Re_k, of about 10, based upon the square root of the permeability as the length scale. However, the critical Reynolds number for transition is not universal and depends on the microstructure of the porous materials. Figure 7.8 shows friction factor, f_K, versus Reynolds number, Re_k, calculated from data taken when we measured the permeability of the regenerator matrix (see Section 9.4.4). Shown also are correlations (Beavers and Sparrow, 1969; Gedeon, 1999) presented for metal screens. It is seen that for metal screens, transition occurs in the Re_k range 10 ~ 100, instead of 1 ~ 10 observed by Ward (1964). One can see also that a line drawn through our data is not too different than the correlation lines.

Above, it is stated that at a crank angle of about 45°, the rms fluctuation suddenly jumps to a high value. When we calculate the velocity within the

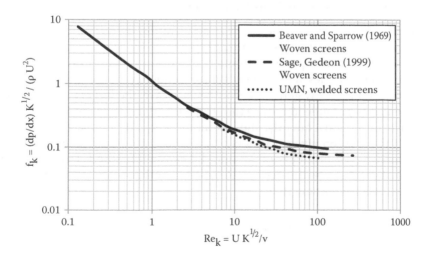

FIGURE 7.8

Transition from the Darcy flow regime to the Forchheimer flow regime in screen mesh porous media. Re_k is based on the square root of the permeability, k, as the length of scale.

matrix at a crank angle of 45° ~ 60°, we get 1.2 ~ 1.5 m/s using the formula: $U = U_{max} \sin(\omega t)$. Here, U_{max} is equal to about 1.7 m/s in the current experiments, and the base-case frequency is 0.4 Hz. The corresponding Reynolds number, Re_k, based on the measured permeability of 1.07×10^{-7} m² is about 25 ~ 30, which falls into the range of Reynolds number predicted from the permeability measurements (see Figure 7.8). Note that the velocities of Figure 7.7 are higher than the value of 1.2 to 1.5 m/sec during the first half of the cycle. This is because the flow is accelerating toward the cooler channel and is at a higher velocity at the measurement location.

7.2.4.1.2 Comments

The maximum Reynolds number, Re_k, during one cycle within the regenerator matrix is only 35. If the critical value for transition is over a range of 10 ~ 100 within metal screens, the flow within the regenerator never reaches a regime that is dominated by eddy transport but is within a transition regime. Thus, the question remains as to whether there is sufficient eddy transport at Re_k of 25 ~ 30 to draw closure to the question about why the heat transfer and the temperature difference are out of phase until about 45° of crank position. How much eddy transport is enough? Because the ratio of eddy transport to laminar transport is several hundred, there may be enough eddy transport to support this theory even if we have barely entered the transition regime.

7.2.4.2 Study with Various Valensi Numbers (Effect of Acceleration on Transition, Laminar to Turbulent)

We attempt to investigate the effect of acceleration on transition by changing Re and Va. The Reynolds number can be fixed by properly adjusting the stroke length and the operating frequency of our oscillatory flow drive. The Valensi number varies with the operating frequency of the drive. With the current oscillatory flow generator (for the base case, the largest stroke length was used), it is possible to reduce the stroke length from 356 mm to 252 mm, or to 178 mm. Consequently, the operating frequency must be increased from 0.4 Hz to 0.565 Hz, or to 0.8 Hz, to maintain the same Re number. The corresponding Va numbers of 3.0 and 4.2 can be obtained. Unfortunately, we cannot measure at the desired Reynolds number with subunity Valensi numbers. Also, we know that:

$$\text{Position, } X = X_{max} \cos(\omega t)$$

$$\text{Velocity, } U = \dot{X} = X_{max} \omega \sin(\omega t)$$

$$\text{Acceleration, } a = \ddot{X} = X_{max} \omega^2 \cos(\omega t)$$

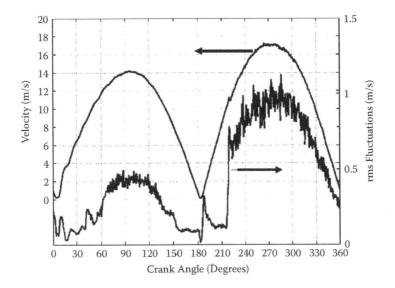

FIGURE 7.9
Instantaneous velocity and rms fluctuation of velocity ($\sqrt{\overline{u'^2}}$), at a radial position aligned with the centerline of a cooler tube and within the cooler-to-regenerator plenum. Taken at $Va = 3.0$.

This means that at the same Reynolds number (the same velocity), a larger Va number attained by increasing the frequency is an indicator of a larger acceleration.

We have taken velocity measurements within the plenum and heat-transfer measurements within the regenerator matrix at the three different Va numbers but with the same Re number. Figures 7.9 and 7.10 show the instantaneous velocity and rms fluctuations within the plenum at Va equal to 3.0 and 4.2. It is seen that rms fluctuations jump to high values at the crank angle of 45 ~ 60°. No systematic difference from the critical crank angle is apparent between these figures, and Figure 7.7, taken at Va equal to 2.1. It appears that transition might be determined by the magnitude of the velocity only and is only weakly dependent upon the flow temporal acceleration, or completely independent of it.

Heat-transfer measurements were also done within the matrix at the different Valensi numbers but with the same Reynolds number. Figure 7.11 shows the instantaneous Nusselt numbers at the three different Valensi numbers. There is a mild trend with Va, showing a lag between heat flux and temperature difference.

7.2.4.2.1 Comments

A question that this comparison raises is, "If transition is dependent only on velocity, why did we not see a sudden change in Nu number during deceleration at the time when the velocity is reduced below the critical velocity observed for acceleration?" This leads us to think that, in fact, the eddies

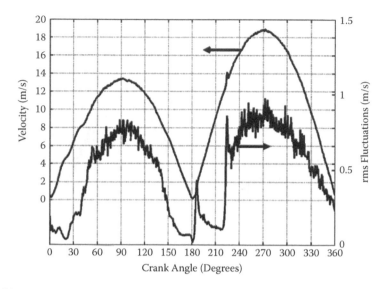

FIGURE 7.10
Instantaneous velocity and rms fluctuation of the velocity ($\sqrt{\overline{u'^2}}$) at the centerline of a cooler tube and within the cooler-to-regenerator plenum. Taken at $Va = 4.2$.

formed inside the matrix during the high-velocity part of the cycle are inclined to persist to lower velocities when the flow is temporally decelerated. While interpreting the rms fluctuation signal of $0° < \theta < 180°$ in Figure 7.9, we must keep in mind that the flow approaching the probe is spatially accelerating toward the cooler channel. Superposed on this spatial acceleration is a

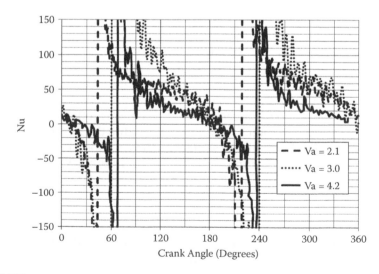

FIGURE 7.11
Instantaneous *Nu* numbers at different *Va* numbers.

temporal acceleration over $0° < \theta < 90°$ and a temporal deceleration over $90° < \theta < 180°$.

7.2.5 Turbulent Transport and Thermal Dispersion within the Porous Matrix

UMN attempted to measure thermal dispersion and flow mixing directly. As the first step, turbulent transport quantities were measured downstream of the regenerator matrix under unidirectional flow conditions using a triple-sensor, hot-wire anemometer probe. The Reynolds shear stress, $\overline{u'v'}$, was directly evaluated. The results showed promise, but the data scatter rendered them rather inconclusive. See Section 7.2.5.1, for the experimental setup, a description of the flow fields, and a comparison of the direct measurements to previous model results.

UMN has experience in measuring time-averaged quantities that represent turbulence at individual points in space within an open flow. However, modeling of transport or dispersion in porous media is usually based on a volumetric averaging technique. See Section 7.2.5.2 for progress in development of a theory that bridges the spatial-average (porous media description) and time-average (measurable) quantities.

Thermal dispersion is also discussed in terms of eddy transport in Section 7.2.5.2.

7.2.5.1 *Turbulent Transport Measurements Downstream of the Regenerator Matrix under Unidirectional Flow*

7.2.5.1.1 *Experiments*

We start with measurements of the eddy transport of momentum downstream of the regenerator under steady, adiabatic flow conditions. The experimental setup (see Figure 7.12, and refer back to the test regenerator

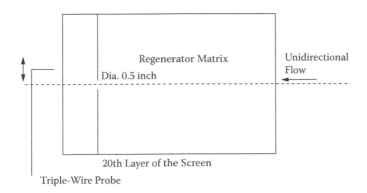

FIGURE 7.12
Experimental setup for turbulent transport measurements.

of Figure 7.1) was similar to that for the permeability measurements (discussed in Chapter 9), except that a triple-sensor, hot-wire probe (TSI 1299BM-20) was used to measure the three velocity components, u, v, and w, for various radial positions, from –1.5 inch to 1.5 inch (about 0.0 which is the radial center of the matrix), using a sampling frequency of 10 kHz. By doing so, the turbulent shear stress, $\overline{u'v'}$, can be evaluated. To create a radial velocity gradient, $\frac{\partial u}{\partial r}$, a plate with a 12.5 mm (0.5 inch) diameter orifice in the middle is inserted into the regenerator at the 20th layer of screen from the exit plane. As a result, radial eddy diffusivity of axial momentum can be obtained as

$$\varepsilon_M = \frac{\overline{u'v'}}{\frac{\partial u}{\partial r}} \tag{7.1}$$

7.2.5.1.2 Data Processing

Figure 7.13 shows a velocity profile at an axial location 12.5 mm (0.5 inch) away from the exit plane of the regenerator. Due to the irregularity of the regenerator matrix, the probe might be located immediately behind a wire or immediately behind a pore. Therefore, the velocity distribution is highly random, which makes it difficult to accurately evaluate local, radial velocity gradients. To obtain smoother velocity gradients and work with larger eddy sizes, the probe is moved farther away from the exit plane.

A color contour plot in Ibrahim et al. (2004a) shows the velocity fields downstream of the regenerator matrix, from 1.6 mm to 25.4 mm (1/8 to 2

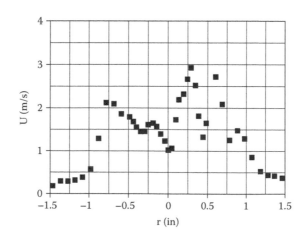

FIGURE 7.13
Velocity distribution at $x = 12.5$ mm (1/2 inch).

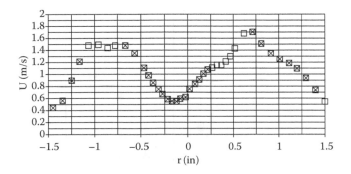

FIGURE 7.14
Velocity distribution at $x = 25.4$ mm (2 inches).

inches) away from the exit plane. Figure 7.14 shows the velocity distribution 25.4 mm (2 inches) downstream of the exit plane. It can be seen that the velocity profile gets much smoother at this point, which ensures an accurate evaluation of the velocity gradient. However, the eddy transport differs from that immediately behind the exit plane. We use these various planes to determine a relationship between the eddy diffusivity and the axial distance from the exit plane, as $\varepsilon_M = \varepsilon_M(x)$, to extrapolate back to the exit plane for evaluation of eddy diffusivity within the matrix. Points with smooth gradients are chosen to evaluate the mean eddy diffusivity, $\overline{\varepsilon_M}$, by:

$$\overline{\varepsilon_M} = \frac{\Sigma \varepsilon_M}{N} \tag{7.2}$$

The parameter N is the number of selected points for data processing. In the case of the axial location 25.4 mm (2 inch) away from the exit plane, N is equal to 29, the number of points indicated by × symbols in Figure 7.14.

To derive a model, we use the Prandtl mixing length hypothesis to compute the eddy diffusivity as the product of a characteristic length (taken to be the hydraulic diameter) and a characteristic velocity (taken to be the in-matrix average velocity),

$$\varepsilon_M = \overline{\lambda} d_n U \tag{7.3}$$

an average diffusivity coefficient is defined as

$$\overline{\lambda} = \frac{\Sigma (\varepsilon_M / d_n U)}{N} \tag{7.4}$$

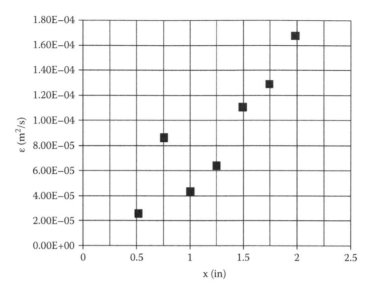

FIGURE 7.15
Eddy diffusivity from Equation (7.3) at various axial locations.

Here, U is the individual local velocity values at the selected points, and d_h is the hydraulic diameter of the regenerator matrix. Figures 7.15 and 7.16 indicate the values of ε_M and $\bar{\lambda}$, respectively, for various axial locations. The value of $\bar{\lambda}$ at $x = 12.5$ mm (1/2 inch) seems to be less than 0.005 but could be in the range $0 < \bar{\lambda} < 0.015$, given the degree of scatter.

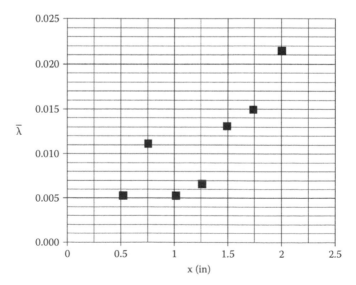

FIGURE 7.16
Average diffusivity coefficients, $\bar{\lambda}$, from Equation (7.4).

7.2.5.1.3 Discussion

The direct measurements of eddy diffusivity are compared to the model derived from our previous jet penetration measurement:

$$\varepsilon_M = \lambda\, d_h U, \quad \text{where } \lambda = 0.15 \sim 0.37 \tag{7.5}$$

The Sage (Gedeon, 1999) one-dimensional simulation program has a model for a similar term, but for thermal dispersion. One can apply Reynolds' analogy, $\varepsilon_M = \varepsilon_H$, and develop an expression for the coefficient of Equation (7.5).

$$\lambda = 1.1(\text{Re Pr})^{-0.37}\beta^{-3.1} - (\text{Re Pr})^{-1.0} \tag{7.6}$$

For the *Re*, *Pr*, and β values of the present study, this Sage model gives $\lambda = 0.15$. Obviously, the model values are much larger than those measured values from the current experiments. Clearly, more remains to be done regarding direct measurements of eddy transport.

7.2.5.2 A Turbulence Model for Thermal Dispersion through Porous Media

7.2.5.2.1 Search for a Mathematical Connection between a Spatially Averaged Dispersion Term (Porous Media Theory) and a Time-Averaged Dispersion Term (Turbulence Theory) (Masuoka and Takatsu, 2002)

In order to include thermal dispersion in the energy equation, a volume average technique has been derived for porous media. For any quantity, W_α, associated with the α-phase (either fluid or solid), a volumetric average, W_α, is:

$$\langle W \rangle_\alpha = \frac{1}{V_\alpha} \int_{V\alpha} W\, dV \tag{7.7}$$

The interval, V_α, is sufficiently large relative to the characteristic length scale of the α-phase to average out the scales of spatial fluctuations (or variations). Therefore, any microscopic quantity, W, can be decomposed into the sum of the volumetric average (macroscopic) quantity <W> and a spatial variation, $\widetilde{W}$:

$$W = \langle W \rangle + \widetilde{W} \tag{7.8}$$

$\widetilde{W}$ is a function of both time and space, and <W> is a function of time, only.

In contrast, when turbulence quantities in free and bounded flows (but not porous media) are of interest, a time-average technique is used to obtain the turbulence field. For any quantity, W, a time-average $\overline{W}$ taken during a time interval Δt is defined as:

$$\overline{W} = \frac{1}{t}\int_{t}^{t+t} W\, dt \tag{7.9}$$

The interval Δt is sufficiently large relative to the characteristic time scale of the fluid to average out the time scales of turbulence. Therefore, this quantity can be defined as the sum of the time-averaged value and the temporal fluctuation:

$$W = \overline{W} + W' \tag{7.10}$$

W' is a function of time and space, and $\overline{W}$ is a function of space, only.

Combining Equations (7.7) and (7.10) allows writing the volumetric average value, $<W>$, as decomposed into the volumetric average of the time average and temporal fluctuation as follows:

$$\langle W \rangle = \frac{1}{V_\alpha} \int_{V_\alpha} W \, dV = \frac{1}{V_\alpha} \int_{V_\alpha} (\overline{W} + W') dV = \langle \overline{W} \rangle + \langle W' \rangle \tag{7.11}$$

$\langle W' \rangle$ may be a function of time but would be small if time and space scales for averaging are large enough. Similarly, the time average value, $\overline{W}$, can be decomposed into the time-average of the volumetric-average and spatial fluctuation as follows:

$$\overline{W} = \frac{1}{t} \int_{t}^{t+t} W \, dt = \frac{1}{t} \int_{t}^{t+t} (\langle W \rangle + \tilde{W}) dt = \overline{\langle W \rangle} + \overline{\tilde{W}} \tag{7.12}$$

Again, $\overline{\tilde{W}}$ may be a function of space but would be small if time and space scales for averaging are large enough.

In a laboratory, UMN has experience in measuring time-average quantities at individual spatial points. This raises a question, "Is it possible to convert a time-averaged value into a volumetric average value?" Applying Equation (7.11) to fields of components of velocity and temperature of fluid phase, we have:

$$\langle u \rangle = \frac{1}{V_f} \int_{V_f} u \, dV = \frac{1}{V_f} \int_{V_f} (\overline{u} + u') dV = \langle \overline{u} \rangle + \langle u' \rangle \tag{7.13}$$

$$\langle T \rangle = \frac{1}{V_f} \int_{V_f} T \, dV = \frac{1}{V_f} \int_{V_f} (\overline{T} + T') dV = \langle \overline{T} \rangle + \langle T' \rangle \tag{7.14}$$

For the spatial fluctuations at individual spatial points:

$$\tilde{u} = u - \langle u \rangle = \overline{u} + u' - (\langle \overline{u} \rangle + \langle u' \rangle) \tag{7.15}$$

$$\tilde{T} = T - \langle T \rangle = \overline{T} + T' - (\langle \overline{T} \rangle + \langle T' \rangle) \tag{7.16}$$

If u' and T' can be substituted with the rms of the velocity, $\sqrt{u'^2}$, and tempe-rature, $\sqrt{T'^2}$, respectively, the spatial fluctuation (or variation) components

$\tilde{u}$ and $\tilde{T}$ at individual spatial points can be evaluated using Equations (7.15) and (7.16). Again, applying the volume-average technique to obtain the volumetric average of the product of $\tilde{u}$ and $\tilde{T}$ and applying a coefficient to account for the correlation between the two:

$$\langle \tilde{u}\tilde{T} \rangle = \frac{1}{V_f} \int_{V_f} \tilde{u}\tilde{T}\,dV = C\sqrt{\overline{u'^2}}\sqrt{\overline{T'^2}} \tag{7.17}$$

This reduces the problem to one of finding the correlation coefficient. It is proposed that this be found with two-point radial spatial correlations measured at the exit plane. More development along this direction is needed.

7.2.5.2.2 Physical Explanation of Turbulent Transport within a Porous Medium (Masuoka and Takatsu, 1996)

Masuoka and Takatsu (1996) studied transport phenomena through a porous medium in terms of vortex transport. Vortices contributing to dispersion through porous media have been categorized into two types: void vortices and pseudo vortices. The pseudo vortices result from flow distortion due to interruption with the solid surface (local separation). They transport the fluid over a large scale. The void vortices are the interstitial vortices formed in the pore between two solid surfaces (shown in Figure 7.17).

They seem to be due to a breakdown of free shear layers in the pore spaces. The characteristic length scales of the pseudo and void vortices are estimated to be the order of the particle diameter and the square root of the permeability

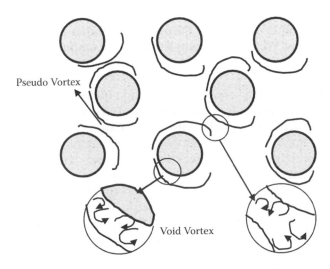

FIGURE 7.17
Schematic model of vortices in packed beds.

(i.e. the pore scale), respectively. Thus, the eddy viscosity is the sum of the eddy viscosity due to the pseudo and that due to the void vortices, as follows:

$$\varepsilon_t = \varepsilon_{t,p} + \varepsilon_{t,v} \qquad (7.18)$$

Further, the ratio between the pseudo and void eddy viscosities (also for thermal diffusivities) is defined as:

$$\gamma = \frac{\varepsilon_{t,p}}{\varepsilon_{t,v}} \qquad (7.19)$$

A model was derived to compute thermal dispersion. The result with $\gamma = 100$ was in agreement with the empirical data of Lage et al. (2002). From this, Masuoka and Takatsu (1996) concluded that the contribution of the void vortex to the thermal dispersion is negligible, and the pseudo vortex is the main contributor to thermal dispersion.

In a laboratory scale, the feature sizes of the porous media are still too small to measure the void and pseudo vortices by direct local measurements. If the void vortex does not have a significant contribution to the thermal dispersion, the thermal dispersion can be determined merely by the measurements of the pseudo vortex. Since the pseudo vortex plays a role of long-distance (large-scale) momentum transport, perhaps it can be measured immediately downstream of the porous medium, instead of within the porous medium.

This short discussion above indicates the direction pursued for direct measurement of turbulent transport.

7.2.6 Measurements of Thermal Dispersion within a Porous Medium near a Solid Wall, and Comparison with Computed Results

7.2.6.1 Summary

The regenerator is a key component of the efficiency of a Stirling cycle machine. Typical regenerators are of sintered fine wires or layers of fine-wire screens. Such porous materials are contained within solid-wall casings. Thermal energy exchange between the regenerator and the casing is important to cycle performance for the matrix, and casing would not have the same axial temperature profile in an actual machine. Exchange from one to the other may allow shunting of thermal energy, reducing cycle efficiency. As reported in Simon et al. (2006), temperature profiles within the near-wall region of the matrix are measured, and thermal energy transport, termed thermal dispersion, is inferred. The data show how the wall affects thermal transport. Transport normal to the mean flow direction is by conduction within the solid and fluid and by advective transport by eddies residing within the matrix. In the near-wall region, both are interrupted from their normal in-core pattern. Solid conduction paths are broken and scales of eddy transport are damped. The near-wall layer typically acts as an insulating layer. This should be considered in design or analysis. Effective thermal conductivity within the core is uniform. In-core transverse thermal effective conductivity values are compared to direct and indirect measurements reported elsewhere and to three-dimensional (3-D) numerical simulation results, computed previously and reported elsewhere. The 3-D CFD model is composed of six cylinders in cross-flow and in a staggered arrangement to match the dimensions and porosity of the porous matrix used in the experiments. The commercial code FLUENT is used to obtain flow and thermal fields. The effective thermal conductivities for the fluid (including thermal dispersion) are computed from the CFD results. See Simon et al. (2006) for a detailed discussion of these experimental and computational results.

7.2.6.2 Conclusion

If there were no effects of the wall, the cross-stream thermal transport would be essentially uniform. Our data show this to be generally equivalent to values found by another indirect measurement, a direct measurement, and computation. However, there are lower values of cross-stream thermal transport near the wall. This is expected because the conduction path at the wall is not continuous. The present results give a model for the radial variation of cross-stream thermal transport near an impermeable wall developed from experiments. This variable transport will significantly affect the heat transfer between the porous medium and the casing wall. Thus, it must be considered in an analysis of a Stirling cycle regenerator.

7.2.7 Computational Fluid Dynamics (CFD): Simulation of UMN Test Rig (CSU)

Two attempts were made to model the whole UMN test rig utilizing 3-D and two-dimensional (2-D) computational models. The model includes the piston, regenerator, cooler, heater, and isolation duct (see Figure 7.3). The 3-D modeling was done with a quadrant of the test rig. After running the 3-D model case and gaining more insight about the flow structure in the test rig, it was decided to shift to 2-D models. The 3-D model was too expensive, computationally, to run.

We utilized the 2-D model to compare results with the UMN rig in two areas: (1) the growth of a jet in the regenerator where the jet had emerged from the cooler, and (2) the effect of the plenum thickness on the jet growth. The jet spread was about the same length as was found experimentally, but the angle of spreading was quite different. This is mainly attributed to the laminar flow model used. In regard to the plenum thickness, our model showed consistent results with the UMN experimental finding. In the following sections, results are presented for the two models. The dimensions of the different components (piston, regenerator, cooler, heater, and isolation duct) of the rig are given in Table 7.1.

7.2.7.1 Three-Dimensional (3-D) Simulation of the UMN Test Rig

We have taken a quadrant of the test rig to be simulated and made use of the two planes of symmetry. After running the 3-D model case and learning more about the flow in the test rig, it was decided to shift to 2-D models. A single simulation took almost 3 weeks (at the time of processing of these cases) for the 3-D model. Below is the theory behind the 2-D regenerator matrix model; 2-D model results are given after the initial discussion of the theory.

7.2.7.2 Theory and Unidirectional Flow Simulation (2-D Regenerator Matrix Modeling, Micro and Macro Scale)

Modeling the regenerator matrix (the main objective of the CFD effort) can be treated generally as modeling of porous media. In the regenerator matrix, unsteady fluid mechanics and heat-transfer characteristics are quite complicated. In this respect, irregular "local" transport phenomena at the pore level are important because basically they result in such macroscopic phenomena as increase of pressure loss and heat-transfer augmentation. Fortunately, the quantities of interest change in a regular fashion with respect to space and time over the pore scale, which led us to a way of macroscopic modeling. Therefore, the general transport equations are used to integrate over a representative elementary volume (REV), which accommodates the fluid and the solid phases within a porous structure. Though the loss of information with respect to the "local" transport phenomena is inevitable with this approach, the integrated quantities, coupled with a set of proper constitutive equations that represent the effects of "local" interactions on the integrated quantities, do provide an effective basis for analyzing transport in porous media.

The REV technique requires information on numerous empirical transport coefficients that should be supplied by well-established experimental findings and analytical solutions for some rather simplified cases. Empirical correlations exist that predict flow and heat-transfer behavior through a variety of porous media, predominantly in granular media. They work well under the condition specified. However the present conditions (see Figure 7.18) go beyond the limitations of the correlations that were developed for packed beds of granular media. The Reynolds number in a wire-screen regenerator matrix is often much larger than those reached in packed beds of granular media because the regenerator is typically much more porous ($50\% < \beta < 98\%$) with a higher possible flow velocity due to the higher porosity. Plus, the form of the regenerator matrix is different than that of the spheres or densely packed granular media. This may cause unpredictable flow patterns to develop and alter the outcome of a model that was based on granular media. These issues can be overcome by taking a new approach to viewing

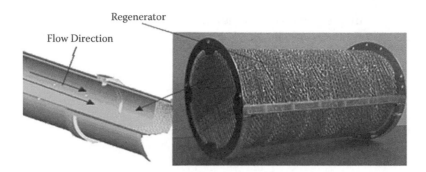

FIGURE 7.18
Wire-screen regenerator.

the regenerator matrix and attempting to directly model the geometry by defining a representative elementary volume (REV) that captures the intricate details of the regenerator structure, a way of microscopic modeling. In this section, we use different microscopic geometries and utilize the CFD data to determine the unknown variables for the process of volume averaging. The solution serves as a closure model for a macroscopic set of equations.

7.2.7.3 Macroscopic Equations Governing Regenerator Matrix Behavior

An incompressible, uniform flow passes through a number of cylinders placed in a staggered array, as shown in Figure 7.19. The periodic (north and south) boundary condition is applied, and considering the limitation of our computer resources, the geometry that consists of a 36-structure unit (9 rows by 4 columns) is taken as the computation domain. One structural unit, taken as a REV, is indicated with black solid lines.

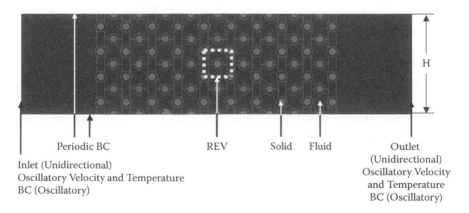

FIGURE 7.19
Geometry for a semi-infinite periodic array of staggered circular cylinders in cross-flow with specified boundary conditions.

The governing equations (continuity, momentum, and energy respectively, as obtained from Chapter 2) for the 2-D microscopic structure are given as follows:

$$\nabla \bullet (u) = 0$$

$$\frac{\partial \rho u}{\partial t} + \nabla \bullet \rho uu = -\nabla p + \nabla \bullet \tau$$

$$\rho_f C_{pf}\left[\frac{\partial T}{\partial t} + (u \bullet \nabla)T\right] = k_f \nabla^2 T :$$

$$\rho_s C_{ps}\frac{\partial T}{\partial t} = k_s \nabla^2 T : solid$$

(7.20)

The boundary conditions are:

At the solid wall

$$u = 0$$

$$T|_s = T|_f$$

$$k_s \frac{\partial T}{\partial n}\bigg|_s = k_f \frac{\partial T}{\partial n}\bigg|_f$$

At the periodic boundaries

$$u\big|_{y=-H/2} = u\big|_{y=H/2}$$

$$\int_{-H/2}^{H/2} v\,dx\big|_{y=-H/2} = \int_{-H/2}^{H/2} v\,dx\big|_{y=H/2}$$

$$T\big|_{y=-H/2} = T\big|_{y=H/2}$$

Integrating the above microscopic equation over a control volume, V, which is much larger than a microscopic (pore structure) characteristic size but much smaller than a macroscopic characteristic size, yields the following volume-averaged equations:

$$\nabla \bullet \langle u \rangle = 0$$

(7.21)

$$\frac{\rho_f}{\beta}\frac{\partial \langle u \rangle}{\partial t} + \frac{\rho_f}{\beta^2}\langle u \rangle \bullet \nabla \langle u \rangle = -\nabla \langle p \rangle^f + \frac{\mu}{\beta}\nabla^2\langle u \rangle + \frac{1}{V}\int_{Aint}(\tau_{ij}\,dA_j - p\,dA_j) - \frac{\rho_f}{\beta}\frac{\partial \langle u_i' u_j' \rangle}{\partial x_j}$$

(7.22)

For the fluid phase:

$$\rho_f C_{pf}\left[\frac{\varepsilon\partial\langle T\rangle^f}{\partial t}+\langle u\rangle\bullet\nabla\langle T\rangle^f\right]=\nabla\bullet\left[k_f\beta\nabla\langle T\rangle^f+\frac{1}{V}\int_{A\,int}k_fT\,dA-\rho_f C_{pf}\langle T'u'\rangle\right]$$

$$+\frac{1}{V}\int_{A\,int}k_f\nabla T\,dA$$

(7.23)

For the solid phase:

$$\rho_s C_s\frac{(1-\beta)\partial\langle T\rangle^s}{\partial t}=\nabla\bullet\left[k_s(1-\beta)\nabla\langle T\rangle^s-\frac{1}{V}\int_{A\,int}k_sT\,dA\right]-\frac{1}{V}\int_{A\,int}k_s\nabla T\,dA$$

(7.24)

Applying the Forchheimer-extended Darcy's law, the volume-averaged momentum equation can be transformed into:

$$\frac{\rho_f}{\beta}\frac{\partial\langle u\rangle}{\partial t}+\frac{\rho_f}{\beta^2}\langle u\rangle\bullet\nabla\langle u\rangle=-\nabla\langle p\rangle^f+\frac{\mu}{\beta}\nabla^2\langle u\rangle-\frac{\mu}{K}|<u>|-\rho_f\frac{C_f}{\sqrt{K}}|<u>|<u>$$

(7.25)

Comparing the above equations with the two energy equations introduced heuristically in the literature by Schlunder (1975), we can obtain:

$$\overline{\overline{k}}_{eff}^f\bullet\nabla\langle T\rangle^f=k_f\beta\nabla\langle T\rangle^f+\frac{1}{V}\int_{A\,int}k_fT\,dA-\rho_f C_{pf}\langle T'u'\rangle$$

(7.26)

$$k_{dis}\bullet\nabla\langle T\rangle=-\rho_f C_{pf}\langle T'u'\rangle$$

(7.27)

$$\nabla\bullet\overline{\overline{k}}_{eff}^s\bullet\nabla\langle T\rangle=k_s(1-\beta)\nabla\langle T\rangle^s-\frac{1}{V}\int_{A\,int}k_sT\,dA^s$$

(7.28)

$$h_{sf}a_{sf}(\langle T\rangle^s-\langle T\rangle^f)=\frac{1}{V}\int_{A\,int}k_s\nabla T\,dA$$

(7.29)

and transform the volume-averaged energy equation into:

Fluid phase:

$$\beta\rho_f C_{pf}\left[\frac{\partial\langle T\rangle^f}{\partial t}+\langle u\rangle\bullet\nabla\langle T\rangle^f\right]=\nabla\bullet\overline{\overline{k}}_{eff}^f\bullet\nabla\langle T\rangle^f+h_{sf}a_{sf}(\langle T\rangle^s-\langle T\rangle^f)$$ (7.30)

Solid phase:

$$\rho_s C_{ps}\frac{(1-\beta)\partial\langle T\rangle^s}{\partial t} = \quad \bullet\,\bar{k}_{eff}^{s}\,\bullet\quad \langle T\rangle^s - h_{sf}a_{sf}(\langle T\rangle^s - \langle T\rangle^f) \qquad (7.31)$$

As mentioned above, several unknown variables appear in the process of volume averaging. To minimize the need for empirical expressions, direct transport simulation of microscopic geometries is used to determine the unknown variables. The microscopic numerical results thus obtained are integrated over a unit to evaluate these unknown variables, which serve as closure models for the macroscopic set of equations. The unknown variables to be determined include permeability, drag coefficient, thermal dispersion tensor inside the fluid, the interfacial heat transfer coefficient between the fluid and the solid, and the static thermal conductivity of the regenerator.

7.2.7.3.1 Unidirectional Flow in Porous Media—Micro Models

The CFD simulations for single cylinders discussed in Section 6.5 are a first step in the microscopic analysis toward analyzing the regenerator matrix. As the number of cylinders increases, the complexity and the computation requirements (memory and central processing unit [CPU] time) increase. Our goal is to compute for a reasonable number of cylinders that will provide adequate representation of the matrix and be computationally affordable. This subcell, in turn, can be used to simulate the entire matrix. A step in this direction is choosing square/circular cylinders in cross-flow and in staggered arrangements, as discussed below.

7.2.7.3.1.1 Turbulent Flow in an Array of Staggered Square Cylinders in Cross-Flow

Figure 7.20 shows the grid structure for an array of staggered square cylinders in cross-flow with inlet and exit plenums. The upper and lower boundaries were chosen as periodic boundary conditions. We investigated a steady, unidirectional, highly turbulent (Re_H: based on the cell height, H, and inlet mean velocity is 113,820) case in order to compare our modeling results with the literature (see, for example, Kuwahara and Nakayama, 2000). We carried out computations using the set of microscopic equations available in the CFD-ACE solver (CFD-ACE User Manual, 1999) for the row of periodic structural units shown in Figure 7.20. Then we applied a user subroutine "flu. dll" to integrate and obtain volume averages: for the whole cell (solid and fluid) for velocity $\langle u \rangle$; and only for the fluid for pressure, turbulence kinetic

FIGURE 7.20
Grid structure for an array of staggered square cylinders in cross-flow with inlet and exit plenums.

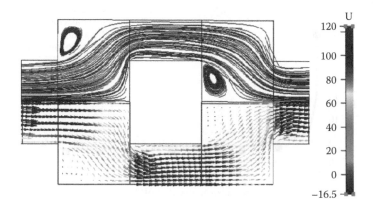

FIGURE 7.21
Velocity vectors (m/s) and streamline contours in the sixth cell of the square arrangement. Porosity = 75%. Cell height = 0.03816 m, and spacing between squares = 0.01808 m. Mean velocity = 50 m/s, Reynolds number (based on mean velocity and cell height) = 113,820. The standard k-ε turbulent flow model is applied.

energy, and dissipation of turbulence kinetic energy, $\langle p \rangle^f$, $\langle k \rangle^f$, and $\langle \varepsilon \rangle^f$, respectively. The solutions were grid-independent and all normalized residuals were brought down to 1×10^{-5}.

Figure 7.21 shows the velocity vectors and streamline contours in a microscopic porous structure with porosity $\beta = 0.75$, and $Re_H = 113,820$. The flow is from west to east. The computation reveals the presence of two large vortices as the fluid enters the cell and as it turns around the solid square. Figures 7.22 and 7.23 show the microscopic field of the turbulence kinetic energy and its dissipation

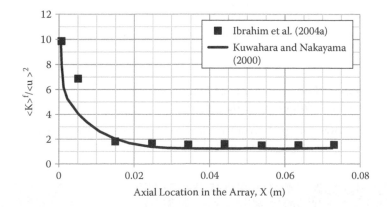

FIGURE 7.22
Turbulence-kinetic-energy comparison between Kuwahara and Nakayama (2000) computational fluid dynamics (CFD) data and the present work. Turbulent flow model for an array of staggered square cylinders in cross-flow. Porosity = 75%. Turbulence kinetic energy variation with x.

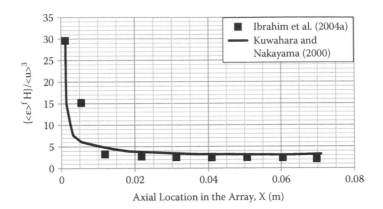

FIGURE 7.23
Dissipation rate of turbulence kinetic energy comparison between the Kuwahara and Nakayama (2000) CFD data and present work—turbulent flow model for an array of a square cylinder in cross-flow. Dissipation rate of turbulence kinetic energy variation with *x*.

rate obtained from integrated results. These results are compared with those of Kuwahara and Nakayama (2000). The inlet turbulence kinetic energy and its dissipation rate are determined using the following correlations:

$$\langle k \rangle^f = 37 \frac{1-\beta}{\sqrt{\beta}} \langle \bar{u} \rangle^2 \tag{7.32}$$

$$\langle \varepsilon \rangle^f = 117 \frac{(1-\beta)^2}{\beta} \frac{\langle \bar{u} \rangle^3}{H} \tag{7.33}$$

A standard $k - \varepsilon$ model was used for our present work. The streamwise decay of averaged turbulence kinetic energy and its dissipation rate, based on microscopic computation, compare well with the values of Kuwahara and Nakayama (2000).

7.2.7.4 Laminar Flow in an Array of Staggered Circular Cylinders in Cross-Flow

Figure 7.24 shows the grid structure for an array of staggered circular cylinders in cross-flow with inlet and exit plenums. The north and south boundaries

FIGURE 7.24
Grid structure for an array of staggered circular cylinders in a cross-flow with inlet and exit plenums.

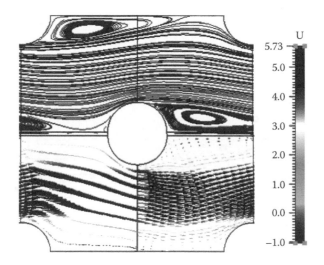

FIGURE 7.25
Velocity vectors (m/s) and streamline contours in the sixth cell of an array of staggered cylinders in a cross-flow. Porosity = 90%. Wire diameter = 0.8133 mm, mean velocity = 2.69 m/s, Reynolds number (based on mean velocity and wire diameter) = 137.7. Laminar flow model.

were chosen as periodic boundary conditions. We investigated a steady, unidirectional, laminar (Re_{Dw}: based on the wire diameter = 137.7) in order to compare our modeling results with the UMN experimental data for unidirectional flow. Again, from the microscopic point of view, we carried out computations using the set of microscopic equations available in the CFD-ACE solver for the row of periodic structural units shown in Figure 7.24. Then we used a subroutine to integrate the microscopic results of flow quantities, mainly $\langle u \rangle$, $\langle p \rangle^f$ over every unit to obtain the macroscopic quantities. Figure 7.25 shows the velocity vectors and streamline contours in a microscopic porous structure with porosity $\beta = 0.9$ obtained at $Re_{Dw} = 137.7$. Again, the computation reveals the presence of two large vortices as the fluid enters the cell and as it turns around the solid cylinder, both are results of the wake behind the cylinder.

A series of computations was carried out for Re_{Dw} from 0.5 to 307. The resulting integrated velocities and pressures are compared with the UMN experimental data in Figure 7.26. The results show agreement between CFD and experiments in the range of mean velocity from 2 to 5 m/s (Re_{Dw} from 100 to 250). A maximum difference was noted (about 25%) at $U = 1.1$ m/s ($Re_{Dw} = 56$). Also, agreement was found between the CFD data and the UMN experimental data for the permeability and friction coefficient (not shown here).

Our microscopic model (both one array and multiarray) was also extensively compared with reference experimental data available for various porosities. Figures 7.27 and 7.28 show microscopic model simulation results compared with data from Tong and London (1957) for steady-state flow through woven-screen matrix with porosity of 0.832 and 0.602, respectively.

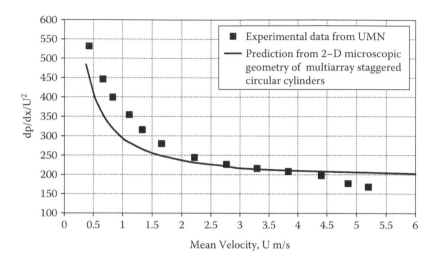

FIGURE 7.26
Comparison between the University of Minnesota experimental data in unidirectional flow and computational fluid dynamics laminar model for an array of staggered circular cylinders in a cross-flow. Porosity = 90%.

Further examination of Figures 7.27 (porosity = 0.832) and 7.28 (porosity = 0.602) indicate that the agreement between our CFD results and the experimental data for friction factor is better for higher porosity. This implies that 2-D modeling will be more satisfactory for higher-porosity matrices where 3-D flow features are less significant.

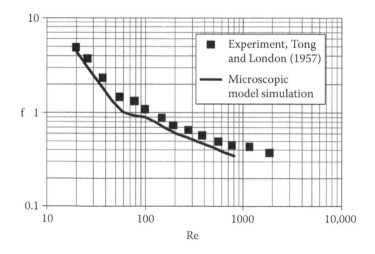

FIGURE 7.27
Comparison between Tong and London (1957) experimental data and microscopic model for multiarray-staggered circular cylinders in a cross-flow. Porosity = 83.2%.

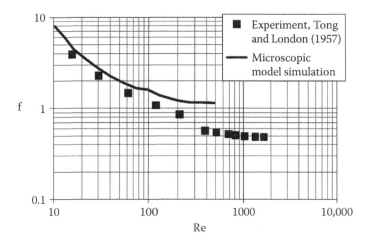

FIGURE 7.28
Comparison between Tong and London (1957) experimental data and microscopic model for multiarray-staggered circular cylinders in a cross-flow. Porosity = 60.2%.

7.2.7.5 Determination of Stagnation Thermal Conductivity

Hsu (1999) proposed a two-energy-equation model for the case of pure conduction in saturated porous media:

$$\rho_f C_{pf} \left[\frac{\beta \partial \langle T \rangle^f}{\partial t} + \langle u \rangle \bullet \ \langle T \rangle^f \right] = \ \bullet [k_{f,\mathit{eff}} \ \langle T \rangle^f + k_{dis} \ \langle T \rangle^f] \quad \text{for fluid} \quad (7.34)$$

$$\rho_s C_{ps} \frac{(1-\beta)\partial \langle T \rangle^s}{\partial t} = \ \bullet [k_{s,\mathit{eff}} \ \langle T \rangle^f] \quad \text{for solid} \quad (7.35)$$

$$k_{f,\mathit{eff}} = \left[\beta + \left(1 - \frac{k_s}{k_f} \right) G \right] k_f \quad (7.35)$$

$$k_{s,\mathit{eff}} = \left[1 - \beta + \left(\frac{k_s}{k_f} - 1 \right) G \right] k_s \quad (7.36)$$

$$G = \frac{\frac{k_{stg}}{k_f} - \beta - (1-\beta)\frac{k_s}{k_f}}{\left(\frac{k_s}{k_f} - 1 \right)^2} \quad (7.37)$$

Now k_{stg} is the only unknown to be determined.

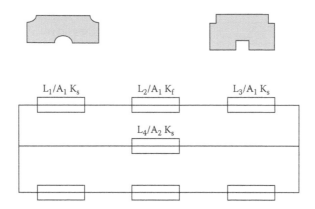

FIGURE 7.29
Equivalent thermal circuits for staggered circular and square cylinders. The results compare well with Hsu's analytical correlation.

Hsu's (1999) analytical expression, based on a 3-D cube model, gives an expression for k_{stg}:

$$\frac{k_{stg}}{k_f} = 1 - (1-\beta)^{2/3} + \frac{(1-\beta)^{2/3}(k_s/k_f)}{(1-(1-\beta)^{1/3})(k_s/k_f)+(1-\beta)^{1/3}} \tag{7.38}$$

The stagnation thermal conductivity is evaluated using equivalent thermal circuits as shown in Figure 7.29.

7.2.7.6 Determination of Thermal Dispersion

We now apply Equation (7.27) to a structural unit. The thermal conductivity due to thermal dispersion in the flow direction can be determined from

$$k_{dis} = \frac{-\frac{\rho_f C_{pf}}{H^2}}{T/H} \int_{H/2}^{H/2} \int_{H/2}^{H/2} (T - \langle T \rangle)(u - \langle u \rangle^f) \, dx \, dy \tag{7.39}$$

The thermal conductivity due to thermal dispersion is determined by feeding the microscopic numerical results into Equation (7.39). The results are plotted against the Peclet number based on wire diameter, Pe_D, in Figure 7.30.

The present results agree with the functional relationship based on a 2-D model by Kuwahara et al. (2001) for the thermal dispersion conductivities at low and high Peclet number ranges, in terms of the Peclet number and porosity, which are:

$$\frac{(k_{dis})}{k_f} = 0.022 \frac{Pe_D^2}{1-\beta} \quad \text{for } (Pe_D < 10) \tag{7.40}$$

$$\frac{(k_{dis})}{k_f} = 2.7 \frac{Pe_D}{\beta^{0.5}} \quad \text{for } (Pe_D > 10) \tag{7.41}$$

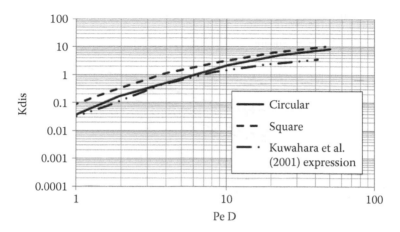

FIGURE 7.30
Effective thermal conductivity due to thermal dispersion versus Peclet number.

From the figure, we found that thermal conductivity due to dispersion over-whelms the other two terms in Equation (7.34) as Peclet number, Pe_D, becomes sufficiently large.

7.2.7.7 Determination of Convective Heat-Transfer Coefficient

We now apply Equation (7.29) to a structural unit. The convective heat trans-fer coefficient in one structural unit (always chose the middle one) can be determined from:

$$h_{sf} = \frac{\frac{1}{V} \int_{Aint} k_s \ TdA}{a_{sf}(\langle T \rangle^s - \langle T \rangle^f)} \tag{7.42}$$

The convective heat transfer coefficient determined by inputting the micro-scopic numerical results into Equation (7.42) is (as Nu) plotted against the Reynolds number based on wire diameter, Re_D, is shown in Figure 7.31.
    The present results are compared against the results of the functional rela-tionship based on the 2-D model by Kuwahara et al. (2000) in Figure 7.31. The convective heat transfer coefficient is given in terms of the Reynolds number and porosity. This functional relationship of Kuwahara et al. (2000) is

$$\frac{h_{sf}D}{k_f} = \left(1 + \frac{4(1-\beta)}{\beta}\right) + 0.5(1-\beta)^{0.5} Re_D^{0.6} Pr^{1/3} \tag{7.43}$$

7.2.7.7.1 CFD-ACE Porous Media Model

We have taken another approach to study the regenerator matrix. This one deals with macroscopic modeling of the matrix. We took the values obtained

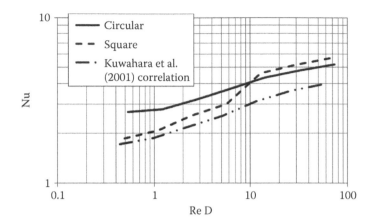

FIGURE 7.31
Comparison between present work and data by Kuwahara et al. (2001). (Data from Kuwahara, F., Shirota, M., and Nakayama, A., 2001, A Numerical Study of Interfacial Convective Heat Transfer Coefficient in Two-Energy Equation Model for Convection in Porous Media, *International Journal of Heat and Mass Transfer*, 44(6), 1153–1159.)

from the experiments for permeability and friction coefficient as input to our CFD-ACE Code.

The CFD model includes 2-D and 3-D configurations. The 2-D model is an axisymmetric pipe unidirectional flow into porous media as in Figure 7.32. The dimensions of the pipe as well as the inlet conditions were taken to be the same as in the experimental set up (see Table 7.1). As for the 3-D model, it is only one quadrant of the cross section of the whole pipe. The inlet boundary condition and properties of the porous media are the same as in the 2-D model. The CFD computations include both laminar and turbulent flow. Figure 7.33 shows a comparison between the UMN experimental data and the CFD results. The CFD simulation results for laminar models (2-D and 3-D) are in agreement with the experimental results. The turbulent CFD simulations are an order of magnitude higher. These data suggest that using the laminar model with permeability and friction coefficient is sufficient to model the porous media.

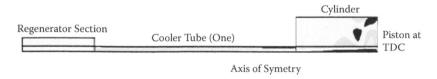

FIGURE 7.32
Two-dimensional computational fluid dynamics model of the University of Minnesota rig.

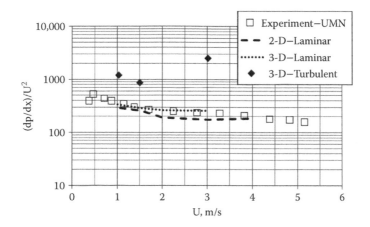

FIGURE 7.33
Comparison between the University of Minnesota experimental data and computational fluid dynamics (laminar, turbulent, two-dimensional, and three-dimensional) unidirectional flow in the regenerator matrix—macroanalysis.

7.2.7.8 2-D Simulation Results of an UMN Test Rig

A 2-D model was developed to simulate the fluid flow in the UMN oscillatory flow test rig. Figure 7.32 shows this 2-D model with the piston at TDC, one cooler tube, and the regenerator section.

We utilized this 2-D model to compare results with UMN test results in two areas: (1) the jet growth as the flow issues from the cooler and enters the regenerator, and (2) the effect of the plenum spacing on the jet growth.

Figure 7.34 shows a comparison of the jet growth between the UMN experimental data and the CFD model. The growth in the UMN case was obtained from temperature measurements and ½ of U_{max} was used for the CFD model. This CFD model was based on a laminar flow field with permeability $= 1.65 \times 10^{-8}$, and plenum width $= 1$ D (D = cooler tube diameter, this is equivalent to $4\delta_{nomial}$, where δ_{nomial}, is the nominal plenum thickness). The figure shows the streamline contours and U velocity vectors at a crank angle of 270° (at this crank angle the maximum velocity of the jet flows into the regenerator matrix). Upon examining the data from the figure, we can see that the model shows a jet spreading over about the same length, but the angle of spreading is quite different. This is mainly attributed to the laminar flow model used.

Figure 7.35 shows the streamline contours and U velocity vectors at a crank angle of 270° for three cases of plenum width: 0, 0.25 D, and 1 D. The figures indicate that the flow field from matrix face to downstream is almost

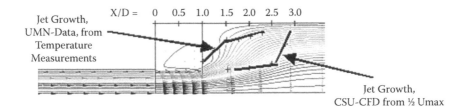

FIGURE 7.34

Comparison of jet growth, University of Minnesota and CSU-CFD. Streamline contours and U velocity vectors, laminar flow, permeability = 1.65×10^{-8}, crank angle = $270°$, plenum width = 1 $D = 4\,\delta_{nominal}$ (D = cooler tube diameter).

Plenum Width = Zero

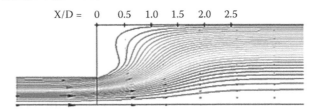

Plenum Width = 0.25 D ($\delta_{nominal}$)

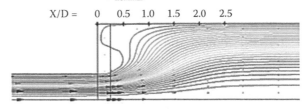

Plenum Width = 1 D ($4\,\delta_{nominal}$)

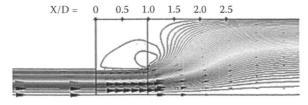

FIGURE 7.35

Streamline contours and U velocity vectors, laminar flow, permeability = 1.65×10^{-8}, crank angle = $270°$, plenum width = 0, 0.25, and 1 D, respectively (D = cooler tube diameter).

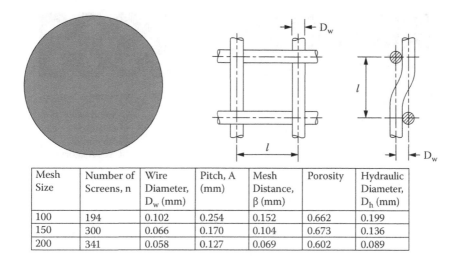

Mesh Size	Number of Screens, n	Wire Diameter, D_w (mm)	Pitch, A (mm)	Mesh Distance, β (mm)	Porosity	Hydraulic Diameter, D_h (mm)
100	194	0.102	0.254	0.152	0.662	0.199
150	300	0.066	0.170	0.104	0.673	0.136
200	341	0.058	0.127	0.069	0.602	0.089

FIGURE 7.36
Type and properties of stainless steel wire screen used in the experiment of Zhao and Chang (1996).

similar for the three cases. This is consistent with the UMN experimental results.

7.2.8 Simulation of Pressure Drop through Porous Media Subjected to Oscillatory Flow

A research paper by Zhao and Cheng (1996), whose experiment was well documented, was useful for comparison with the CSU CFD results. The type and properties of the regenerator used in this experiment are depicted in Figure 7.36. (The 100-mesh size screen data was used for the CFD simulation.)

7.2.8.1 The Microscopic Model and Similarity Parameters

For this case, we constructed a model of 5 (rows) × 19 (columns) of cylinders in cross-flow, in staggered arrangements as shown in Figure 7.37. The model matched the Zhao and Cheng (1996) experiment by choosing of the same porosity, wire diameter, and hydraulic diameter as for the 100-mesh wire and the similarity parameters: A_R and Va.

7.2.8.2 Results and Discussion

The CFD runs were made for porosity, $\beta = 0.662$; $A_R = 615$, 843, and 1143; and $Va = 0.01005$, 0.03770, and 0.05529.

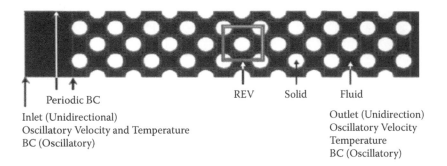

Periodic BC

Inlet (Unidirectional)
Oscillatory Velocity and Temperature
BC (Oscillatory)

REV   Solid   Fluid

Outlet (Unidirection)
Oscillatory Velocity
Temperature
BC (Oscillatory)

FIGURE 7.37
Microscopic geometry and boundary conditions.

The comparison shown in Figure 7.38 indicates that the CFD model gave the same trend in dp/dx amplitude as the experimental data, as the *Va* increases. However, there is a phase angle difference between the CFD and experiment which increases with *Va*.

Figure 7.39 shows another comparison between Zhao and Cheng (1996) data and the present model. In this case, the *Va* was kept = 0.04524 while the dimensionless fluid displacement was varied. Again the comparisons are good, except for a phase difference between the CFD and the data that increase as the dimensionless fluid displacement increases.

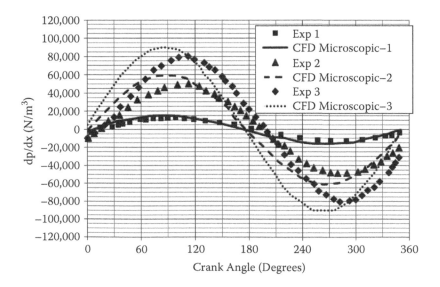

FIGURE 7.38
Variations of pressure drop per unit length as a function of crank (phase) angle, C.A., for $(Ao)_{Dh}$ = 843.38, $(Re_w)_{Dh}$ = 0.01005, 0.03770, and 0.05529, computational fluid dynamics and experiment (Zhao and Cheng, 1996, *Cryogenics* 36, 333–341).

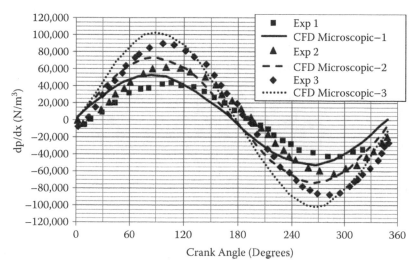

FIGURE 7.39

Comparison of pressure drop per unit length as a function of crank (phase) angle, C.A., for $(Re_w)_{Dh} = 0.04524$, $(Ao)_{Dh} = 614.73, 843.38$ and 1143.25, computational fluid dynamics and experiment (Zhao and Cheng, 1996).

7.2.9 Simulation of Thermal Field inside Porous Media Subjected to Oscillatory Flow

Obtaining the unsteady heat-transfer coefficient between the flow and the screen (solid) as the oscillatory flow goes through the regenerator is our focus in this section. Based on the temperature measurements (from UMN) near the cooler side of the regenerator, the jets spread in the region close to the cooler but are mixed very well by about 15 screens deep into the matrix. The CFD results confirm that the internal flow temperatures are not functions of the radial position. By comparing the instantaneous Nu numbers at different radial and axial locations, it could be concluded that generally the heat-transfer coefficient does not change from location to location within the portion of the regenerator matrix where the flow is thermally fully developed. Also note that transport within the regenerator is sufficient (due to high porosity) to keep the velocity and temperature uniform at any time. Omitting the first term on the right side of Equation (7.31) yields:

$$\rho_s C_s \frac{(1-\beta)\partial\langle T\rangle^s}{\partial t} = -h_{sf} a_{sf}(\langle T\rangle^s - \langle T\rangle^f) \tag{7.44}$$

$$\frac{h_{sf} D_h}{k_f} = \frac{\rho_s C_s D_h \frac{(1-\beta)\partial\langle T\rangle^s}{\partial t}}{a_{sf} k_f(\langle T\rangle^f - \langle T\rangle^s)} \tag{7.45}$$

Thus, from Equation (7.45), the unsteady heat-transfer coefficient can be calculated based on the volume averaged wire temperature and fluid-solid temperature difference during one cycle.

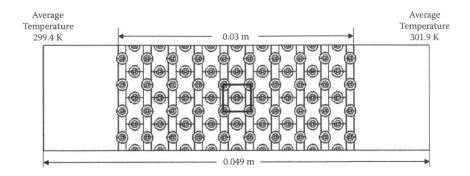

FIGURE 7.40
The computational domain (699 × 150 grids) for a 9 rows × 19 columns matrix with dimensions and thermal boundary conditions.

Figure 7.40 shows our 2-D model for simulating the regenerator matrix. It is composed of a cross-flow over 9 (rows) × 19 (columns) cylinders in a staggered arrangement. Both the velocities and temperatures at the two ends (east and west) were taken as sinusoidal in agreement with the UMN test rig conditions. The north and south boundaries were taken as periodic. We extracted the results from the REV, which is shown as a black box in Figure 7.44.

The test section design parameters chosen here are Reynolds number (Re_{max}) = 800 and Valensi number = 2.1 based on maximum bulk mean velocity (U_{max}) and the hydraulic diameter of the screen matrix (D_h). Figure 7.41 shows the REV, dimensions for the wire, spacing, and hydraulic diameter. Also shown are the different expressions for relating these quantities.

Figures 7.42, 7.43, and 7.44 show the CFD calculated pressure, U-velocity, and temperature contours, respectively, at a crank angle of 90°. It can be seen from the U-velocity contours that the flow behind each cylinder is streamlined because we are using a laminar flow model (little or no mixing). The same is true for the thermal field shown in Figure 7.44.

Figure 7.45 shows the solid (wire) temperature versus crank angle for five different wires (central and four neighboring ones, see corner picture). The two wires in the far west coincide and are at colder temperatures, while the ones on the far east also coincide, but at a higher temperature. The central wire appears in the middle. The data from the UMN are also shown. The CFD results are in phase angle agreement with the UMN data but show a smaller amplitude.

Figure 7.46 shows the solid (wire) and fluid temperatures versus crank angle from the UMN data and CFD results. Two fluid temperatures (from CFD) are shown, one on the diagonal west of the wire and the other on the east (see corner picture). The two fluid temperatures (from CFD) are in agreement with each other but differ from UMN data both in phase angle and amplitude.

Figure 7.47 shows the friction factor versus the crank angle for both CFD and Sage results. The agreement is very good.

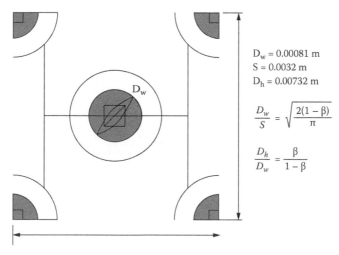

$D_w = 0.00081$ m
$S = 0.0032$ m
$D_h = 0.00732$ m

$$\frac{D_w}{S} = \sqrt{\frac{2(1 - \beta)}{\pi}}$$

$$\frac{D_h}{D_w} = \frac{\beta}{1 - \beta}$$

FIGURE 7.41
One computational cell (representative elementary volume) with dimensions and definitions.

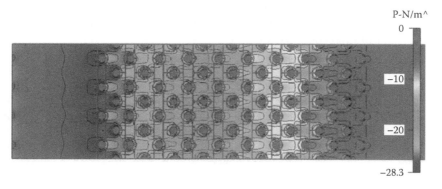

$P\text{-N/m}^\wedge$

FIGURE 7.42
Pressure (*Pa*) contours at 90° crank angle.

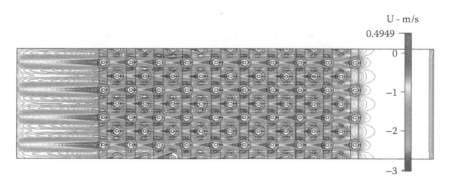

U - m/s

FIGURE 7.43
U-Velocity (m/s) contours at 90° crank angle.

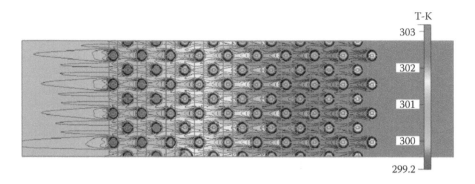

FIGURE 7.44
Temperature (K) contours at 90° crank angle.

The Nusselt numbers determined by inputting the microscopic CFD-ACE numerical results into Equation (7.45) are plotted in Figure 7.48 against the crank angle during one cycle. Figure 7.48 shows a comparison of the present work with the UMN experimental data for cases running under the same condition. Also, a correlation from Sage for modeling woven screens is plotted as an additional comparison. It is:

$$Nu = (1 + 0.99 \, Re^{0.66} \, Pr^{0.66}) \tag{7.46}$$

The comparison of the instantaneous friction factor from the CFD and Sage, Gedeon (1999) model are shown to be very good. As for the instantaneous Nu, the comparison between the present calculations and the UMN experimental results are shown to be in agreement during the deceleration

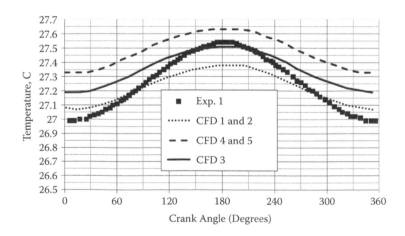

FIGURE 7.45
Results for the solid temperature, C, versus crank angle, University of Minnesota data and computational fluid dynamics.

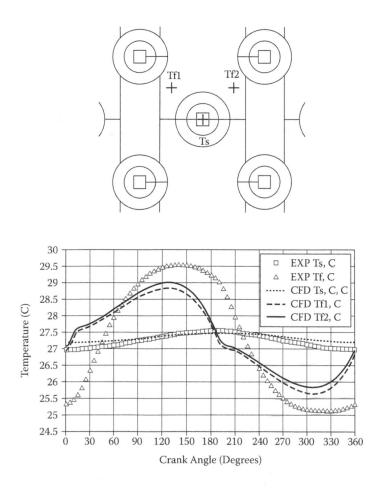

FIGURE 7.46
Results for the solid and fluid temperatures, C, versus crank angle, University of Minnesota data and computational fluid dynamics.

part of a cycle from 90° to 180° and from 270° to 360°. However, there are disagreements, such as follows: (1) Our CFD calculations showed differences of about 40% when compared with the average heat transfer coefficient over the cycle; (2) underprediction during the acceleration portion of the cycle; and (3) about 50° phase angle difference. These differences might be attributed to the geometry chosen for the model being 2-D axisymmetric compared to 3-D, or the temperature boundary conditions (taken uniform for the incoming fluid) which might be unrealistic. Additional efforts were made, including modifying the temperature boundary condition; modeling more wires, both in the streamwise and cross-stream directions; and model- ing it as a 3-D geometry. The first and second attempts did not produce any significantly different results. The 3-D approach is more expensive computationally and will be explored in the future under funding from other sources.

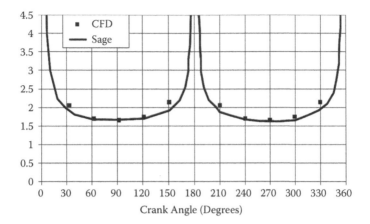

FIGURE 7.47
Friction factor versus crank angle, Sage and computational fluid dynamics.

We also modeled the same experiment utilizing the macroscopic porous media model available in CFD-ACE+. These macroscopic results are discussed in Section 7.2.10 below and by Ibrahim et al. (2002, 2003).

7.2.10 Validation of the CFD-ACE+ Porous Media Model

In this section, we show results of an additional modeling effort for the regenerator matrix done utilizing the macroscopic porous media model in CFD-

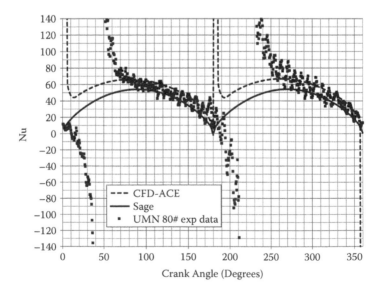

FIGURE 7.48
Nu number versus crank angle (degrees) for oscillatory flow; comparison among University of Minnesota experimental data, Sage and computational fluid dynamics.

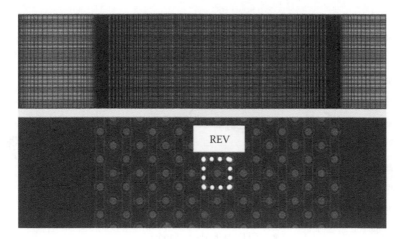

FIGURE 7.49
Geometries of macroscopic model (upper) and microscopic model (lower).

ACE+. Figure 7.49 shows the computational domain used for the two cases (i.e., microscopic and macroscopic). The upper grid is for the macroscopic porous media model, and the lower one shows a schematic of the microscopic model (as shown earlier).

Figure 7.50 shows a comparison between the CFD results for macro- and microscopic scale modeling for the pressure drop. The models agree, with

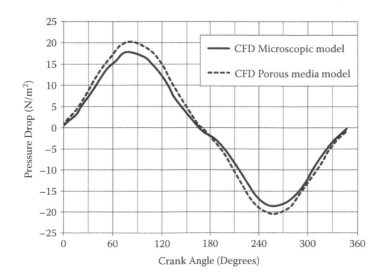

FIGURE 7.50
Comparison of CFD-ACE+ macroscopic porous media model pressure drop calculation with microscopic model results. Porosity = 0.9.

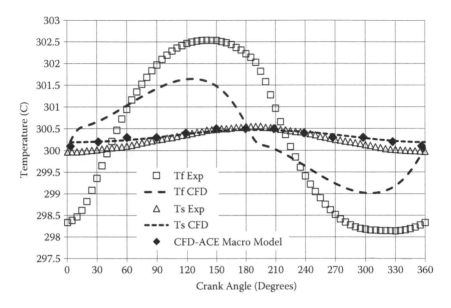

FIGURE 7.51

Comparison of CFD-ACE+ macroscopic porous media model thermal field calculation with microscopic model and University of Minnesota experimental results.

the macroscopic-scale model showing slightly higher values. Notice that there are no experimental data here to compare against.

Figure 7.51 shows the solid and fluid temperature variations with the crank angle. Shown in the figure are data from the UMN experiments and CFD results from the macro- and microscopic scale modeling. The microscopic results were discussed earlier. The macroscopic model assumes a gas temperature that is identical to that of the solid, which is unacceptable if accurate regenerator enthalpy flux losses are desired, as discussed earlier.

8

Segmented-Involute-Foil Regenerator—Actual Scale

8.1 Introduction

The goal of the National Aeronautics and Space Administration (NASA) microfabricated regenerator project (see Ibrahim et al., 2007, 2009a) was to develop a new regenerator of high durability as well as high efficiency using emerging microfabrication technology. In addition to the benefit to Stirling engine space-power technology, such regenerator development would also benefit Stirling cycle coolers and NASA's many cryocooler-enabled missions. This project was conducted in three phases, I, II, and III. Phases I and II were conducted by Cleveland State University (or CSU, the lead institution), the University of Minnesota (UMN), Sunpower Inc. (Athens, Ohio), Gedeon Associates (Athens, Ohio), Infinia Corporation (Kennewick, Washington), and International Mezzo Technologies (Mezzo) (Baton Rouge, Louisiana) (see Ibrahim et al., 2007), while Phase III was conducted by CSU, Sunpower Inc., Gedeon Associates, and Mezzo (see Ibrahim et al., 2009a).

In Phase I of this project, a microscale regenerator design was developed based on state-of-the-art analytic and computational tools. For this design, a 6% to 9% engine-efficiency improvement was projected. A manufacturing process was identified, and a vendor (Mezzo) was selected to apply it. Mezzo completed electric discharge machining (EDM) tools for fabricating layers of the chosen involute-foil microregenerator design, based on the team's specifications. They were ready to begin producing regenerator layers (annular portions of disks) by the end of Phase I. Also, a large-scale mock-up (LSMU) involute-foil regenerator was designed and fabrication had begun by the end of Phase I. Computational fluid dynamics (CFD) analysis for different geometries was employed to model the fluid flow and heat transfer under both steady and oscillatory-flow conditions. The effects of surface roughness were included (Ibrahim et al., 2004b). Several geometries: lenticular, parallel plates (equally/nonequally spaced), staggered parallel plates (equally/nonequally spaced), and three-dimensional (3-D) involute foils were studied via CFD. CFD modeling was also applied to both the microscale involute-foil regenerator and to the LSMU model of it.

The Phase II report (Ibrahim et al., 2007) of this project covered in detail the preliminary design process that was used for adapting a microfabricated regenerator to a Sunpower FTB (frequency-test-bed) Stirling engine/alternator (Wood et al., 2005). The FTB Stirlings produce about 80 to 90 W of electrical power with a heat input of 220 W and are the direct ancestors of the advanced Stirlings now under development by Sunpower and NASA Glenn Research Center (GRC) for future NASA space missions. They were originally designed for random-fiber regenerators. During Phase II, several tasks were completed. The team developed a preliminary microfabricated regenerator design based on its similarity to a parallel-plate structure; analyzed radiation losses down the void part of the regenerator; analyzed thermal conduction losses in the solid part of the regenerator, using closed form as well as two-dimensional computational analysis; built a prototype microfabricated regenerator for use in the NASA/Sunpower oscillatory flow test rig; tested that regenerator and derived design correlations for heat transfer and pressure drop; and performed system modeling of a FTB engine with a microfabricated regenerator using the Sage simulation software (Gedeon, 1999, 2009, 2010); this was done first using a theoretical parallel-plate correlation for heat transfer and pressure drop, then with the correlations derived from actual test data.

During the Phase III effort, a new nickel segmented-involute-foil regenerator was microfabricated and was tested in a Sunpower Inc., FTB Stirling engine/alternator. Testing in the FTB Stirling produced about the same efficiency as testing with the original random-fiber regenerator. But the high thermal conductivity of the prototype nickel regenerator was responsible for significant performance degradation. An efficiency improvement (by a 1.04 factor, according to computer predictions) could have been achieved if the regenerator was made from a low-conductivity material. Also, the FTB Stirling was not reoptimized to take full advantage of the microfabricated regenerator's low flow resistance; thus, the efficiency would likely have been even higher had the FTB been completely reoptimized. This chapter discusses the regenerator microfabrication process, testing of the regenerator in the Stirling FTB, and the supporting analysis. Results of the pretest CFD modeling of the effects of the regenerator-test-configuration diffusers (located at each end of the regenerator) are included in Appendix L. The chapter also includes recommendations for accomplishing further development of involute-foil regenerators from a higher-temperature material than nickel. (The NASA space-power Stirlings operate at a hot-end temperature of ~850°C, above the practical temperature range for use of nickel.)

8.2 Selecting a Microfabricated Regenerator Design

Currently, Stirling convertor regenerators are usually made of woven screens or random fibers. These types of structures suffer from the following features: locally nonuniform flows; local variations in porosity which result in local mismatches in flow channels that contribute to axial thermal transport; high flow friction combined with considerable thermal dispersion, a thermal loss mechanism that causes an increase in apparent axial thermal conduction; wire screens that require long assembly times which tends to increase their cost; and for space engines, an assurance that no fibers of the matrix will eventually work loose and damage vital convertor parts during the mission.

Research efforts thus far have shown that attractive features for effecting high fluid-to-matrix heat transfer with low pressure drop are a matrix in which the heat-transfer surface is smooth, the flow acceleration rates are controlled, flow separation is minimized, and passages are provided to allow radial mass flow for a more uniform distribution when the inlet flow or the in-channel characteristics are not radially uniform. It is thought that properly designed microfabricated regular geometries could not only reduce pressure drop, maintain high heat transfer, and allow some flow redistribution when needed, but could show improved regenerator durability for long missions.

8.2.1 Problems with Current Generation Random Fibers

In designing a new regenerator, it is first useful to review the reliability and performance issues relating to the current generation of random-fiber regenerators typically used in Stirling engines for space power.

Because a random-fiber structure relies on a network of sintered contact points between fibers, it is possible that there will be "loose" fibers that will migrate out of the regenerator into other parts of the convertor where they may damage close-tolerance seals or clog gas-bearing ports. This is a concern, not a known problem. It is difficult to "prove" anything about a random structure without testing.

Corrosion of thin materials in hot helium environments is another potential problem. No long-term studies document the durability of random-fiber materials at high temperatures in a helium environment. A NASA GRC study, Bowman (2003), documented significant wire corrosion in 316L stainless steel wires under such conditions with significant debris generation. The trace levels of oxidizing contaminants in a helium environment may actually be worse than higher concentrations because they may prevent the formation of a protective oxide layer on the wire surfaces. NASA is currently investigating the replacement of 316L stainless steel with ceramics or certain oxidation-resistant metals, such as nickel aluminide (NiAl), which can resist oxidation at high temperatures.

Besides the reliability issues, random fiber regenerators come nowhere close to the ideal regenerator performance in terms of heat transfer and flow resistance. Based on a figure of merit defined later in Section 8.2.3, random-fiber regenerators (and woven-screen regenerators, for that matter) are worse by about a factor of 4 compared to theoretical "best-possible" regenerator performance. One reason for poor performance is the presence of wakes, eddies, stagnation zones, and nonuniform flow passages within a random-fiber

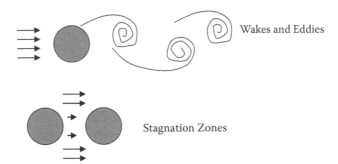

FIGURE 8.1
Flow features that reduce random-fiber regenerator performance.

structure, as illustrated in Figure 8.1. These flow features lead to increased flow resistance and a loss mechanism known as thermal dispersion.

8.2.2 New Microfabricated Regenerator Helium Flow and Structural Requirements

In the new regenerator, the matrix should consist of precisely oriented microscale (25 to 100 µm) features. In general, these features should be uniformly spaced with a relatively low surface roughness and produce low local streamwise acceleration and deceleration rates and minimal flow separation. These properties are key factors to achieving high heat-transfer rates with low pressure drops. Passages should provide some radial flow opportunity to give a more uniform distribution when the inlet flow or the in-channel characteristics are not radially uniform. The overall solid volume fraction should be small to reduce axial thermal conduction losses. There is a minimum amount of solid material necessary to achieve adequate solid heat capacity relative to the helium, but this is typically 10% or less of the volume for modern small Stirling engine applications. Also for minimal axial conduction, it is preferred to have a tortuous solid conduction path through the regenerator in the axial direction.

It is important that the matrix resist deformation under mechanical and thermal stresses. Mechanical stresses arise from the method used to hold the regenerator in its canister. Thermal stresses arise from heating the regenerator from room temperature to as high as 850°C at the hot end. If there is nonuniform heating or differing rates of thermal expansion in the regenerator canister, the regenerator passages must be robust enough to withstand it.

The regenerator in the space-power engine (such as that in Figure 1.1) must operate for at least 14 years, so it must also be durable. Mechanical shock and vibration can occur during launch and during normal operation.

8.2.3 Performance Characterization

Preliminary evaluation of regenerator performance can be expressed with a figure of merit that roughly measures the "heat transfer performance per unit flow resistance penalty." A high figure of merit indicates good performance. As an extension to the conventional figures of merit, Backhaus and Swift (2001) and Ruhlich and Quack (1999), Gedeon Associates developed a new figure of merit, which includes thermal dispersion, defined in Equation (8.1) (Gedeon, 2003a). (Also, see Appendix H for implications of this figure of merit in actual Stirling engines, as explained by Gedeon Associates.)

$$F_M = \frac{1}{f_r \left(\frac{R_e P_r}{4 N_u} + \frac{N_k}{R_e P_r} \right)} \tag{8.1}$$

In Equation (8.1), f_r is the friction factor, Pr is the Prandtl number, Re is the Reynolds number, N_u is the Nusselt number, and N_k is an "enhanced conductivity ratio" defined as the effective axial conductivity divided by the molecular conductivity of the heat-transfer fluid. The effective conductivity includes conduction in the solid, conduction in the fluid, and eddy transport in the fluid. For uniform flow channels under steady, fully developed, laminar flow conditions, N_k is near unity. The friction factor, f_r, and the Nusselt number, N_u, are given in Sage (Gedeon, 1999, 2009, 2010) for several regenerators of interest. For example, for the parallel plate regenerator, $f_r = 96/Re$ and $N_u = 8.23$.

Figure 8.2 presents the figures of merit for several regenerator designs as a function of Reynolds number (assuming $Pr = 0.7$). Figure 8.2 shows that the figures of merit are near their maximum values when the mean Reynolds number of the regenerator is on the order of 100, which is also ≈ the Reynolds number expected in an operating modern small Stirling engine (of the type

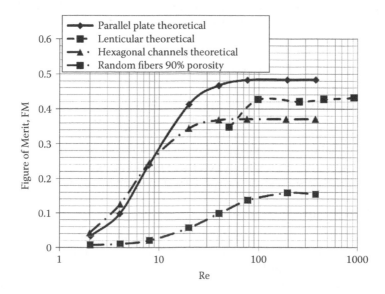

FIGURE 8.2
Figure of merit for various matrices constructed from stainless steel.

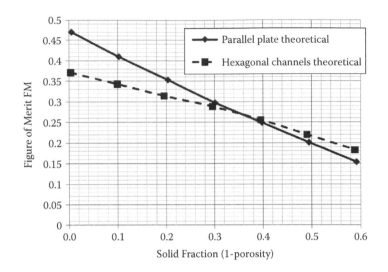

FIGURE 8.3
Continuous channel regenerators: figure-of-merit degradation versus wall thickness at mean
$Re = 62$.

shown in Figure 1.1). The regenerator geometries having the three highest figures of merit in Figure 8.2 are discussed in more detail below.

Solid-mode thermal conduction is a loss that will reduce the performance of a regenerator matrix below the theoretical predictions shown in Figure 8.2. The Sage Stirling-cycle modeling software, Gedeon (1999, 2009, 2010) was used to estimate the degradation due to solid conduction for continuous-channel stainless steel regenerators, such as parallel plates or hexagonal channel regenerators. The degradation is shown in Figure 8.3. The figure of merit decreases linearly as the solid fraction increases.

For stacked-segment regenerators, the solid-mode thermal conduction can be reduced by gaps or offsets between segments. These gaps will also provide radial flow opportunities that can lead to a more uniform flow distribution. However, the temperature difference across such thermal breaks must be small compared to the solid-to-helium temperature difference in a regenerator that uses a high-conductivity gas like helium, for if not, the gas will bridge the conduction gap. For this reason, thin solid layers are preferred in segmented matrices.

8.2.4 Micro Feature Tolerance

To determine the tolerance to channel width variations of a microfabricated regenerator, the Sage software was used to model parallel-plate regenerators, Gedeon (2003a, 2003b) and Gedeon (2004a, 2004b), with some channels of a particular width and the others of another, slightly smaller, width. These calculations indicated that DC (direct current, i.e., nonzero mean) flow circulation will be generated by nonuniform spacing in parallel-plate regenerators,

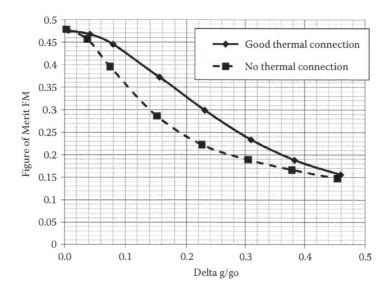

FIGURE 8.4
Parallel plate regenerators: figure-of-merit degradation versus flow gap variation at mean $Re = 62$.

reducing overall regenerator performance. Figure 8.4 shows figure of merit degradation as a function of relative gap variation at mean $Re = 62$, one for the case where the two different-sized regenerator channels are adjacent to each other (good thermal contact) and one for the case where the two regenerator channels of different size are far from each other (no thermal contact). "Delta-g/g_0" represents the fractional amount that one channel is larger than another. As revealed by Figure 8.4, the "no thermal contact" case has a lower figure of merit. This is because the thermal contact between the two parts can suppress the temperature skewing effect of the DC flow. Figure 8.4 also indicates that the figure of merit will be reduced significantly by increasing the gap variation, whether for the "good connection" case or the "no thermal connection" case. The main reason for the decrease in figure of merit with an increase in gap variation is the higher cycle-averaged DC enthalpy flow.

The results shown in Figure 8.4 can be applied to the general case of any channel-type regenerator, such as a honeycomb type. In that case, hydraulic-diameter variation would replace gap variation, but the resulting curves should be quite similar. If the gap (or hydraulic diameter) variation is limited to ±10% for the microfabricated regenerator, the figure of merit will be at least 0.4, according to Figure 8.4.

The above helps explain why wrapped-foil regenerators have not performed well in Stirling engines. The main reason is the difficulty in maintaining gap/structural integrity when subjected to larger temperature and stress gradients.

8.2.5 Surface Roughness

The flow-passage walls should be smooth to approach the theoretical laminar flow assumed in theoretical regenerator modeling. The challenge is to determine a maximum roughness limit to ensure that the channel can be regarded as smooth. Many studies of the maximum relative roughness for smooth tube behavior have been carried out. Schlichting (1979) presented results from experiments on rough pipes with relative roughness values, k_s/R (k_s is the sand grain roughness, and R is the radius of the pipe), from 1/500 to 1/15 which had been carried out by Nikuradse (1933). Nikuradse's experiments showed that pipes with maximum relative roughness, k_s/D, of 0.033 behave as though they were smooth, in the laminar regime. Samoilenko and Preger (1966) (reported by Idelchick, 1986) developed an equation to calculate the onset of roughness (or departure from the laminar 64/Re line) as the friction factor rises, which is valid for rough tubes with $0.06 > \varepsilon/D > 0.007$, where ε is absolute roughness height, Re_0 is the Reynolds number value at which the friction factor departs from the Hagen-Poiseuille theory, rising to higher values. Apparently, when $\varepsilon/D < 0.007$, it is presumed that Re_0 is large enough that transition to turbulence occurs before transition to rough-laminar behavior is observed. Bucci et al. (2003) presented data of water flow in stainless steel capillary tubes which agreed somewhat with Equation (8.2).

$$Re_0 = 754 \exp\left(\frac{0.0065}{\varepsilon/D}\right) \tag{8.2}$$

Equation (8.2) indicates that $Re_0 = 936$ for a relative roughness of 0.03. Because the Reynolds number of the proposed regenerator is approximately 100, channels with $\varepsilon/D < 0.033$ are considered to be smooth.

For noncircular channels, the hydraulic diameter, d_h, is treated as the diameter, D, of circular tubes. The hydraulic diameter is defined as

$$D_h = 4\left(\frac{A}{P_{wet}}\right) \tag{8.3}$$

where A and P_{wet} are the area and the wetted perimeter of the noncircular channels, respectively. For the microfabricated regenerator, the hydraulic diameter is around 200 μm. Because the maximum relative roughness is 0.033, microfabricated surfaces with a maximum absolute roughness height of approximately 7 μm can be treated as smooth surfaces.

8.2.6 Regenerator Design Concepts

8.2.6.1 Early Concept 1: Lenticular Arrays

We began our regenerator research with a "lenticular array" concept that is shown in Figure 8.5. The name comes from the cross-section view at the top

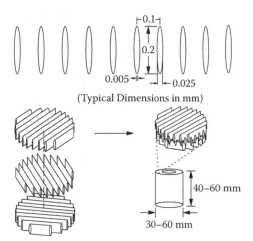

FIGURE 8.5
A three-dimensional lenticular array geometry.

showing "lentil" or lens-shaped elements. Previously published computational analysis (Ruhlich and Quack, 1999) showed that the two-dimensional lenticular array structure produced a good figure of merit (see Figure 8.2). One problem is how to hold the structural elements in alignment. Our solution was to nest together layers of lenticular arrays as shown in Figure 8.5. Every layer is composed of several parallel lenticular rows. Each layer is rotated relative to the adjacent layers (by 60°), and each layer penetrates halfway into the next. In addition to being structurally strong, the lenticular shape array minimizes flow separation, and the almost uniform cross section prevents significant flow acceleration and deceleration. This array also allows flow redistribution among channels, thus accommodating redistribution of the entry flow.

It was found that the lenticular array geometry could not be made by current technology, so researchers moved on to other concepts.

8.2.6.2 Early Concept 2: Honeycomb Structures

A honeycomb structure would consist of a network of hexagonal channels as shown in Figure 8.6. The illustrated wall thickness of the honeycomb matrix is 16 μm, and the space between two parallel walls is 260 μm. A complete regenerator could consist of a single sheet of hexagonal channels (i.e., 60 mm long) or a stack of honeycomb "slices." Stacking with offset would allow flow redistribution among channels, thus accommodating redistribution of the entry flow.

The main problems with honeycomb structures are that they are difficult to make with the required passage uniformity, and their theoretical performance is not as good as that of parallel-plate structures. Mezzo has fabricated honeycomb structures via LiGA/EDM for intended use in Stirling coolers (see Figure 8.17 and discussion).

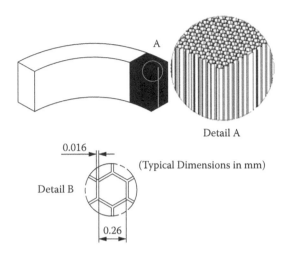

Detail A

0.016

(Typical Dimensions in mm)

Detail B

0.26

FIGURE 8.6
Honeycomb geometry.

8.2.6.3 Involute-Curved Parallel Plate Structures

Achieving a parallel plate structure that performs up to its theoretical expectations has been the Holy Grail of microfabricated regenerator research. Many researchers have tried to produce parallel-plate structures by winding thin metal foils around cylindrical forms, with the layers spaced by some sort of spacing elements. Such regenerators have never performed up to expectations, presumably because of poor uniformity of layer spacing. The required gap between layers is on the order of 100 microns, which is very difficult to maintain within the required 10% tolerance (see Section 8.2.4) both during initial assembly and after the effects of thermal distortions. For the microfabricated regenerator, we rejected wrapped foil because of its historical record.

We also rejected a variation of the foil-type regenerator consisting of stacked flat foil elements (see Backhaus and Swift, 2001). Flat elements have the structural problem that if you constrain their length while subjecting them to thermal expansion they buckle. That is, they deflect one way or the other at random with the ratio of lateral deflection to lengthwise shortening approaching infinity. Any buckling of a regenerator foil element is bad because it affects the spacing between layers. Flat elements also cannot be arranged in an axisymmetric way, leading to potential problems with flow uniformity.

We believe that a parallel-plate regenerator can only succeed with curved elements. The great advantage of curved elements is structural. If a curved element has constrained endpoints while subjected to thermal expansion, its curvature changes slightly in a regular way to accommodate the new arc length. It does not buckle. Therefore, the spacing between layers is not affected much, and the effect on spacing is predictable.

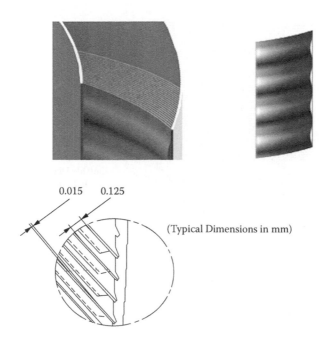

0.015   0.125

(Typical Dimensions in mm)

FIGURE 8.7
Involute-foil geometry.

The type of curved parallel plate structure selected for further study consists of precisely aligned, thin, curved plates as shown in Figure 8.7. The illustrated wall thickness is 15 μm, and the space between two parallel walls is 125 μm. Similar to the honeycomb regenerator, a complete regenerator could consist of a single-sheet parallel-plate structure (i.e., 60 mm long) or a stack of parallel plate "slices." Again, stacking could be with offset to allow flow redistribution among channels.

The foil elements follow involute curves. An "involute of a circle" is what you get when you unwind a string from a circular cylinder. It is the curve traced out if you keep the string tight and attach a pencil to the end of the string. A family of such curves spaced by constant rotational angle increments (shortening the string by increments) has the property that the normal distance between curves (the gap) is constant along the curve. So, involute curves are a way to pack foil with the desirable property of uniform gap.

The main problem with the illustrated involute foil regenerator, in Figure 8.7, was that no one could make it.

8.2.6.4 Selected Concept: Concentric Rings of Involute-Curved Elements

Our selected design for the new manufactured regenerator was a variation of the above involute foil idea as illustrated in Figure 8.8 through Figure 8.11. A batch-mode EDM process, discussed in detail later (Section 8.3.4), can be

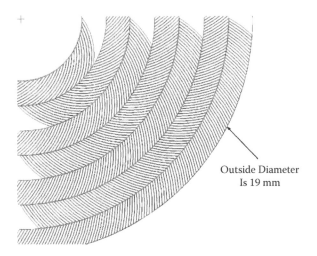

Outside Diameter
Is 19 mm

FIGURE 8.8
Initial conceptual geometry of concentric involute rings.

used to make it. To increase structural integrity, the curved "foil" elements are designed to be relatively short and packaged within concentric rings. The ratio of foil element length to spacing gap is an important design criterion. As illustrated, the ratio is on the order of 10, which is a rule-of-thumb criterion for parallel-plate flow.

FIGURE 8.9
Solid model for concentric involute rings.

FIGURE 8.10
Stacking arrangement for concentric involute rings.

The concentric ring approach adapts well to the thin-annular regenerator canisters used for Stirling convertors. In that case, there might be only three or so concentric rings of elements between the inner and outer walls. For testing purposes, the geometry could be extended to almost completely fill a cylindrical canister, as was required for the oscillating-flow testing planned for Phase III. Note that there is a progressive change of the radial angle of the foil elements in successive rings. This is not expected to result in any significant flow nonuniformity, because the normal gap between plate elements is always the same, and the flow direction is predominantly in the axial direction.

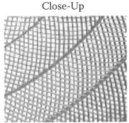

Close-Up

FIGURE 8.11
Two-disk end view for concentric involute rings.

Figure 8.8 shows a pattern for concentric rings of involute-curved elements. 19 mm outer diameter (OD) corresponds to NASA/Sunpower oscillating-flow regenerator test facility sample dimensions.

Figure 8.9 shows a solid model of a test-configuration regenerator disk. Key dimensions are 86 micron channels, 14 micron walls, 1 mm ring spacing, 0.5 mm thick. Approximately 40 disks would be required to complete a full test regenerator.

Figure 8.10 shows flipping successive disks alternates involute direction allowing flow distribution between disks and interrupting solid conduction path. No rotational alignment is required during assembly. There was a slight further evolution in the design of the involute-foil disks before testing occurred. Figure 8.11 shows an end view of two final disks stacked together showing the pattern made by successive curved involute elements. This final design is also illustrated in Figure 8.11 and discussed in Section 8.2.6.4. Figure 8.11 shows the two types of disks (final design); they were alternated in the stack to promote good uniform axial (average) flow.

8.2.7 Predicted Benefits to a Stirling Engine: 6% to 9% Power Increase for the Same Heat Input

The figure of merit of Equation (8.1) measures only the relative performance of one regenerator against another, considering only the regenerator. To evaluate the expected benefit to the overall Stirling engines requires system modeling.

The approach we took was to use the Sage computer simulation (Gedeon, 1999, 2009, 2010) to optimize two Stirling convertors, one with a random-fiber regenerator and the other with an ideal-foil regenerator, everything else being the same. To make the results as relevant as possible to NASA, the models are similar to those used to design the advanced Stirling convertor (ASC) (see Wood et al., 2005). This is a machine with 230 W thermal input (design specification) and roughly 100 W pressure volume (PV)-power output.

Essentially, Sage optimized all of the ASC machine dimensions subject to a number of constraints in order to maximize PV power output for a given heat input. The baseline simulation model for the random-fiber regenerator matrix is grounded in heat-transfer and flow-friction correlations that came from actual test results using the NASA/Sunpower regenerator test rig. The ideal-foil regenerator model is based on theoretical correlations for developed laminar flow between parallel plates. The model includes the conduction loss down the solid part of a stainless-steel ideal-foil matrix. The solid conduction was reduced by a factor of 10 to account for a possible tortuous conduction path in the microfabricated structure compared to actual parallel plates. In other words, the model took advantage of all of the good things about parallel plate regenerators (high heat transfer and low flow resistance) and discounted the bad thing (solid conduction).

TABLE 8.1

Dimensions of Optimized Ideal-Foil Regenerator

Canister inner diameter (ID)	36.9 mm
Canister outer diameter (OD)	43.6 mm
Canister length	133 mm
Gap between layers	0.137 mm
Foil thickness	0.018 mm

The result of the two optimizations was that PV power output increased by 8.6% for the ideal foil regenerator. That and other regenerator-related numbers of interest are summarized in Tables 8.1 and 8.2.

In Table 8.2 available energy (AE) losses are essentially lost PV powers according to reversible heat-engine theory. Note that the most dramatic improvement for the ideal foil regenerator was the reduction in AE loss to gas conduction (thermal dispersion) although most of the other AE losses also decreased. It is interesting that the reduction of total AE losses for the ideal foil regenerator (5.6 W) was less than the resulting increase in PV power output (9.4 W). This means that Sage's optimizer was able to take advantage of loss reductions in other areas made possible by the more efficient regenerator. For example, the ideal foil regenerator turned out to be 48% longer than the random-fiber regenerator, thereby also lengthening the regenerator pressure wall, displacer cylinder, and displacer shell, and reducing their conduction losses. However, there may be structural reasons (displacer support issues) that would argue against increasing the regenerator length by that amount.

TABLE 8.2

Regenerator Comparison

	Random Fiber	Ideal Foil
Hydraulic diameter (micron)	345	274
Mean Reynolds number	54	52
Valensi number	1.1	0.68
AE loss to flow friction (W)	5.76	3.75
AE loss to heat transfer (W)	5.28	4.77
AE loss to gas conduction (W)	3.38	0.22
AE loss to solid conduction (W)	0.10	0.17[a]
Total regenerator AE loss (W)	14.5	8.9 ($\Delta = -5.6$)
PV power output (W)	109.2	118.6 ($\Delta = +9.4$)

[a] With foil solid conduction multiplier *Fmult* = 0.1 to account for expected tortuous solid flow path in microfabricated structure. *Fmult* is a Sage computer code (Gedeon, 1999, 2009, 2010) input parameter, used here for adjustment of solid conduction.

Note: AE, available energy.

Because of concerns over increased regenerator length, we ran another comparison optimization with the regenerator length constrained to 70 mm. To be even more conservative, we removed the 0.1 solid conduction multiplier and assumed an uninterrupted solid conduction path for 15 micron thick stainless-steel foil. Under these conditions, the resulting benefit of the foil regenerator dropped a bit but still showed 6.6% more power available for the same heat input.

8.3 Manufacturing Processes Considered and Manufacturing Vendor Selection

Manufacturing vendor selection entailed identifying manufacturing processes and vendors and evaluating vendor process capabilities. The manufacturing vendor needed to be capable of producing regenerators that met the geometry, size, and tolerance requirements outlined in Section 8.2. Through a survey of the microfabrication and rapid prototyping industries and research groups, several vendors were identified and contacted. All vendors were provided with a document that described several regenerator designs (see Appendices F and G). Vendors were then contacted, and their responses to specific questions related to process capabilities, experience, and cost were tabulated in a selection matrix. This matrix was the basis for down-selecting to two vendors from 20. A request for proposals was issued to the two vendors. The proposals were evaluated by the entire team and International Mezzo Technologies was selected to be the manufacturing vendor. Details of discussions with potential vendors are summarized in Ibrahim (2004a).

Mezzo proposed to fabricate a regenerator that consisted of stacked metal layers. An involute cell pattern was to be formed through EDM into each 300 to 500 micron thick layer (in concept, similar to Figure 8.9). The EDM tools were to be created through a microfabrication process, LiGA, a combination of lithography and electroplating.

8.3.1 Overview of Manufacturing Processes Considered

Several fabrication processes were identified as potentially suitable for producing microscale regenerators: (1) extrusion/powder metallurgy (Tuchinskiy and Loutfy, 1999); (2) LiGA (a combination of x-ray lithography and electrodeposition) (Sandia, 2004); (3) LiGA combined with EDM (Takahata and Gianchandani, 2002); (4) microfoil lamination (Paul and Terhaar, 2000); (5) EFAB (electrochemical fabrication) (Cohen et al., 1999); (6) LENS (laser engineered net shape) (Atwood et al., 1998); (7) SLS (selective laser sintering) (Deckard and Beaman, 1988); and (8) three-dimensional (3-D) printing (Sachs et al., 1992).

Several of these processes, SLS, LENS, and 3-D printing, although promising, could not then meet microscale feature size and surface requirements. In these processes, a 3-D part is built by sintering layers of metal or ceramic particles. The particle size was 30 to 50 microns, and thus, the surface roughness and feature size requirements could not be achieved.

The remaining processes (steps 1 through 5 above) were evaluated relative to the design and functional criteria presented in Section 8.2—that is, (1) metals are desirable; (2) aligned parallel plates show the highest figure of merit; (3) micron-scale features must be produced; (4) tolerance must be in the micron scale; and (5) desired roughness is less than 7 microns. Detailed descriptions of these processes are provided in the following paragraphs. A comparison of these processes based on the above selection criteria, (1) through (5), is shown in Table 8.3.

Cellular materials with a high aspect ratio (i.e., long length compared to the cell dimensions) can be produced through a combination of powder metallurgy and extrusion. In this process (Figure 8.12), bimaterial rods are compacted and extruded to form an oriented cellular structure. The bimaterial rod consists of an outer metallic or ceramic powder layer surrounding a sacrificial core. After compaction and extrusion, the outer powder layer is sintered, and the filler material is removed. The wall thickness and porosity are dictated by the powder geometry and limits of the sintering process. Powders as fine as 15 to 20 microns in diameter are used in the process and at least two "powder" layers are required to create the walls. Also, the powder geometry limits the wall density to no higher than 95%. This process, then, is capable of producing honeycomb structures with a 30 to 50 micron thick wall. The roughness of the final part will be on the order of the powder size. Because this is an extrusion process, the overall length of the regenerator can be achieved by a single array of honeycomb cells (no "slices" are required). One of the disadvantages of the process is the difficulty in obtaining microscale tolerance and uniformity of the cell openings and wall thicknesses.

In the LiGA process, a combination of x-ray lithography and electrodeposition is used to create metallic structures on a silicon wafer. A photo resist (PR), usually poly methyl meth-acrylate (PMMA), is deposited on the wafer and exposed to an x-ray light source. X-rays are required to expose the thick PR layer because their short wavelengths allow very fine patterns to be resolved. The exposed material is then removed and the remaining PR/wafer is electroplated. Finally, the unexposed PR is removed, leaving a metal part that can be used as an EDM tool, or a mold. These parts and tools can have fairly high aspect ratios, in the range of 10 to 50 (Kalpakjian and Schmidt, 2003). Researchers have used LiGA-created EDM tools to manufacture metallic parts with microscale features. Figure 8.13 shows an EDM electrode developed by Sandia (2004), which was made by LiGA. The tool features shown are approximately 150 μm high and 50 μm wide. For microregenerator applications, individual regenerator slices could be fabricated by the LiGA process and stacked to form the 6 cm tall regenerator. Alternately, EDM tools could be

TABLE 8.3

Comparison of Various Manufacturing Techniques

Process	Materials	Regenerator Design	Feature Size Capability (Shown)	Estimated Tolerance (Including Repeatability/Uniformity)	Estimated Roughness
Powder metallurgy and extrusion to produce CMOS (cellular materials with oriented structure)	Cu, Al, Ni, Ti, steel, stainless (also ceramics)	Honeycomb Possibly single component up to 10 cm tall	Powders 15–20 µm diameter, two to three layers; walls 30–50 µm, no limit on pore size; porous walls to 95%	Micron range, see Figure 8.12	Medium/high
LiGA technology with electric discharge machining (EDM) tool (use synchrotron)	Any metal that can be EDM (includes stainless)	Honeycomb Disks 100–300 µm thick	Approximately 15 µm walls, aspect ratio range from 10:1–50:1	Submicron range (no taper)	Low
LiGA technology, do not use synchrotron, so this limits aspect ratio and leads to tapered walls	Metals such as nickel and copper	Honeycomb Disks 100–300 µm thick	Aspect ratio limit of 20:1 (based on light source), tapered walls	Submicron range, but taper could pose problem, sides will not be straight	Low
Foil lamination	NiAl (metal aluminides) preferred—to avoid warpage; also stainless steel, copper, and aluminum	Aligned channels	Current heat exchanger designs have a ratio of 40:1, ratio will also be limited by foil thickness	Existing designs show some "waviness" in plate surface	Low
EFAB: micron level buildup with sacrificial layers and metal deposition	Metals that can be sputtered: copper, aluminum, gold	Lenticular design or honeycomb	Build up in 6 µm layers (so shape will have 6 micron steps)	Micron/submicron	Medium/high (6 micron steps)

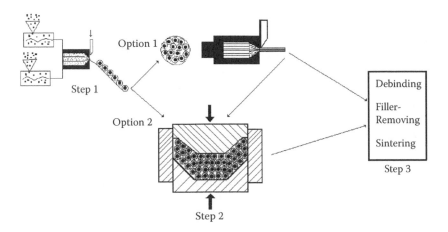

FIGURE 8.12
Combined powder metallurgy and extrusion process. Process diagram is from Tuchinsky and Loutfy (1999).

created in a combined LiGA/EDM process. Then, metallic regenerator "slices" could be produced by EDM. The geometry of parts or tools fabricated by the LiGA process are limited only to planar shapes. Thus, either the honeycomb or the parallel plate geometry can be achieved with a LiGA process.

A combination of LiGA and EDM can achieve a smooth surface finish. The surface roughness resulting from the LiGA process is less than 50 nm

FIGURE 8.13
Electric discharge machining (EDM) electrode to create screens. (From Sandia, 2004, www.mst. sandia. gov/technologies/meso-machining.html.1400_ext/1400_ext_MesoMachining.htm.Sandia.)

(Kalpakjian and Schmidt, 2003). Researchers at Ritsumeikan University (Ritsumei, Japan) (2004) fabricated high aspect ratio electrostatic microactuators using the LiGA process and found the side wall roughness was 23.1 nanometers. The EDM process, however, produces parts with roughness values varying from very rough to very smooth, depending on the metal removal rate. High rates produce a very rough finish, and low rates processed with an orbiting electrode can provide very fine surface finishes. Surfaces with a roughness of 0.3 μm can be achieved in EDM fabrication (Kalpakjian and Schmidt, 2003). Because the LiGA process produces negligible roughness, a combination of the LiGA and EDM processes will result in a surface roughness of approximately 0.3 μm, meeting the requirements for surface finish.

Microfoil lamination has been used to create microscale heat exchangers with aspect ratios as high as 40:1 (Paul and Terhaar, 2000). The lamination process begins with precision laser cutting of 25 to 250 μm thick foils. These foils are stacked, aligned, and (reactive) diffusion bonded to form microchannels (Figure 8.14). The tolerance in aligning the foils is approximately 5 μm. The surface finish, governed by the laser-processing step, is expected to be within the desired roughness of 7 μm. Either the honeycomb or the aligned plate array can be produced by this approach. The challenge in fabricating the regenerator from laminated foil sheets is in precisely aligning a stack of microchannels that is 6 cm high.

EFAB (electrochemical fabrication) technology is an additive microfabrication process based on multilayer selective electrodeposition of metals, including nickel, copper, silver, gold, platinum, and stainless steel. The minimum feature size on a plane which is normal to its height is 5 to 10 μm. Up to a 6 micron thick layer can be deposited (as compared to a 1 to 2 micron thick layer deposited by traditional microelectromechanical systems [MEMS] deposition processes) and any 3-D shape, such as an array of lenticular-shaped elements, can be fabricated by depositing sacrificial layers. The process, however, is time consuming, as multiple sacrificial and structural layers

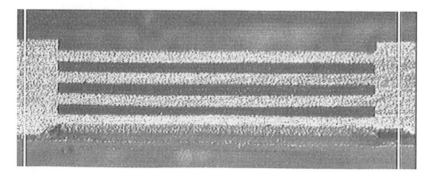

FIGURE 8.14
Laminated foil sheets with approximate dimensions 3.8 mm wide, 16 mm long (into the page), 100-μm wall, and 100-μm gap.

are required to produce a complex geometry. Thus, although EFAB is capable of fabricating all of the three proposed regenerator geometries discussed in Section 8.2.6, the time required to produce a single regenerator could be more than 6 months (after the process details have been established). Additional details regarding manufacturing process capabilities and regenerator design criteria can be found in Sun et al. (2004).

8.3.2 Background of the Mezzo Microfabrication Process and Initial, Phase I, Development

8.3.2.1 LiGA-EDM Background

The word LiGA is an acronym for three German words that, when translated, stand for lithography, electroforming, and molding. In the lithography step, a mask with the desired geometric features is placed on top of a resist such as PMMA or SU-8. The resist is then exposed to radiation or light. After the exposure, the resist is developed and the exposed volumes of the resist are dissolved, resulting in very smooth high-aspect-ratio structures with nearly vertical side walls. In the electroforming step, the voids created during the lithography step are electroplated to fill them with metal. The remaining photo-resist material is then removed leaving a metal structure with the geometric pattern of the mask. This metal structure is the molding tool, which can be used as an electrode in EDM to form the final microfabricated component. Using LiGA-fabricated electrodes in conjunction with EDM is a process that has been used at Mezzo and referred to as LiGA-EDM. A schematic of this process can be seen in Figure 8.15.

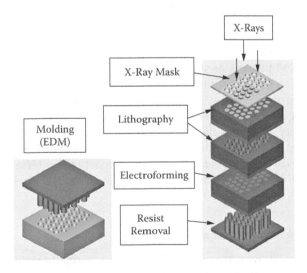

FIGURE 8.15
Schematic of LiGA-EDM (electric discharge machining) manufacturing process.

The LiGA-EDM manufacturing process was chosen for fabrication of the new regenerator. The LiGA technology has been well established for many years, while EDM technology began over 50 years ago and has been developed in recent years to allow for high precision and control over the process. EDM removes material using an electric spark and can be applied to any conductive material. In this technology, a high-frequency pulsed AC or DC current is applied through an electrode or wire to a material, which melts and vaporizes the surface of the workpiece. The electrode never comes into contact with the piece but instead discharges its current through an insulating dielectric fluid across a very small spark gap. This process has been utilized in materials that are too hard or when the geometry required cannot be machined using conventional methods. EDM technology has progressed to allow machines to individually machine holes as small as 25 microns but is limited to making one hole at a time. LiGA-EDM allows the process to form more complex geometries and to simultaneously form tens of thousands of holes with one machining operation and a single electrode. Through this process, it is possible to fabricate a variety of geometries in almost any metal and some ceramics.

By using x-ray lithography, high-aspect-ratio (5 to 35) microstructures can be formed, which allows for taller features. This is important because the EDM process consumes the tool as it forms each individual disc. When taller microstructures are used, more regenerator discs can be formed with a single tool. The process also allows for the perimeter of the individual discs to be formed in parallel with the formation of the microholes, thus removing the problems associated with alignment of the microholes with the perimeter of the device. The perimeter is made by first including a wall in the EDM tool, which surrounds the array of microposts. When the wall is plunged into the metal foil, it cuts the perimeter of the disc. Once the cut is made, the disc is ready to be inserted into the stack. This process also allows for multiple discs to be formed simultaneously as shown in Figure 8.16.

The LiGA-EDM process as well as other micro EDM processes have been used at Mezzo to form similar microstructures in erbium metal and tungsten carbide. Figure 8.17 shows some micrographs of both of these cases.

These micrographs illustrate that the LiGA-EDM process can be applied to a wide variety of materials and can theoretically be used to form these microstructures in any conductive material. With its ability to fabricate microstructures at high aspect ratios out of a variety of materials, it appeared that the LiGA-EDM method could produce the geometries needed to create a new type of Stirling engine regenerator.

8.3.2.2 EDM Electrode Design

For this project, it was necessary to develop a regenerator pattern that matched both the designs of the project and the requirements of the LiGA-EDM fabrication process. This pattern was based on the involute design that

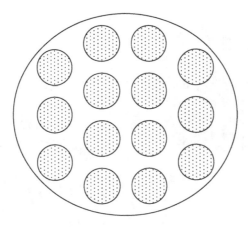

FIGURE 8.16
Batch process wafer layout.

was the focus of this project, and was used to produce a mask that was used to filter ultraviolet (UV) light for lithography. An early design is shown in Figure 8.18.

This UV mask was completed and an x-ray mask was fabricated from this pattern using UV lithography. This is the first step in the LiGA-EDM process for producing Stirling regenerator discs of the desired geometry. The dimensions of this mask are listed in Table 8.4 along with the desired dimensions of the final regenerator part.

The deviations in dimensions between the mask and final disc are due to the EDM process creating features that are always larger then the corresponding section of the EDM electrode. This will result in a product that has a smaller OD and a larger ID than the original UV mask.

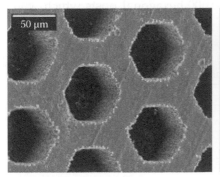

FIGURE 8.17
(Left) Erbium formed using LiGA-EDM. (Right) Tungsten carbide formed using micro-EDM.

FIGURE 8.18
Ultraviolet (UV) mask layout.

8.3.2.3 EDM Electrode Fabrication

Before electrodes were manufactured from the involute pattern, EDM electrodes made from previous regenerator patterns were used to begin the research effort. Mezzo had previous experience with this technology in manufacturing microchannel regenerators for cryocoolers, but that project involved forming microgeometries in erbium metal instead of the high-temperature materials used in Stirling engines. This initial electrode design consisted of an array of over 10,000 hexagonal posts which would

TABLE 8.4

Ultraviolet (UV) Mask Dimensions

Dimension	UV Mask	Final Disc
Regenerator ring outer diameter	19.2 mm	19.05 mm
Regenerator ring inner diameter	4.9 mm	5.05 mm
Slot width	40 microns	86 microns
Spacing between slots	60 microns	14 microns
Spacing between discs	66 microns	20 microns

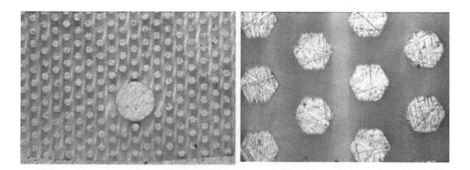

FIGURE 8.19
Micrographs of EDM electrodes for honeycomb regenerator.

be used to form a honeycomb-pattern regenerator. Each electrode hexagon had a diameter of 55 microns from point to point on the hexagon diameter. This pattern allowed Mezzo to begin performing EDM work in stainless steel before the involute pattern electrodes were completed. This hexagon pattern actually makes it more difficult to manufacture the electrodes and perform the EDM work due to the large number of features and the spacing between each feature. Micrographs of the hexagon electrodes can be seen in Figure 8.19.

While tests were being conducted with the hexagon electrodes, the involute electrodes were fabricated out of both nickel and copper to use in future research. Scanning electron microscope (SEM) micrographs show some of the features of these electrodes in Figure 8.20. Now that the involute electrodes had been fabricated, future EDM work could use these electrodes and preliminary regenerator discs could be fabricated with the final EDM step.

FIGURE 8.20
Scanning electron microscope (SEM) micrographs of electric discharge machining (EDM) electrodes for involute pattern.

FIGURE 8.21
(Left) EDMed stainless steel. (Right) Ni electrode.

8.3.2.4 Preliminary EDM Results

Results of using the LiGA-EDM technology to form microgeometries in stainless steel were obtained by using a hexagon electrode in a 420 stainless steel. SEM micrographs of one example of steel and Ni electrodes used in these initial tests can be seen in Figure 8.21.

The micrograph on the left shows a fairly large area containing a number of microholes that were produced in stainless steel. Because the electrodes used in these initial results were not of the best quality, the defects that were present in the electrode transferred to the stainless steel. This was expected to be avoided when electrodes were produced specifically for this application. Close-up micrographs of the features produced in the stainless steel can be seen in Figure 8.22.

One parameter that varies with different EDM settings and materials is the overcut. The overcut is the distance that the individual sparks travel from the electrode to the workpiece. The larger this gap is, the larger the resulting

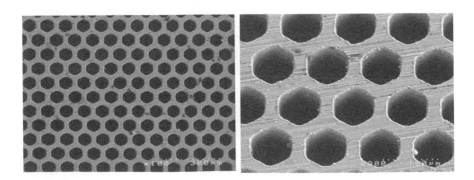

FIGURE 8.22
Close-ups of EDMed stainless steel.

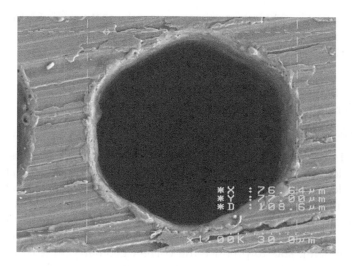

FIGURE 8.23
Close-up of EDMed hole showing dimensions.

hole would be for the same size electrode. This gap increases with increased current and reduced frequency. In order to characterize what this gap was for the current setting, a micrograph was taken of one of the holes and is shown in Figure 8.23.

This micrograph showed that the hole diameter was approximately 25 to 30 microns greater than the original electrode, resulting in an overcut of about 12 to 15 microns. The involute design allows for an overcut of up to 23 microns and will enable increasing the settings of the EDM to get faster machining times and less electrode wear.

The EDM settings used to make these features were set to minimize the surface roughness and obtain better feature definition. The surface was characterized so that during future refinement of the EDM settings, the resulting surface properties could be directly compared and the optimal settings would be established.

8.3.2.5 Mezzo Microfabrication Process Conclusions at the End of Phase I of the NASA Contract

The LiGA-EDM process development, during Phase I, showed that it was capable of forming thousands of microstructures in stainless steel, and it had been shown previously to work for more exotic metals and some ceramics. Some electrodes for EDM purposes were fabricated to produce an advanced regenerator for a Stirling engine; it was thought that continued development of the process should eventually provide advanced regenerators for both Stirling engines and cryocoolers that were already in production. With the combination of using more sensitive x-ray resists, UV lithography tools, and further refinements of the EDM process, this method could possibly be developed into a more economical process for manufacturing Stirling regenerators.

8.3.3 Continued Development of the Mezzo Microfabrication Process during Phase II

8.3.3.1 Phase II Development

The initial research plan involved Mezzo using the LiGA process to fabricate well-defined, high-aspect-ratio EDM tools. These LiGA-fabricated EDM tools would then be used to make the micromachined regenerator parts from materials with the desired high-temperature properties and low thermal conductivity (stainless steel, Inconel, etc.). EDM tools were fabricated via LiGA, and efforts to EDM parts from stainless steel showed initial promise in terms of being able to produce the correct geometry, at least at shallow depths. But the process was very slow, tool wear rate was high, and it became apparent that the probability of fabricating the desired stainless-steel regenerator using LiGA-EDM with the available funding was low.

To fabricate the regenerator on schedule, Mezzo changed its manufacturing approach. The standard LiGA process was used to directly produce individual nickel regenerator components which were then assembled, and subsequently tested at Sunpower. By changing its fabrication strategy, Mezzo was able to provide the regenerator for the project, and Sunpower was able to experimentally verify that the involute-foil layered regenerator geometry provided high performance. This change in plans also supported the desire to move the oscillating-flow rig testing from Phase III (year 3) to Phase II (year 2)—because available Phase III funding for this and other NRA contracts was in jeopardy. For details of Phase II, see Ibrahim et al. (2007).

8.3.3.2 Fabricated Regenerator

The process described in Ibrahim et al. (2007) was used to fabricate the regenerator tested in this project. Micrographs of typical parts are shown in Figure 8.24a,b,c,d. The nickel webs are approximately 15 μm in width, and arranged in an involute pattern (Figure 8.24a,b). The thickness of each disk is approximately 250 μm. Figure 8.24c shows a single involute-foil slipped onto the stacking fixture. Figure 8.24d shows a single disk leaning against the outer housing of the regenerator. Figure 8.25 shows the final regenerator that was tested.

8.3.3.3 Microfabrication Process Conclusions at the End of Phase II

In Phase II, the LiGA micromachining process was used to fabricate a regenerator that was tested and found to provide very good performance in the

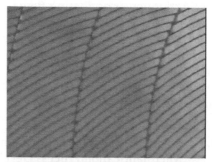

(a) Nickel Webs of Involute   (b) Lower Magnified View of Involute Pattern

(c) Disk Stacked onto Fixture   (d) Disk Leaning against Outer Housing

FIGURE 8.24

Different magnified views of regenerator disks: (a) nickel webs of involute, (b) lower magnified view of involute pattern, (c) disk stacked onto fixture, and (d) disk leaning against outer housing.

FIGURE 8.25

Assembled regenerator (stack of 42 disks).

NASA/Sunpower oscillating-flow test rig (see Section 8.4.1.4 for test results). Also, the manufacturing approach of using LiGA-fabricated EDM tools to fabricate regenerator parts seemed initially to offer little potential due to the extremely low material removal rate. Although the regenerator tested provided good performance, LiGA and LiGA-EDM process optimization could result in a better product. Potential improvements in the process are discussed in Ibrahim et al. (2007).

8.3.4 Engine Regenerator Fabrication (Phase III of Mezzo Microfabrication Process Development)

This section focuses on the contributions of Mezzo in fabrication of a regenerator for testing in a Stirling engine, during the Phase III effort. The standard LiGA process was used to directly produce individual nickel regenerator components. The first regenerator disks suffered from three key defects: underplating, some defects in the regenerator ribs, and contamination of the flow passages from wire EDM.

The goal of Phase III of this project was to build a regenerator for testing in an actual Stirling engine, which was free of the defects seen in the oscillating-flow rig regenerator. The underplating problems, which were caused by high-energy x-ray scattering, were eliminated by changing the substrate material from stainless steel to glass. With this simple material change, the x-rays were able to pass directly through the substrate. The defects in the ribs were corrected by carefully fabricating the x-ray mask with very tight process control. Finally, the contamination problems arising from the use of wire EDM for planarization was corrected by changing to a polishing process.

By changing its fabrication strategy, Mezzo was able to provide the regenerator for the project, and Sunpower was able to experimentally determine the regenerator performance. During the fabrication process, Mezzo developed several advanced processes for fabrication of the second regenerator. This section provides a summary of the revised manufacturing process.

8.4 Analysis, Assembly, and Oscillating-Flow Rig Testing of the Segmented-Involute-Foil Regenerator

8.4.1 Prototype Regenerator Testing in the Oscillating-Flow Test Rig

8.4.1.1 Physical Description

The regenerator sample tested in the NASA/Sunpower oscillating-flow test rig (see Appendix A for a description of the rig) consisted of a stack of 42 involute-foil disks (layers). The CAD rendering in Figure 8.26 shows progressive magnifications of a typical disk viewed from the front.

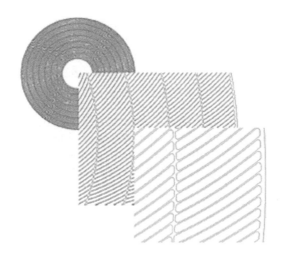

FIGURE 8.26
Involute-foil geometry.

The next CAD rendering (Figure 8.27) shows a typical single-flow channel in the matrix. It is from an early solid model and does not correspond exactly to the final matrix geometry, but it is useful to illustrate some of the important dimensions that define the geometry.

L_c is the flow channel length (disk thickness), W is the channel width (arc length of web), g is the channel flow gap (normal distance across), and s is the basic involute element spacing (gap + web thickness).

From the CAD drawings, it is possible to calculate the representative hydraulic diameter for the entire regenerator matrix like this: A unit cell of the matrix consists of one seven-ring disk and one eight-ring disk. The hydraulic diameter for such a two-disk cell is a reasonable approximation for the entire matrix. There are seven different basic channel shapes in the seven-ring disk and eight in the eight-ring disk. For each basic channel

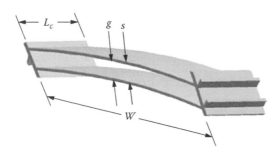

FIGURE 8.27
Channel geometry.

shape, we use available CAD tools to measure individual flow area and wetted perimeter. The total flow area A_T for the two-disk cell is the sum of the areas of each individual channel element multiplied by the number of occurrences per disk. The same is done to compute the total wetted perimeter W_T. The number of channel element occurrences per disk is just its involute generating circle diameter divided by the 100 micron involute spacings (see Appendix G). The final representative hydraulic diameter is $4A_T/W_T$ .The final hydraulic diameter calculated this way is $D_h = 162$ microns.

The above value agrees within 0.2% of the value computed by UMN for the large-scale mock-up (4.872 mm) when scaled up by the scale factor 30. For parallel plates, the hydraulic diameter would be exactly twice the flow gap ($2g$) which is more like 170 microns.

For overall matrix porosity, the value reported by Mezzo was used: $\beta = 0.8384$. This value was based on physical measurements of mass and the known material density for nickel. As reported by Mezzo, it agrees quite well with the theoretical average porosity for a two-disk cell of 0.8299. There is only a 1% porosity discrepancy in the direction that suggests that the flow channel gaps are slightly wider than expected (by about 1%). From this observation, one might expect the actual hydraulic diameter to be larger by about the same amount, on average, which would bring it to about 163.7 microns (D_h scales with flow area if wetted perimeter remains about the same).

Another parameter of interest is the ratio of channel flow length L_c (disk thickness) to hydraulic diameter which Cleveland State University investigated with CFD modeling (see Section 8.5). For the current batch of disks, the mean thickness is 238 microns (42 disks totaling 10 mm), so $L_c/D_h = 1.47$.

Still another parameter of some importance is the channel aspect ratio, or ratio of channel width to hydraulic diameter. The weighted average flow channel width for the two types of disks is about 1200 microns, so the average channel aspect ratio is $W/D_h = 7.4$.

In addition to the above parameters, the way the disks are stacked on top of each other is important. The current scheme of alternating involute directional orientation and staggered ring walls is easy to describe but difficult to quantify in terms of any simple numerical ratio.

8.4.1.2 Test Canister Design

The regenerator disks are housed within the test canister depicted in Figure 8.28 for testing in the NASA/Sunpower oscillating-flow rig.

8.4.1.3 Regenerator Verification from SEM Images

Scanning electron microscope (SEM) images of the involute-foil regenerator disks revealed excellent spacing uniformity of the flow channels and a generally smooth surface finish but also some defects. SEM micrographs were taken with the assistance of NASA Glenn Research Center (GRC).

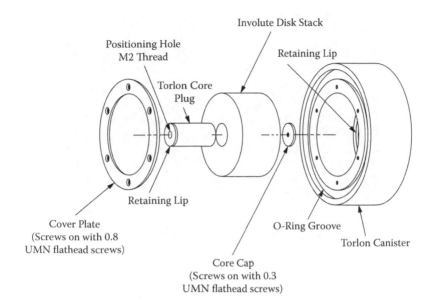

FIGURE 8.28
Testing canister geometry.

8.4.1.3.1 The Specimen

The sample photographed was the first one tested in the NASA/Sunpower oscillating-flow test rig. The matrix was not taken apart. The purpose was not to do an exhaustive survey of every one of the 42 disks inside but only to take a look at what was visible from the two ends of the assembled regenerator canister.

8.4.1.3.2 Spacing Uniformity

One objective was to assess the uniformity of spacing of the foil elements in the matrix. It was previously established that the spacing uniformity should be within ±10% to avoid significant adverse impact on the figure of merit (Appendix B). It was found that the Mezzo matrix is significantly better than this, as can be seen in the following images. The first image (Figure 8.29a) is a low-magnification view that gives one the qualitative impression that the spacing is uniform. The second image (Figure 8.29b) shows actual measurements with the Adobe Acrobat "measurement" tool used to conclude that the spacing uniformity is within ±2% in a small region of the matrix. The spacing may actually be more uniform than this because there was a significant error in the accuracy with which the estimate of the normal direction across the channel and the location of the channel edges was done with the measurement crosshairs.

The measurements are hard to read on the photograph, but the values were saved and examined using Microsoft® Excel's statistical functions; the standard deviation of foil spacing was computed to be 1.6% of the average spacing, for the 12 measurements.

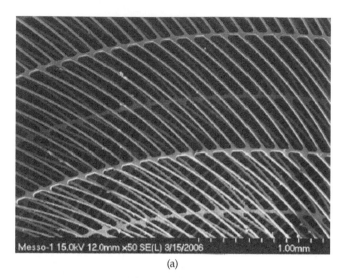

(a)

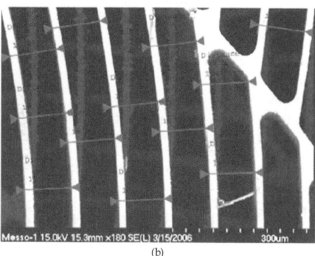

(b)

FIGURE 8.29
(a) Face-on view at low magnification showing overall good foil spacing uniformity. (b) Adobe Acrobat measurement tool in action measuring localized foil spacing in a close-up image.

8.4.1.3.3 Perspective View

Figures 8.30 and 8.31 were obtained by tilting the sample by 25 degrees. Figure 8.30 gives a good idea of the 3-D matrix structure and also shows a very smooth surface finish on the walls of the flow channels. Figure 8.31 shows a magnified view of a channel wall. We estimated the "roughness" height to be less than 1 micron, compared to an 85 micron channel gap. The roughness mainly consists of occasional pores and shallow grooves or steps, parallel to the flow direction.

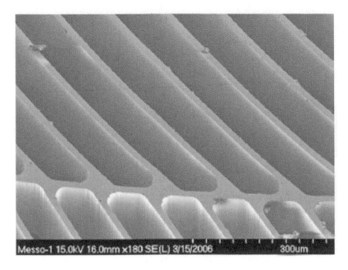

FIGURE 8.30
A 25° tilted view showing three-dimensional (3-D) structure near an annular ring with involute webs above and below. The dust visible on the surface is probably from testing or handling.

8.4.1.3.4 Defects

The matrix was not without its flaws. The image in Figure 8.32 shows significant spatter-like debris on the face of the second disk below the surface of the regenerator—at one location (microscope operators tend to focus on the defects). The debris comes from EDM removal of the disk from the plating substrate. The top disk also shows some evidence of this debris on its

FIGURE 8.31
Detail of flow channel surface roughness.

FIGURE 8.32
Visible debris attached to the face of the second disk within the sample, visible below the top disk.

lower surface. Other images show similar debris including small spherical particles of what appears to be nickel (preliminary SEM assay).

Eliminating the debris and edge roughness would be an important objective in further development. Roughness will have some effect (documented in the oscillatory-flow tests) on the helium entering the flow channels and might pose a contamination problem if any of the material works loose.

Another "flaw" in this particular assembly is that the angular involute orientations are the same in the top two disks instead of opposite (crossed), as they are supposed to be. Subsequent inspection by Mezzo showed that there was a more or less random involute orientation throughout the matrix, although the two types of disks were correctly alternated. This stacking problem was corrected and the properly stacked regenerator was also tested.

There is another sort of defect (see Figure 8.33) that appears to be associated with a flaw in the photolithographic mask used to expose the photo-resist material. Occasionally, one sees notches extending the full disk thickness, as shown in Figure 8.33. This particular notch extends a bit less than halfway through the web thickness.

The presence of such defects might put an effective lower limit on the thinness of the webs before a significant number are cut completely through. In fact, it may explain why some webs were cut completely through in some cases.

8.4.1.3.5 Recommendations

In spite of the flaws, the test results for this matrix were very encouraging. However, some of the flaws are of concern for long-term reliability, so work on quality control issues is needed for future matrices. Based on these photographs, the two main concerns are spatter-like debris near the ends of the

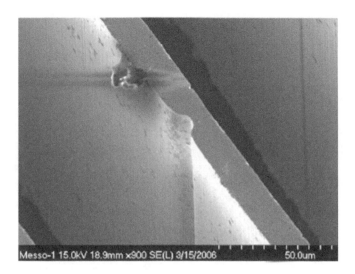

FIGURE 8.33
Notch defect extending for the full length of an involute web.

flow channels and notch defects extending completely through or mostly through channel walls.

8.4.1.4 Oscillating-Flow Rig Test Results

The results of testing in the NASA/Sunpower oscillating-flow test rig were very promising and are summarized along with results from several other regenerator types in Figure 8.34.

The microfabricated regenerator has a figure of merit substantially higher than the other regenerator types, including the 90% random-fiber regenerator that is roughly what is being used in the current generation of space-power Stirling engines.

The figure of merit, given in Equation (8.1), is a first-cut measure of overall regenerator performance. It is inversely proportional to the product of regenerator pumping loss, W_p, thermal loss, Q_t, and the square of regenerator mean flow area, A_f (see details in Appendix H, by Gedeon Associates), as follows:

$$F_m \propto \frac{1}{W_p Q_t A_f^2}$$

A_f tends to be constrained by power density (void volume), so when comparing regenerators in similar engines it may be ignored.

8.4.1.4.1 New 96% Porosity Random-Fiber Data

In Figure 8.34 the figure of merit for 96% porosity random fibers is based on the latest oscillating-flow-rig test data with improved accuracy compared to data

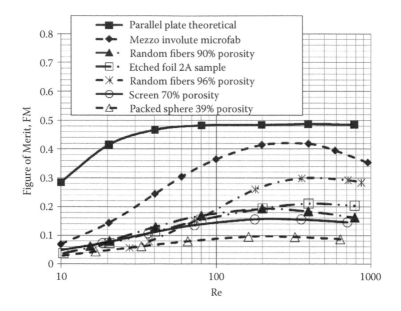

FIGURE 8.34
Figures of merit for various matrices.

previously reported. Based on the earlier data, it appeared that the figure of merit for 96% porosity random fibers was actually higher than that of the microfabricated regenerator at some Reynolds numbers. That is no longer the case, and the microfabricated regenerator now has a figure of merit substantially higher than that for the 96% porosity random fibers. The microfabricated regenerator now ranks as the best regenerator ever tested in the NASA/Sunpower test rig.

8.4.1.4.2 Unexplored Dimensionless Ratios

The pressure-drop and heat-transfer correlations discussed below (Equations 8.4 through 8.7) are for a particular regenerator matrix. Because the correlations are in terms of dimensionless quantities (like Reynolds number), they also apply to geometrically similar matrices. What does that mean?

Aside from the essential geometric property of the involute channels, namely that they consist of uniform-gap planar flow passages, and the chosen stacking geometry, the important dimensionless specifications are porosity β, the ratio of flow channel length to hydraulic diameter, L_c/D_h, and the aspect ratio, W/D_h.[*] Of these, probably L_c/D_h is most important because it affects the degree to which flow is fully developed within any given flow channel at a given Reynolds number. Porosity is probably of

---

[*] Because the porosity is a relatively good measure of the ratio of flow gap to element spacing, g/s, and the fill factor $(1-\beta)$ is a relatively good measure of the ratio of web thickness to element spacing $(1-g/s)$, there is no need to separately use g/s or $(1-g/s)$ to characterize the matrix.

lesser importance because it does not directly affect the nature of the flow channels. But it does affect the thickness of the webs between flow channels and, therefore, the way the flow enters the channels in distributing from one disk to the next. The aspect ratio is probably not too critical so long it is not significantly lower than ~10.

One should beware of applying the correlations below to microfabricated involutes with significantly different porosity, L_c/D_h or W/D_h. But if one does, it is expected that the present correlations are conservative when applied to regenerators with higher porosity or higher L_c/D_h or higher W/D_h. In either of those cases, the flow would then be closer to the ideal parallel plate flow, and the figure of merit should increase somewhat.

8.4.1.4.3 Restacked Canister

Because the original test regenerator was found to have been incorrectly assembled (random spiral orientation) after it had been tested in the oscillating-flow test rig, it was decided that the spiral orientation was important enough to warrant retesting the regenerator. Mezzo restacked the regenerator using the same disks, but this time with disks both correctly sequenced and with the spiral direction reversed at each disk transition, according to the original plan. It was then tested in the oscillating-flow rig and found that the overall figure of merit changed slightly compared to the original testing, as shown in Figure 8.35.

Figure 8.35 shows that the correct stacking produces a slightly better figure of merit at high Reynolds numbers and slightly worse values at low Reynolds numbers. As will be seen below, the friction factors for the two cases are

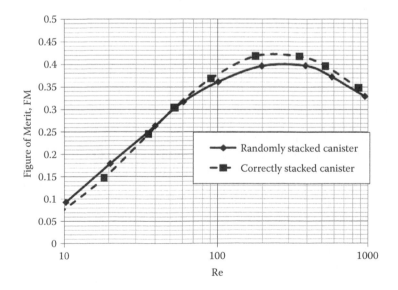

FIGURE 8.35
Figure of merit values, random and correctly stacked canister.

TABLE 8.5

Values for the Originally and Correctly Stacked Regenerators

	Random Stacking	Correct Stacking
Mean pressure (bar)	50.0	50.0
Piston amplitude (mm)	4.001	4.000
Coolant flow rate (g/s)	6.161	5.712
Coolant ΔT (°C)	2.149	2.264
Heat rejection (W)	**55.39**	**54.08**
Tmean regenerator hot end (°C)	449.2	450.5
Tmean regenerator cold end (°C)	340.1	339.2
ΔT regenerator (°C)	**109.1**	**111.3**

almost identical, so the main reasons for the differences are thermal losses, with the restacked regenerator producing more thermal loss at low Reynolds numbers and less at high Reynolds numbers. The increased thermal loss at low Reynolds numbers is probably a result of improved measurement accuracy as a result of new rig operating procedures, as explained below. The reduced thermal loss at high Reynolds numbers appears to be real.

To get a better feel for what is going on, two data points near the peak of the figure of merit curve at Reynolds numbers around 400 were taken, one data point for the original regenerator and one for the restacked canister. Both correspond to tests with 50-bar nitrogen.

As shown in Table 8.5, the two data points are nearly identical in all respects except that the restacked canister has about 2.4% lower thermal loss (heat rejection to cooler) for a 2% larger regenerator temperature difference. Assuming thermal loss scales in proportion to temperature difference, this implies that the restacked regenerator would produce about 4.5% lower thermal loss for the same temperature difference. This is consistent with the figure of merit calculation that shows about a 5% improvement for the restacked regenerator over the originally stacked regenerator at a Reynolds number of 400.

At low Reynolds numbers, where heat rejection is smaller, the thermal differences between the two regenerators are likely due to changes in test-rig operating procedures. For the restacked regenerator, a "single ramp-up" procedure designed to minimize difficulties with long-term thermal drift was adopted. Under this procedure, the operator increases the piston amplitude in 1 mm increments from 0 to 10 mm, waiting about 30 minutes for the rig to equilibrate after each change. Under the previous procedure, the operator ramped piston amplitude up and down twice over the range 0 to 10 mm, with only about 8 minutes equilibration time between changes. There was also a new procedure for measuring the baseline cooler heat rejection due to static thermal conduction. The rig was operated at 5 mm piston amplitude (midrange) for 2 hours then allowed to sit for 30 minutes at zero piston amplitude before logging the baseline static thermal conduction data points. Previously, static thermal conduction was averaged from zero-amplitude data points logged several times during the test without waiting as long for the rig to settle down to thermal equilibrium. The new procedure is considered to produce more accurate results at the low Reynolds number end of the experimental range.

8.4.1.4.4 Friction-Factor Correlations

The Darcy friction factors for the original and restacked tests are:

$$f = \frac{120.9}{R_e} + 0.362R_e^{-0.056} \quad \text{(original stacking)} \tag{8.4}$$

and

$$f = \frac{117.3}{R_e} + 0.380R_e^{-0.053} \quad \text{(correct stacking)} \tag{8.5}$$

Plotted as functions of Reynolds number, there is hardly any difference (correct stacking about 0.9% lower on average). The following plot (Figure 8.36) focuses on a small range of Reynolds numbers to better resolve the two curves. If it is plotted over the full range from 10 to 1000, then the two

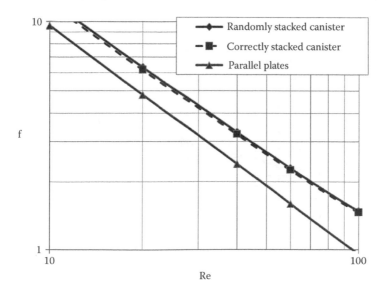

FIGURE 8.36
Darcy friction factors, f, as functions of Reynolds number, Re, for the originally and correctly stacked regenerators.

TABLE 8.6

Dimensionless Groups for the Pressure Drop Tests

Peak R_e (Reynolds number) range	3.4–1190
V_a range (Valensi number)	0.11–3.8
δ/L range (tidal amplitude ratio)	0.13–1.3

curves become indistinguishable. The range of key dimensionless groups for these tests was as listed in Table 8.6.

8.4.1.4.5 Heat-Transfer Correlations: Simultaneous N_u and N_k

The correlations for simultaneous Nusselt number and enhanced conductivity (thermal dispersion) ratio derived for this matrix are:

$$N_u = 1 + 1.99\, P_e^{0.358} \quad N_k = 1 + 1.314\, P_e^{0.358} \quad \text{(random stacking)} \tag{8.6}$$

and

$$N_u = 1 + 1.97\, P_e^{0.374} \quad N_k = 1 + 2.519\, P_e^{0.347} \quad \text{(correct stacking)} \tag{8.7}$$

$P_e = R_e P_r$ is the Peclet number. N_u and N_k are plotted individually below for the case $P_r = 0.7$ (see Figures 8.37 and 8.38).

As a reminder, N_u and N_k are designed to be used together in a model like Sage (Gedeon (1999, 2009, 2010), where the Nusselt number may be understood

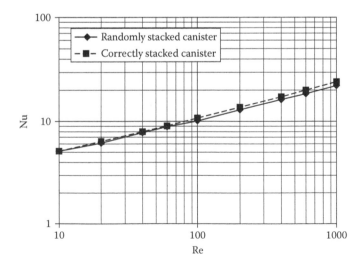

FIGURE 8.37
Mean Nusselt number N_u values as functions of Reynolds number, *Re*, for the originally and correctly stacked regenerators.

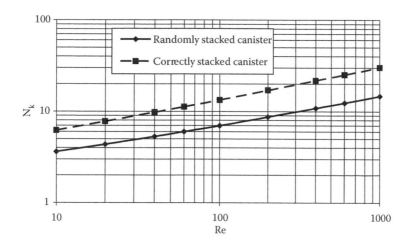

FIGURE 8.38
Mean enhanced thermal conductivity (thermal dispersion) ratio, N_k, values as functions of Reynolds number, Re, for the originally and correctly stacked regenerators.

as based on section mean temperature rather than velocity-weighted (bulk) temperature. Under this assumption, N_k compensates for any discrepancy in enthalpy flow compared to using a bulk-temperature approach.

Comparison of results for the two regenerators, in Figure 8.37, shows that the N_u values are very close to one another except for a slight, but significant, increase at high Reynolds numbers for the correctly stacked case. The N_k values in Figure 8.38 are quite different, probably due to larger measured thermal losses at low Reynolds numbers attributed to lower baseline thermal losses measured under the new rig operating procedures. The combined effects of N_u and N_k together are best seen in the plot of figure of merit in Figure 8.35.

The range of key dimensionless groups for these tests is as presented in Table 8.7.

8.4.1.4.6 Parallel-Plate Nusselt Number Comparison

Figure 8.39 compares two Nusselt-number (N_u and N_{ue}) plots derived for the microfabricated involutes against the theoretical Nusselt number ($N_u = 8.23$) for fully developed flow between parallel plates under the uniform heat flux boundary condition.

TABLE 8.7

Test Parameter Ranges

Peak R_e range (Reynolds number)	2.6–930
V_a range (Valensi number)	0.064–2.4
δ/L range (tidal amplitude ratio)	0.17–1.8

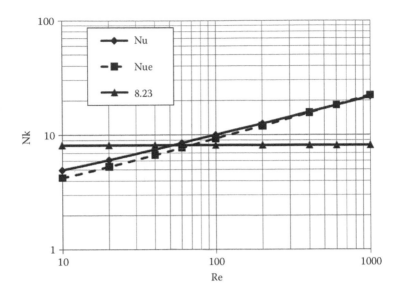

FIGURE 8.39
Nusselt-number values as functions of Reynolds number, Re, compared to the fully developed channel value.

The curve labeled "Nu" is the N_u part of the simultaneous N_u, N_k correlation. The curve labeled "Nue" is an effective N_{ue} correlation, derived under the assumption that $N_k = 1$. The two derived Nusselt numbers are close to each other but far from the theoretical Nusselt number. At high Reynolds numbers, the derived values are higher than the theoretical value, which makes sense because the flow in the microfabricated flow channels is more and more like developing flow with increasing Reynolds numbers, and Nusselt numbers are known to be higher in developing flow (as documented by the CSU see Section 8.5). At low Reynolds numbers, the derived values are lower than the theoretical value. This may be due to the effects of solid (nickel) thermal conduction within the individual regenerator disks which is of some significance at low Reynolds numbers (see Section 8.4.2.3).

8.4.2 Analytical Support to Prepare for Testing in a Frequency Test Bed (FTB) Stirling Engine

8.4.2.1 Adapting the Sunpower FTB for a Microfab Regenerator

During Phase II, the eventual goal of installing a microfabricated regenerator into the Sunpower FTB (frequency test bed) Stirling engine for testing was considered. Various options of increasing difficulty and cost required to adapt the engine to an involute-foil microfabricated regenerator with expected increasing levels of performance were identified.

8.4.2.1.1 Background

The Sunpower FTB Stirling convertor (engine plus linear alternator) was a prototype for an advanced Stirling convertor (ASC) under development for NASA. The FTB was a good choice for trying out a microfabricated regenerator, because it was available and operates at a relatively low temperature (650°C) compared to the ASC (850°C). Parts were easier to make, and operation did not require special high-temperature considerations. The goal was to adapt the FTB to use a microfabricated regenerator with minimal changes.

During Phase I of the CSU-NRA contract, some estimates were made of the performance advantage for a space-power engine using a microfabricated regenerator (see Section 8.2.7). During those studies, there was the luxury of completely optimizing the engine to take advantage of the new regenerator. With the FTB, that freedom was not available. A mostly fixed engine was available, allowing a swap-out of the random-fiber regenerator for a microfabricated regenerator and maybe a few other changes.

Here are the things that Sunpower was willing to change, listed in order of increasing difficulty and cost:

1. Use either of two available heater heads, allowing two different regenerator lengths.
2. Make a new heater head allowing for an increased regenerator length, with the same outer diameter and a new acceptor heat-exchanger insert with increased flow resistance (to increase pressure drop for displacer tuning reasons).
3. Make a new heater head as above, but also with a different outer diameter to accommodate a "thinner" regenerator.

Sunpower requested adherence to the following constraints:

1. Keep the current displacer rod diameter so as to be able to use the existing piston/cylinder assembly for the new regenerator and thereby eliminate that experimental uncertainty.
2. Restrict the regenerator length to no more than 60 mm to avoid displacer cantilever support problems.

8.4.2.1.2 Sage One-Dimensional (1-D) Code Model

This Sage code (Gedeon, 1999, 2009, 2010) modeling was done prior to any actual testing of the involute-foil test regenerator, so the regenerator is modeled as a simple foil-type regenerator. The material was stainless steel, and foil element thickness was fixed at 15 microns. The flow gap (between involute elements) was allowed to float as an optimized variable. The solid conduction empirical multiplier *Kmult* was set to 1 as a conservative estimate pending a better understanding of the advantages of interrupting the solid conduction path.

In order to maintain the displacer phase angle with a fixed rod, the overall flow resistance of the rejector + regenerator + acceptor must be maintained. The lower flow resistance of the microfabricated regenerator compared to the random fiber regenerator it replaces could not be used to full advantage. The model maintained flow resistance indirectly by imposing the constraint that the excess displacer drive power (*Wdis*) be zero. The displacer was constrained by a "displacer driver" component, and *Wdis* is the power it requires to move the displacer at the desired amplitude and phase. To satisfy this constraint, the model optimized the regenerator flow gap and, sometimes, the acceptor flow passage dimension. When the acceptor flow passage dimension was optimized, there was enough slack to also maximize efficiency. Otherwise, the regenerator flow gap was used only to meet the tuning constraint without regard to efficiency.

8.4.2.1.3 Performance Estimates

The simulated performances for the microfabricated regenerator with the above progressive FTB accommodations are given in Table 8.8, compared to the baseline random fiber performance (via the ratio of the improved microfab efficiency to that of the baseline).

The final efficiency value falls in the range of the 6% to 9% efficiency gain projected earlier (see Section 8.2.7). Most of the efficiency gain is achieved by making a new heater head, slightly longer but of the same diameter. The added bother of reducing heater head diameter hardly increases efficiency, though it does increase power level somewhat.

The higher efficiency values of the second two microfabrication cases (Table 8.8) are a result of the acceptor (heater) taking the burden of providing the additional pumping dissipation to maintain the displacer phase

TABLE 8.8

Simulated Performance Results

	W_{pv} (W)	Q_{in} (W)	PV Efficiency	Efficiency/ Baseline
Baseline random-fiber regenerator	128.0	303.7	0.422	—
Microfab—existing head, Lregen = short[a]	123.8	292.1	0.424	1.005
Microfab—existing head, Lregen = long[a]	113.2	260.8	0.434	1.028
Microfab—new head, same outer diameter (OD), Lregen = 60 mm, smaller acceptor passages	101.7	227.5	0.447	1.059
Microfab—new head, smaller OD, Lregen = 60 mm, smaller acceptor passages	110.7	247.2	0.448	1.062

[a] Short and long regenerators refer to two different dimensions allowed for design and not available for publication.

TABLE 8.9

Flow-Gap Values

	Flow Gap (microns)
Microfab—existing head, Lregen = short[a]	52.3
Microfab—existing head, Lregen = long[a]	58.0
Microfab—new head, same outer diameter (OD), Lregen = 60 mm, smaller acceptor passages	92.7
Microfab—new head, smaller OD, Lregen = 60 mm, smaller acceptor passages	91.6

[a] Short and long regenerators refer to two different dimensions allowed for design and not available for publication.

angle, freeing the regenerator gap to optimize for efficiency. The acceptor is the best place to put the added dissipation, because at the hot temperature it should, in theory, be partially recoverable. The extra dissipation subtracts from the available PV power, at-least somewhat, and it would be better to reduce the displacer rod diameter instead. But that is not allowed in the current FTB and will have to wait until the future when a space-power engine/alternator might be designed from the ground up to employ a microfabricated regenerator.

8.4.2.1.4 Regenerator Dimensions

The regenerator flow gaps for the microfab cases of Table 8.8 are given in Table 8.9.

The second two have much larger gaps (presumably much easier to make) as a result of the acceptor providing the additional damping to maintain the displacer phase angle. The overall regenerator dimensions are not included in the table to avoid revealing any proprietary FTB dimensions.

8.4.2.1.5 Recommendations

Converting an FTB to use a microfabricated regenerator is feasible, and it appears that the best option is the next to the last—a new longer heater head with the same OD and a redesigned acceptor insert. It remained to be worked out exactly how to hold the new regenerator in place and what measures will be required at either end to distribute the flow from the acceptor and rejector.

8.4.2.2 Radiation-Loss Theoretical Analysis

During the Phase I final review, an evaluation of the effects of radiation through the regenerator was requested. This was done by a simplified theoretical analysis of radiation down a long thin tube (and later, a CSU CFD analysis). A long thin tube overestimates radiation through a stack of involute-foil disks, because there is a clear sight path down the whole length of the tube. In the actual involute-foil stack, it is impossible to see any light passing through it when held up to a bright light source.

The conclusion is that the radiation heat flow down the regenerator would be very small. Near the cold end about 6×10^{-4} of 18 W, or about 10 mW. Near the hot end about 1×10^{-2} of 18 W, or about 200 mW. In terms of the time average enthalpy flux (13 W) and the solid thermal conduction (7 W), the radiation loss is smaller by two orders of magnitude. For detailed derivation of the radiation heat transfer loss see Appendix C.

8.4.2.3 Solid Thermal Conduction in Segmented Foil Regenerators (Derivation by Gedeon Associates)

One benefit of our stacked-disk design was that it interrupts the solid thermal conduction path from one end to the other. This section explores this statement in detail and finds that the truth is more complicated and depends on disk thickness, solid thermal conductivity, and also the properties and Reynolds number of the gas flowing through it.

An analysis of segmented foil regenerators shows a complicated reality. In one extreme, coupling between the regenerator gas and solid bridges the contact resistance between segments, producing solid conduction in individual segments approaching that of a continuous foil regenerator. In the other extreme, high thermal conduction within each segment produces a stair-step solid temperature distribution with distinct temperature gaps between segments, increasing the net enthalpy flow down the regenerator. In either case, solid conduction shows up as a regenerator thermal loss. The Mezzo regenerator lies closer to the second extreme, which is likely the cause of the reduced figure of merit at low Reynolds numbers in recent testing. This raises concerns about using high-conductivity materials, such as the pure metals (nickel, gold, and platinum) favored by the current electroplating fabrication process (LiGA). If such materials are used, the optimal regenerator disks need to be shorter and have thinner walls than disks made of lower conductivity materials.

8.4.2.3.1 Average Solid Conduction

Equation (8.8) is a simple approximation for the average solid conduction in a segmented foil regenerator (symbol definitions for this section are shown in Table 8.10):

$$\frac{\bar{Q}_s}{Q_{s0}} \approx \frac{1}{1 + 2\frac{(1-\beta)^k}{\beta}\frac{s}{k}\frac{1}{F}} \tag{8.8}$$

where F is the not-quite-so-simple factor:

$$F = \langle R_e P_r \rangle \frac{L_c}{D_h}\left(1 - e^{-\frac{2N_u}{\langle R_e P_r \rangle}\frac{L_c}{D_h}}\right)^2 \tag{8.9}$$

TABLE 8.10

Symbol Definitions for Section 8.4.2.3

$\bar{Q}_s$	Spatial average solid conduction for segmented foil regenerator
Q_{s0}	Conduction for continuous foil regenerator with same solid area
β	Regenerator porosity (void fraction)
D_h	Hydraulic diameter (twice flow gap)
L_c	Foil segment length (disk thickness of Mezzo design)
k_s	Solid thermal conductivity
k	Gas thermal conductivity
$<R_eP_r>$	Time averaged RePr product
R_e	Reynolds number based on hydraulic diameter
P_r	Prandtl number
N_u	Nusselt number (hD_h/k)

The physics behind this approximation is that heat-transfer coupling between the gas and solid tends to short-circuit the contact resistance between solid segments, reducing the temperature gap between segment endpoints. This imposes temperature gradients within the segments and, therefore, thermal conduction. The thermal conduction varies along the segment length, but the average value is of interest. The details are derived in the following subsections.

The point in looking at average solid conduction is that when it is high—approaching that of a continuous-foil matrix—then any contact resistance between segments is doing no good. Is that the case for the Mezzo regenerator? Table 8.11 shows some key values for that regenerator.

For these values, the average solid conduction ratio of Equation (8.8) reduces to Equation (8.10):

$$\frac{\bar{Q}_s}{Q_{s0}} \approx \frac{1}{1 + \frac{180}{F}} \tag{8.10}$$

The factor F depends mainly on Reynolds number and relative segment length L_c/D_h and is plotted in Figure 8.40 for the case of Prandtl number, $Pr = 0.7$ and Nusselt number, $N_u = 8.23$ (developed flow between parallel plates, constant heat flux boundary condition).

TABLE 8.11

Key Values for the Mezzo
Involute-Foil Regenerator

β	0.84
L_c/D_h	$250/2(85) = 1.5$
k_s/k	$86/0.18 = 480$

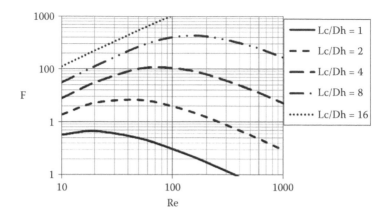

FIGURE 8.40
Factor F of Equation (8.9) as a function of Reynolds number at various segment lengths L_c/D_h, for $Pr = 0.7$ and $Nu = 8.23$.

As relative segment length L_c/D_h increases, F grows quickly so that solid conduction approaches that of a continuous-foil regenerator. For the Mezzo regenerator with $L_c/D_h = 1.5$, the F factor peaks at about 15, at a Reynolds number of about 30, which means that solid conduction is never more than about 8% of continuous foil conduction, according to Equation (8.10). In other words, the Mezzo regenerator segments never have more than about 8% the regenerator-average temperature gradient. Each segment is relatively isothermal. The high contact resistance blocks most of the solid thermal conduction.

8.4.2.3.2 Price of Localized Solid Conduction

But isothermal regenerator segments are not without their cost. A stair-step solid temperature profile (piecewise constant) results in a minimum enthalpy flow per unit flow area of :

$$h_{min} = -c_P \rho \langle u \rangle \ T_g \tag{8.11}$$

where $\langle u \rangle$ is the time-averaged flow velocity (section-mean, absolute value), and ΔT_g is the solid temperature difference between successive regenerator segments (see Figure 8.41). This is easy to understand by considering a regenerator cross section between segments (disks) where the overall regenerator temperature is increasing in the positive direction. For positive flow, the gas temperature is always lower than the solid temperature of the negative segment because it is heating up overall. For negative flow, the gas temperature is always higher than the solid temperature of the positive segment. So, the time-average difference in gas temperatures passing through the gap must be at least ΔT_g. The minimum enthalpy flow can be put in dimensionless form by dividing by molecular gas conduction $-k \ dT/dx$. A key observation

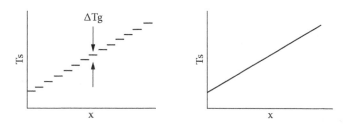

FIGURE 8.41
Extreme limiting cases for solid temperature distribution in a segmented foil regenerator.

is that the overall regenerator temperature gradient dT/dx is just $\Delta T_g/L_c$ for a stair-step temperature distribution. After a bit of simplification, the minimum enthalpy flow for a stair-step temperature distribution reduces to

$$h_{\min} = -k\frac{dT}{dx}\langle R_e P_r \rangle \frac{L_c}{D_h} \tag{8.12}$$

Compare this to the following continuous-regenerator enthalpy flow of Equation (8.12) (Gedeon and Wood, 1996):

$$h = -k\frac{dT}{dx}\left\langle \frac{(R_e P_r)^2}{4N_u} \right\rangle \tag{8.13}$$

At low Reynolds numbers, the minimum enthalpy flow $h_{\min}$ dominates h while the reverse is the case at high Reynolds numbers. The Reynolds number at which the two are the same is found by equating $h_{\min}$ to h and solving. That critical Reynolds number, assuming constant Nusselt and Prandtl numbers, is

$$\langle R_e \rangle_c = \frac{4N_u}{P_r}\frac{L_c}{D_h} \tag{8.14}$$

Whenever a Reynolds number is higher than this, the effect of the stair-step temperature distribution will be of diminishing significance. For the Mezzo regenerator, $<R_e>_c$ is about 50, where the value of $h_{\min}$ from Equation (8.12) is about 50 times higher than the helium static conduction loss. If the regenerator were continuous foil, the solid conduction loss would be a factor k_s/k $(1-\beta)/\beta$ higher than helium static conduction (conductivity ratio times area ratio) or about 91 times higher. The segment-localized solid conduction produces an adverse helium enthalpy flow that is nearly as bad as if the solid conduction passed unimpeded by contact resistance through the entire regenerator.

This coupling of localized solid conduction to helium enthalpy flow likely explains why the experimental test results for the Mezzo regenerator showed

TABLE 8.12

Solid Conduction

Material	Thermal Conductivity Near Room Temperature[a] (W/m-°C)	Ratio Relative to 316 Stainless Steel
316 stainless steel	16	1
Platinum	73	4.6
Nickel	86	5.4
Gold	300	19

[a] From 1985 Material Selector Handbook, Materials Engineering.

a suspiciously low Nusselt number below $R_e \approx 75$ (as in Figure 8.39). The derived Nusselt number was forced to account for the h_{min} enthalpy flow based on a regenerator model that did not assume a stair-step temperature distribution. High solid conduction was at the root of the problem, in spite of the high contact resistance between regenerator disks. Had the solid conduction been lower, then the temperature gaps would have been smaller, and h_{min} would have been smaller.

What might be done about this problem? The obvious solution is to make the regenerator disks from lower-conductivity material. The ratio of solid-to-gas conductivity k_s/k is rather high for a nickel regenerator, or about five times that for a stainless-steel regenerator. As a rule of thumb, alloys have lower thermal conductivity than pure metals. Table 8.12 gives some idea of the solid conduction losses of pure-metal regenerators compared to a stainless-steel regenerator.

Let's not be too pessimistic here. The good figure of merit measured for the nickel regenerator includes the effects of solid conduction loss. But the involute-foil would do even better in the future with a lower conductivity material or thinner walls, especially at lower Reynolds numbers where the temperature-gap effect dominates. The Mezzo regenerator figure of merit peaked at a Reynolds number of about 200, whereas most Stirling engine optimizations performed in the past prefer Reynolds number amplitudes on the order of 100 or lower.

8.4.2.3.3 Solid Conduction Derivation

The solid temperature in a segmented-foil regenerator lies somewhere between the two extremes illustrated in Figure 8.41 between a piecewise constant stair-step distribution and a continuous distribution. The first case corresponds to static conduction in vacuum with very high contact resistance between segments. There is a temperature gap ΔT_g between segments. In the second case, the gas is in very good thermal contact with the solid and has sufficient heat capacity to dump any required amount of heat in the segment entry regions so as to eliminate the temperature discontinuities.

The reality is somewhere between the two. That reality looks something like the illustrations in Figure 8.42 that show the gas and solid temperature

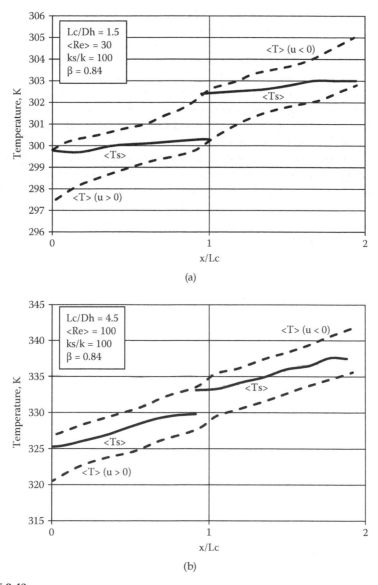

FIGURE 8.42
Time-average solid- and gas-temperature distributions for the middle two segments of a six-segment foil regenerator. (Top) 250 micron long (shorter) segments; (bottom) 750 micron long (longer) segments.

distributions (time averages) for the middle two segments of a six-segment (shorter) segmented regenerator. A 1-D Sage-code simulation, as discussed in detail later, generated the solutions. Gas temperatures are averaged over the positive and negative flow half-cycles individually. Without time averaging, the temporal temperature variations (due to finite solid heat capacity) would appear significant at this scale, but they occur roughly 90° out of phase with the mass flow rate, so they do not affect the net enthalpy flow. The upper illustration shows a solution for a segmented regenerator with the same properties as the involute-foil regenerator (250 microns thick), except with stainless-steel solid material in order to better show the solid temperature variation. The lower illustration is for a stainless-steel regenerator with three times longer segments, corresponding to 750 micron thick disks, and a higher Reynolds number, where the solid temperature gradient increases to about 60% of the regenerator average.

In any case, the average solid conduction within a segment is proportional to the average solid temperature gradient, which may be written $(\Delta T_m - \Delta T_g)/L_c$, where ΔT_m is the temperature difference between neighboring segment centers, ΔT_g is the temperature gap between neighboring segment endpoints, and L_c is the segment length. The average solid conduction is then the product of solid conductivity k_s, solid cross-sectional area A_s, and the average temperature gradient:

$$\bar{Q}_s = -k_s A_s \frac{T_m - T_g}{L_c} \tag{8.15}$$

The solid temperature gap ΔT_g depends on the overall energy balance that can be roughly formulated in terms of a control volume between the beginning of a regenerator segment (left) and its middle (right), as shown in Figure 8.43.

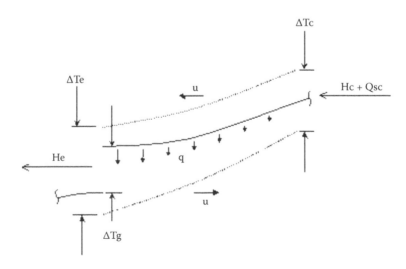

FIGURE 8.43
Energy balance between segment end (left) and center (right), bounded by vertical dotted lines. The solid curve represents time-average solid temperature. The dotted curves above and below represent gas temperatures time averaged separately for positive and negative flow directions.

The energy balance is formulated in terms of the time-average enthalpy flow H carried by the gas and the axial solid conduction Q_s. At the left boundary (segment endpoint), these values are H_e and zero, respectively (zero solid conduction because of presumed high contact resistance between segments). At the right boundary (midsegment), the values are H_c and Q_{sc}. The half-segment energy balance may therefore be written as in the following:

$$H_e = H_c + Q_{sc} \qquad (8.16)$$

The heat transfer q between gas and solid, shown in Figure 8.43, is not part of the energy balance because both gas and solid are included together and what leaves one enters the other.

The essential idea is that the intersegment temperature gap ΔT_g results in an increased net enthalpy flow at the segment entrance compared to the equilibrium value deep within the segment after the gas has had time to drop its extra heat to the solid. For a long segment, the increased enthalpy flow is $-cp<\dot{m}> \ T_g$, where $<\dot{m}>$ is the time-averaged absolute mass flow rate. For a short segment, the increase is not as much. In a section of any length, the difference in net gas enthalpy flow between the segment end and its midpoint is approximately:

$$H_e - H_c \approx -c_p \langle \dot{m} \rangle \ T_g (1 - e^{-A/2})^2 \qquad (8.17)$$

where the extra factor $(1-e^{-A/2})$ is derived in Section 8.4.2.3.7, with A given by Equation (8.23).

The dumping of heat from the gas to the solid results in solid conduction that varies smoothly from zero at the segment endpoint to a maximum at the midpoint. If the segment is not too long, then it is reasonable to assume that the average solid conduction is halfway between the endpoint and midpoint values. In other words,

$$Q_{sc} = 2\bar{Q}_s \qquad (8.18)$$

In terms of the above simplifications, the energy balance can be written as in Equation (8.19):

$$-c_p \langle \dot{m} \rangle \ T_g (1 - e^{-A/2})^2 \approx 2\bar{Q}_s \qquad (8.19)$$

Solving for ΔT_g and substituting into Equation (8.15) gives the average solid heat flux in the form:

$$\bar{Q}_s \approx k_s A_s \frac{T_m}{L_c} - 2 \frac{k_s A_s}{L_c} \frac{1}{c_p \langle \dot{m} \rangle (1 - e^{-A/2})^2} \bar{Q}_s \qquad (8.20)$$

The first term on the right is just the conduction Q_{s0} for a continuous foil with the same solid cross section and boundary temperatures. The second term contains several factors that can be arranged into more standard dimensionless groups. Solving for $\bar{Q}_s/Q_{s0}$, the result is Equation (8.8), given near the beginning of this section.

The average segmented regenerator conduction approaches the continuous foil conduction as the term $2\frac{(1-\beta)}{\beta}\frac{k_s}{k}\frac{1}{F}$ of Equation (8.8) approaches zero, which can happen for a number of reasons, for example, high porosity β (thin foil), low solid conductivity k_s, long segment length L_c, or low Reynolds number. On the other hand, the average segmented regenerator conduction approaches zero as that same term approaches infinity, which happens for the opposite reasons.

8.4.2.3.4 Sage Models

The illustrations in Figure 8.42 were the result of modeling individual segments of a foil regenerator using the Sage code. It would be very awkward and time consuming to model a total regenerator this way (consisting of hundreds of segments), but it is not too difficult to model, say, a six-segment regenerator in order to get an idea of what is going on. The Sage model illustrated in the following Figure 8.44 does just that.

This Sage model is equivalent to the NASA/Sunpower regenerator test rig containing a rather short, six-disk, involute-foil regenerator. Each canister of the model contains a simulation of a single involute-foil regenerator disk, identical to one fabricated by Mezzo except made of stainless steel instead of nickel. The reason for stainless steel is because its lower thermal conductivity results in a higher solid temperature variation that shows up better on plots. The regenerator segments are interconnected by gas flows

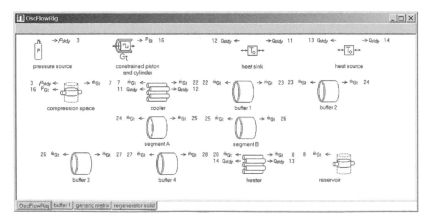

FIGURE 8.44
Sage model diagram for six-segment regenerator.

(including gas thermal conduction continuity), but the solid domains are not connected, corresponding to high contact resistance between disks.

Middle segments A and B, in Figure 8.44, are the segments of main interest (whose solutions are plotted in Figure 8.42) with the "buffer" segments on each side included to give some opportunity for the temperature solutions to develop to their equilibrium (spatially periodic) values.

8.4.2.3.5 Baseline Sage Model

The working gas is helium at 25-bar charge pressure, and the driving piston motion is sinusoidal with amplitude adjusted to produce an average Reynolds number of 30 in the regenerator segments. This Reynolds number gives the largest F factor according to Figure 8.40. The endpoint boundary temperatures are adjusted so the temperature difference between segments is about 2.5° corresponding to about what it would be in a realistic Stirling-engine regenerator (e.g., 800°C temperature difference over 320 disks). The segment temperatures range from about 293°K to 308°K, so the model corresponds to a short section near the cold end of the regenerator.

The Nusselt number in each segment is set to the constant value 8.23, corresponding to developed laminar flow between parallel plates (constant heat-flux boundary condition). This is the most unrealistic part of the model because it neglects the increased Nusselt number in the developing-flow region, which is arguably the entire section. But the intent of the model is only to get a rough idea of what is going on for purposes of validating the simplified theory derived in this section. It does not make sense to use the Nusselt number derived from recent testing of the Mezzo regenerator, because that Nusselt number applies to the regenerator as a whole and not to individual segments of the detailed model.

8.4.2.3.6 Longer Segments

The baseline model is not very satisfying from an academic viewpoint because there is not much solid temperature variation within segments. This can be remedied by increasing the segment length by a factor of 3, producing a ratio $L_c/D_h = 4.5$. To keep the same overall temperature gradient, the temperature difference between segments is also increased by a factor of 3. By increasing the average Reynolds number to 100 (increasing charge pressure and stroke), the F factor of Figure 8.40 is roughly at its maximum value of about 100, and the average temperature of the solid should be about 70% of the regenerator average according to Equation (8.8). That is not too far off from the 60% of the Sage solution shown in the lower illustration of Figure 8.42.

8.4.2.3.7 Steady Heat-Transfer Solution

At the heart of the preceding solid conduction derivation is the energy balance in a single segment of the regenerator. That energy balance is the result of a simplified steady-state solution for heat transfer between a solid

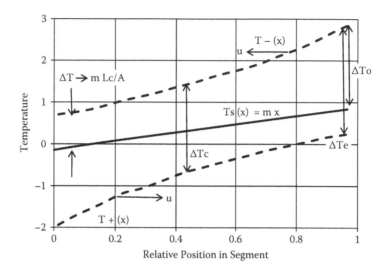

FIGURE 8.45
Steady-flow gas-temperature solutions for regenerator solid segment with linear temperature distribution.

segment of length L_c with a time-invariant linear temperature distribution $T_s(x) = mx$ and two steady gas streams of mass velocities $\pm\rho u$, as illustrated in Figure 8.45.

The governing equation, (8.21), is a simplified energy equation for incompressible, steady flow, considering only heat transfer to the solid:

$$\frac{dT}{dx} = -\frac{hs}{c_p \rho u}(T - T_x) \tag{8.21}$$

In Figure 8.45, the gas temperature solution for positive directed flow is denoted T_+, and the solution for negative flow T_-. The boundary condition is that there is a temperature difference ΔT_0 between the gas and solid at the negative end for T_+ and the positive end for T_-.

Of interest is the temperature difference $T_- - T_+$ at either entrance, denoted ΔT_e, compared to the temperature difference at the segment midpoint, denoted ΔT_c. That temperature difference determines the amount of heat transfer to or from the solid between the segment end and midpoint. Skipping all the details, the final result is:

$$T_e - T_c = \left(T_0 + \frac{mL_c}{A}\right)\left(1 - e^{-A/2}\right)^2 \tag{8.22}$$

where A is

$$A = \frac{4N_u}{R_e P_r} \frac{L_c}{D_h} \tag{8.23}$$

The asymptotic temperature difference $T - T_s$ for a long section is just $\pm mL_c/A$, depending on the flow direction. In that case, the first factor on the right in Equation (8.22) is the total gas temperature change from section entrance to exit, which is the same as the solid temperature difference ΔT_g between regenerator segments. Making this approximation regardless of segment length results in the following approximation, Equation (8.24), for the change in net gas enthalpy flux (per unit flow area) between the ends and middle of a regenerator segment:

$$h_e - h_c \approx -c_p \rho \, |u| \; T_g \left(1 - e^{-A/2} \right)^2 \tag{8.24}$$

This is essentially the approximation used in energy balance Equation (8.17). In making the leap from steady flow to sinusoidal flow, the factor $|u|$ appears in Equation (8.17) in the form of the time-average absolute mass flow rate, the time variation of the factor $(1-e^{-A/2})$ is ignored, and A is evaluated at the mean Reynolds number.

8.5 CFD Results for the Segmented-Involute-Foil Regenerator

8.5.1 Description of the Actual-Size Regenerator and Its Large-Scale Mock-Up

The actual-size regenerator (see Figure 8.46a,b) consists of a stack (42 layers) of involute-foil disks (or annular portions of disks) that have been microfabricated by the LiGA processes. It has a low resistance to flow because it has a reduced number of separation regions, compared to wire screen and random fiber. The resulting figure of merit (proportional to the ratio of heat transfer to pressure drop) has proven superior to the currently used random-fiber and wire-screen regenerator structures.

The large-scale mock-up (LSMU) testing was based on dimensions 30 times actual size, and test conditions dynamically similar to those expected in an engine (see Figure 8.47). The LSMU was fabricated and tested at UMN.

Figure 8.48 (similar to Figure 8.26) shows a progressively exploded flow direction view of one of the involute-foil regenerator disks. On the second

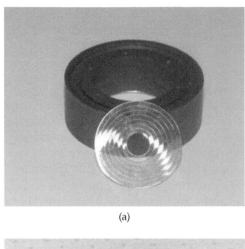

(a)

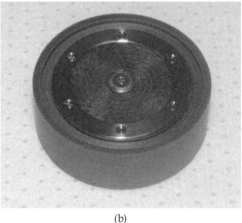

(b)

FIGURE 8.46
(a) A disk leaning against outer housing of the actual-size involute-foil regenerator. (b) Assembled actual-size involute-foil regenerator (stack of 42 disks).

zoom, the channels can be seen more clearly. The disk portrayed in Figure 8.48 has six ribs. There is a second type of disk that has seven ribs. This enables the staggering of the ribs in the stack, to reduce axial conduction and improve axial channeling of the flow.

Figure 8.49 shows a 3-D view of only one channel of one layer with dimension labels. It can be seen in this figure that the walls of the channel are curved with an involute-foil profile. Table 8.13 shows involute-foil channel dimensions and the segment or layer length (thickness, in the table). Note that the channel "width," W, given in Table 8.13 is a nominal value; as can be seen in the middle view of Figure 8.48, W varies somewhat from the inner to the outer ring of each layer of the actual-size involute-foil segments (this

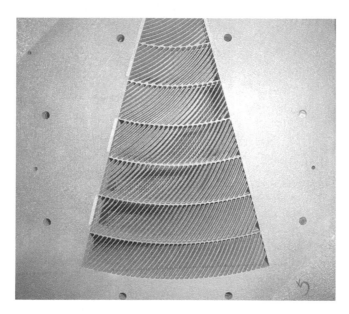

FIGURE 8.47
A picture of five aligned large-scale mock-up (LSMU) plates.

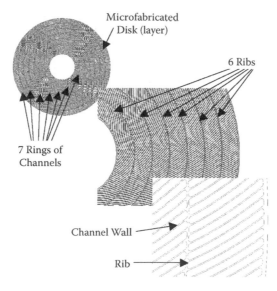

FIGURE 8.48
Exploded view of the microfabricated disk.

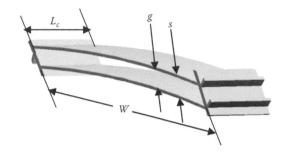

FIGURE 8.49
Segmented-involute-foil channel, of one layer.

variation of W is also true for the large-scale segmented-involute-foils of the LSMU—as can be seen in Figure 8.47).

This design has several potential advantages compared to existing designs (such as random-fiber and wire-screen matrices) as demonstrated by Sage (Gedeon (1999, 2009, 2010), and finite element analysis (FEA) done earlier (Ibrahim et al., 2007). The actual microfabricated hardware and test results showed that this design has improved ratio of heat transfer to pressure drop (i.e., high figure of merit), low axial conduction due to minimal contact area between disks, better reproducibility and control over geometric parameters, and high structural integrity and durability.

It was a challenge to analyze this geometry for fluid flow and heat transfer via CFD so as to facilitate the comparison between this geometry and other regenerator geometries. The first difficulty is the complex 3-D geometry, as depicted in Figure 8.46b. The second is the oscillatory nature of the flow as it occurs in the Stirling engine.

A description will be given for the modeling setup used to analyze the problem. This includes the grid generation for the computational domain and the matrix under investigation. Results of the numerical investigation for

TABLE 8.13

Involute-Foil Channel Dimensions

Dimension	Unit	Value
Gap, g	Micron, 10^{-6} m	86
Gap+wall, s	Micron	100
Wall thickness, s-g	Micron	14
Channel width, W	Micron	1000
Disk (layer, segment) thickness, L_c	Micron	265
Porosity		0.838
Hydraulic diameter, Dh = 4* wetted area/wetted perimeter	Micron	162

the different parameters of the proposed problem will be presented followed by the conclusions.

8.5.2 CFD Computational Domain

It was decided early on that it would not be feasible from a microscopic computational point of view to model the whole regenerator. Therefore, it was necessary to look for simplifications, which usually come in the form of symmetries and boundary-condition approximations. Accordingly, we took several progressive simplifying steps in analyzing the involute-foil matrix via CFD. Below is a brief description of each of the CFD models used, from more complex to simpler models. The FLUENT (2005) commercial code was utilized in this investigation.

8.5.2.1 Model (1), 3-D Involute Channels

Employing the radial and angular periodicity of the geometry, we were able to build a model based on one sector (with 8.87°) and apply periodic boundary conditions as shown in Figure 8.50. Figure 8.51 shows an enlarged area from the middle of the sector of Figure 8.50. In Figure 8.51, one can see the channel walls, the angle formed between channel walls (81° in this middle ring) in two successive layers and how the involute-foil profile of the wall deviates from a flat wall (by about 2°). This figure suggests possible further simplification of the geometry by approximating the involute-foil profile with a straight line and the angle between successive layer walls with a right angle.

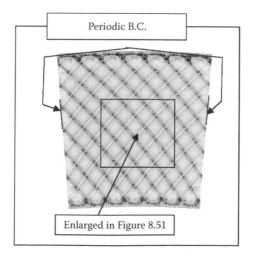

FIGURE 8.50
Periodic sector from Figure 8.48, showing two layers, with computational grids.

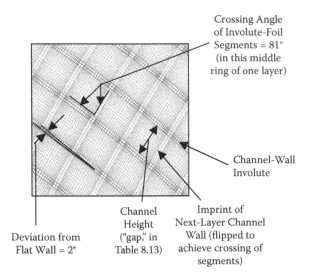

FIGURE 8.51
Enlarged area in the middle of the periodic sector.

The regenerator is a stack of just two types of alternating layers. Thus, the repeating unit is composed of two layers. One can use the flow output of one repeating unit as the input to the next one. Figure 8.52 shows an isometric view of three successive layers (or more precisely, four boundaries of these three layers). Because the frontal area occupied by the circular ribs is very small in comparison to the rest of the frontal area, a simplification was made to line up the ribs from disk to disk. In the real geometry the ribs are staggered from disk to disk. The orientation (crossing angle) of the channel walls was not significantly altered by the alignment of the ribs. As shown in Figure 8.52, the ribs of two successive layers are aligned, but the channel walls are still approximately perpendicular from layer to layer. Aligning the ribs enables a bounded domain in the radial direction for both layers that forms a repeating unit.

As mentioned above, the minimum thickness of the repeating unit has to be the thickness of two layers. However, the interface between two layers is a geometric discontinuity. The exit (velocities and temperatures profiles) of one repeating unit would be used as a boundary inlet to the next one. It is better to have no geometric discontinuities at boundary inlets and outlets. Therefore, the selected repeating unit consisted of half the thickness of one layer, followed by a full-thickness layer, and ending in another half thickness of the next layer. So, a half-layer thickness was used at the entry and exit. Figure 8.53 shows this arrangement.

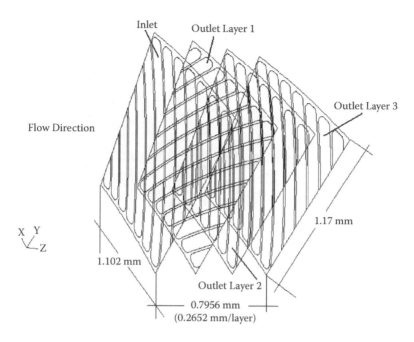

FIGURE 8.52
Isometric view of four of the boundaries of three successive layers.

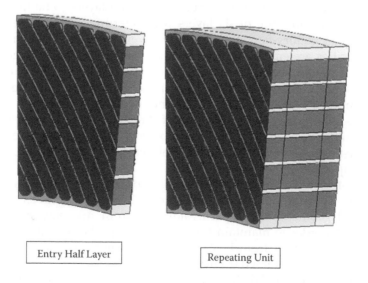

FIGURE 8.53
Model (1), three-dimensional involute-foil layers computational domain, entry unit, and repeating unit. The half-full, half computational domain unit can be repeated periodically until the required stack height is achieved.

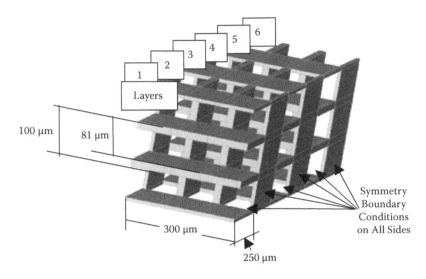

FIGURE 8.54
Model (2), three-dimensional straight-channel layers computational domain, for six layers.

8.5.2.2 Model (2), 3-D Straight Channels

Flow direction periodicity works only for steady-state modeling. The transient simulation for the case studied requires oscillatory (alternating-direction, zero-mean) flow, requiring a stack of several layers to be included in the domain. A minimum of six layers was determined to be adequate to capture the oscillatory-flow phenomena. However, even if radial and angular periodicities are employed, the grid size would still be too large for the available computational capability. Further simplifications, as indicated in Figure 8.54, must be used. If the foil-crossing angle from one layer to the next is approximated to 90° (instead of 81°), the involute-foil profile is approximated as straight, and the round ends of the channels are neglected, then one can build a manageable grid. This is expected to capture most of the 3-D oscillatory-flow phenomena of the microfabricated design. Figure 8.54 shows such a computational domain.

Table 8.14 shows dimensions of the computational domain shown in Figure 8.54. In order to maintain the same hydraulic diameter as the actual involute-foil geometry, the gap was adjusted to 81 microns, from 86 (see Table 8.13). In order to maintain the same spacing, the wall thickness was adjusted to 19 microns, from 14. Another adjustment was made to the layer thickness that was decreased from 265 microns to 250 microns. This was done to better match the actual disks that were fabricated for experimental testing, because the original design called for 265 micron thick disks, but the manufacturer (Mezzo) fabricated 250 micron thick disks. The resulting porosity, as shown in Table 8.14, was 0.81 instead of the actual regenerator's 0.84.

TABLE 8.14

Three-Dimensional (3-D) Straight Channel Computational
Domain Dimensions

Dimension	Unit	Value
Gap, g	Micron, 10^{-6} m	81
Gap+wall, s	Micron	100
Wall thickness, s-g	Micron	19
Channel width, W	Micron	300, symmetry
Disk (layer) thickness, L_c	Micron	250
Porosity		0.81
Hydraulic diameter, $D_h = 2g$	Micron	162

8.5.2.3 Model (3), 2-D Computational Domain

Further simplifications were also made, for part of the CFD study, by using
a 2-D computational domain. For the cases studied, the 2-D domain consisted of a single parallel-plate channel and solid walls with six successive
sections. There was no variation in flow geometry upon exiting one section
and entering the next. However, by changing the solid-interface settings, one
could set various values for the thermal contact resistance (TCR) between
wall sections. This was expected to capture the interruption in the wall thermal conduction obtained by alternating the orientation of the channel walls
(from one layer to the next) in the 3-D domain. Figure 8.55 shows such a 2-D
computational domain. This 2-D domain allowed for quick parametric studies and finding trends that could be later confirmed in 3-D with fewer runs.

8.5.2.4 Code Validation for All Models

As presented above, three different models were identified for dealing with
the geometry under investigation. Model (1) is 3-D for involute-foil layers;
Model (2) is 3-D straight-channel layers; Model (3) is 2-D straight channels.
They represent different levels of compromise between the actual problem
and the resources required to model the problem. Table 8.15 lists the three
different models, the cell count employed in each, and the studies conducted
utilizing each one.

In Model (1) over 5 million cells were used for steady flow runs and $Re =$
50, 94, 183, 449, 1005, and 2213. The solid material applied is stainless steel,

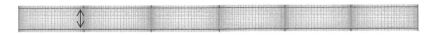

FIGURE 8.55
Model (3), two-dimensional computational domain.

TABLE 8.15

Total Cell Count, Conditions, *Re* Number Range for the Different Models Examined

Model	Total Cell Count	Conditions Examined	Re Number Range Examined		Data (from Literature) against Which the Model Was Validated
			Steady Flow	Oscillatory Flow	
Model (1), (3-D)	5,408,640	* Solid material: Stainless Steel * TCR between Layers: 0	$Re = 50, 94, 183, 449, 1005$ & 2213	NA	University of Minnesota (UMN) data, Equation (8.27) (Ibrahim et al., 2007), Gedeon Equation (8.28) (Ibrahim et al., 2007; and Kays and London, 1964), (see Figure 8.72)
Model (2), (3-D)	3,650,400	* Solid material: stainless steel and nickel * TCR between layers: 0 and ∞	$Re = 50$	$Re_{max} = 50$ $Re_\omega = 0.229$	*For steady state:* Equation (8.25) (Shah, 1978) and Equation (8.26) (Stephan, 1959), see Figures 8.64 and 8.65. Models (2) and (3) match very well in the first layer. *For oscillatory flow:* Equations (8.28) and (8.29) (Ibrahim et al., 2007), see Figures 8.46 and 8.47. Model (2) matches Equation (8.28) (Ibrahim et al., 2007) very well
Model (3), (2-D)	15,600	* Solid material: stainless steel and nickel * TCR between layers: 0 and ∞	$Re = 50, 150,$ and 1000	$Re_{max} = 50$ and 150, $Re_\omega = 0.229$ and 0.687	*For steady state:* Equation (8.25) (Shah, 1978) and Equation (8.26) (Stephan, 1959), see Figures 8.64, 8.65, 8.69, and 8.70 *For oscillatory flow:* Equations (8.28) and (8.29), see Figures 8.60, 8.61, 8.66, and 8.67

Note: TCR, thermal contact resistance between layers.

and zero thermal contact resistance was applied between layers. There are no data, to the best of our knowledge, available in the open literature for the variation of f or Nu along the flow direction. The only f and N_u data available are for the whole stack (UMN Data, Equation 8.27; Gedeon, Equation 8.28 (Ibrahim et al., 2007; Kays and London, 1964). Therefore, those data were utilized to validate the model (see Figure 8.72). Another way of validating this model is to compare its results with 2-D data (Model 3) only in the first layer (see Figures 8.69 and 8.70).

In Model (2), over 3.5 million cells were used for steady ($Re = 50$) and oscillatory flow ($Re_{max} = 50$ and $Re_\omega = 0.229$). The solid materials applied are stainless steel and nickel. Zero and infinite thermal contact resistances were applied between layers. Again for those 3-D cases, there are no data, to the best of our knowledge, available in the open literature for the variation of f and Nu along the flow direction. The only f and N_u data available are for the whole stack. Therefore, for oscillatory flow, Equations (8.28) and (8.29) (Ibrahim et al., 2007), were utilized (see Figures 8.66 and 8.67). The model matches Equation (8.28) (Ibrahim et al., 2007) very well. As for the steady-state cases, we utilized Equation (8.25) (Shah, 1978), and Equation (8.26) (Stephan, 1959) (see Figures 8.64 and 8.65). Because those equations (8.25 and 8.26) are obtained for 2-D geometries, we looked for validation for the first layer only, as seen in Figures 8.64 and 8.65.

In Model (3), 15,600 cells were used for steady ($Re = 50$, 150, and 1000) and oscillatory flow ($Re_{max} = 50$ and 150 and $Re_\omega = 0.229$ and 0.687). The solid materials applied are stainless steel and nickel. Zero and infinite thermal contact resistances were applied between layers. For the steady-state cases, Equation (8.25) (Shah, 1978) and Equation (8.26) (Stephan, 1959) were utilized for the code validation (see Figures 8.64, 8.65, 8.69, and 8.70). As for the oscillatory flow, Equations (8.28) and (8.29) (Ibrahim et al., 2007) were used (see Figures 8.60, 8.61, 8.66, and 8.67).

The way we built the computational grids (grid independence) for the three models is from the 2-D up to the 3-D. Therefore, a grid-independence study was conducted (see Section 8.5.4) for the 2-D model (Model 3) with number of cells 15,600. Then the 3-D models were extended to over 3.6 million cells for Model (2) and to over 5.4 million cells for Model (1).

8.5.2.5 Boundary Conditions

The problem under investigation involves 3-D (Models 1 and 2) and 2-D (Model 3) computational models under steady and oscillatory flow conditions. The 3-D models represent islands within the larger regenerator stack-up, the other boundaries (boundaries that are not inlets or outlets) are either symmetry or periodic boundary conditions. In the case of the 8.87° slice, see Section 8.5.2.1 for more detail, the thermal contact resistance between the solid layers (in Models 2 and 3) were chosen as zero or infinity.

TABLE 8.16

Base Case for Oscillatory-Flow Conditions

Valensi number, Re_ω	0.22885
Maximum Reynolds number, Re_{max}	49.78
Frequency, Hz	27.98
Hydraulic diameter, m	0.000162
Max mass flux, kg/m2-s	6.17215
Cold end solid boundary condition	Adiabatic
Hot end solid boundary condition	Adiabatic
Inlet fluid temperature, cold end, °K	293.1
Inlet fluid temperature, hot end, °K	310.2
Mean pressure, Pa	2500000
Mean, maximum velocity, m/s	1.5488

8.5.2.5.1 Steady-State Runs

For steady-state runs, the solid temperature was kept constant at 673°K while the fluid entered the channel at 660°K. Uniform inlet velocity was selected at the west boundary, and outlet pressure was chosen at the east boundary.

8.5.2.5.2 Oscillatory-Flow Runs

The running conditions for the base case examined in the oscillatory-flow study are shown in Table 8.16.

8.5.3 Friction Factor and Nusselt Number Correlations

The correlations discussed below were used in this CFD study.

8.5.3.1 Steady-State Correlations

8.5.3.1.1 Parallel Plate Friction-Factor and Heat-Transfer Correlations

In order to compare the steady-flow CFD results to the literature, the following correlations were selected from Shah and London (1978). The Fanning friction-factor correlation of Equation (8.25) is attributed to Shah (1978) and applies to laminar, hydrodynamically developing flow, the flow regime for the case studied:

$$f_F = \frac{1}{Re}\left(\frac{3.44}{(x^+)^{1/2}} + \frac{24 + \frac{0.674}{4x^+} - \frac{3.44}{(x^+)^{1/2}}}{1 + 0.000029(x^+)^{-2}} \right) \tag{8.25}$$

For steady heat transfer, the correlation of Equation (8.26) was selected from Shah and London (1978) and is attributed to Stephan (1959). This correlation applies to laminar simultaneously (thermally and hydrodynamically) developing flow which is the flow and thermal regime for the case studied.

This correlation is valid for a constant wall temperature and Prandtl numbers between 0.1 and 1000:

$$\text{Nu\_m} = 7.55 + \frac{0.024(x^*)^{-1.14}}{1 + 0.0358(x^*)^{-0.64}\,\text{Pr}^{0.17}} \tag{8.26}$$

8.5.3.1.2 Large-Scale Mock-Up (LSMU) Darcy Friction Factor

A study of the momentum equation noted that for engine-representative Valensi and Reynolds numbers, the transient term is unimportant and pressure-drop measurements can be taken in steady, unidirectional flow. Such measurements led to the following friction factor correlation for the LSMU (Ibrahim et al., 2007):

$$f = \frac{153}{\text{Re}} + 0.127\,\text{Re}^{0.01} \tag{8.27}$$

8.5.3.2 Oscillatory Flow—Actual Scale

As for the oscillatory-flow cases, the following correlations, of Equations (8.28) and (8.29) for involute-foil friction factor and heat transfer are attributed to Gedeon Associates (Ibrahim et al., 2007). These correlations were obtained from involute-foil experimental data. The experiments were done at Sunpower Inc., on a NASA/Sunpower oscillating-flow test rig equipped with a microfabricated involute-foil regenerator. This regenerator had 42 disks in its stack (see Figure 8.46). The material used for the disks was nickel. The friction factor determined for oscillatory flow condition was

$$f_D = \frac{117.3}{\text{Re}} + 0.38\,\text{Re}^{-0.053} \tag{8.28}$$

The range of key dimensionless groups for these tests are given in Table 8.17. Heat transfer under oscillatory flow conditions is given by

$$\text{Nu\_m} = 1 + 1.97\,\text{Pe}^{0.374} \tag{8.29}$$

The range of key dimensionless groups for these tests are given in Table 8.18.

TABLE 8.17

Dimensionless Groups for the Pressure-Drop Tests

Peak R_e Range	3.4–1190
V_a Range	0.11–3.8
δ/L Range	0.13–1.3

TABLE 8.18

Test Parameter Ranges

Peak R_e Range	2.6–930
V_a Range	0.064–2.4
δ/L Range	0.17–1.8

8.5.4 CFD Grid-Independence Test and Code Validation

Three computational domains were identified as good candidates for modeling the problem at hand. One was a 2-D domain (Model 3), and the other two were 3-D domains (Models 1 and 2). The actual geometry (see Figure 8.49) and range of *Re* (see Tables 8.17 and 8.18) indicate that a laminar, thermally, and hydrodynamically developing flow, will adequately represent this case under investigation. As for the data available for code validation, the following were identified: (1) steady flow, 2-D geometry, and local f (Equation 8.25) (Shah, 1978) and N_u (Equation 8.26) (Stephan, 1959) as functions of the axial flow location; (2) steady flow, 3-D geometry (LSMU) and mean values for f (Equation 8.27) (Ibrahim et al., 2007); and (3) oscillatory flow, 3-D geometry (actual scale), and mean values for f (Equation 8.28) (Ibrahim et al., 2007), and mean values for N_u (Equation 8.29) (Ibrahim et al., 2007). Therefore, a strategy was developed to obtain a grid independence test for the 2-D geometry (Model 3). This geometry was used in the 3-D models (1 and 2). Thus, the availability of local f and N_u for the 2-D geometry will ensure a good grid selection for the CFD experiments. Then the 3-D geometry (as an extension for that grid) can be validated using mean f and N_u for steady and oscillatory flow conditions.

The 2-D computational domains (Model 3) with four different grid sizes were chosen. The number of grids (in-the-*x*-direction × in-the-*y*-direction, per layer) are 20 × 10, 30 × 20, 50 × 20, and 100 × 40. These are summarized in Table 8.19. Figure 8.56 shows the grid for the 20 × 10 only.

The above four grid sizes were tested and the results were plotted for friction factor as a function of dimensionless length, *x*+, for *Re* = 150 (see Figure 8.57). Similarly, the results were plotted for mean Nusselt number, *Nu_m*, as a function of a different dimensionless length, *x**—also for *Re* = 150

TABLE 8.19

Summary of Grids Tested in Grid-Independence Study

Grids/ Axial-Segment	Number of Cells between Plates	Vertical Grid-Spacing Ratio	Number of Cells along Axial Segment (Layer)	Horizontal or Axial (Segment) Spacing Ratio
20 × 10	10	1.15	20	1, uniform
30 × 20	20	1.15	30	1, uniform
50 × 20	20	1.1	50	1.1
100 × 40	40	1.1	100	1.05

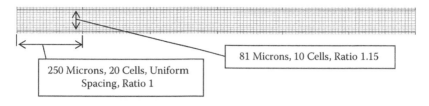

250 Microns, 20 Cells, Uniform
Spacing, Ratio 1

81 Microns, 10 Cells, Ratio 1.15

FIGURE 8.56
The 20 × 10 two-dimensional grid.

(see Figure 8.58). The results are poor for the smallest grid (20 × 10), while little gain in accuracy is achieved by moving from a grid size of 50 × 20 to 100 × 40. The 50 × 20 grid size was eventually selected as the best compromise between accuracy and computing resources, to be used following the completion of the grid-independence study. For more details see Danila (2006).

8.5.5 Results of 2-D and 3-D Modeling

The CFD data will be presented starting with 2-D, Model (3), where extensive computations were conducted followed by 3-D straight channel, Model (2), and finally 3-D involute channel, Model (1).

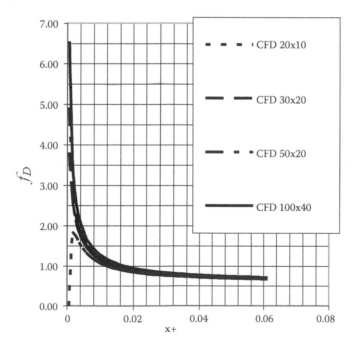

FIGURE 8.57
Grid independence study: Darcy friction-factors, f_D, as functions of dimensionless length, x^+, at Reynolds number, $Re = 150$ (for grids/segment of 20 × 10, 30 × 20, 50 × 20, and 100 × 40, horizontal × vertical).

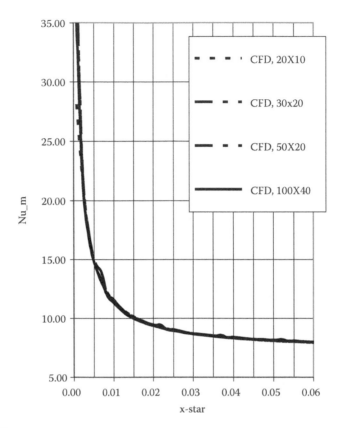

FIGURE 8.58

Grid independence study: Mean Nusselt numbers, Nu_m, as functions of dimensionless length, x^*, at Reynolds number, $Re = 150$ (for grids/segment of 20×10, 30×20, 50×20, and 100×40, horizontal × vertical).

8.5.5.1 Results of Model (3), 2-D CFD Simulations of Involute-Foil Layers

Steady state was examined first at $Re = 50$ in order to compare with available correlations. The comparison was good especially at distances farther away from the entrance. Then oscillatory-flow cases were conducted to examine the effects (on friction-factor and Nusselt number) of changing (1) the thermal contact resistance between the six layers, (2) the oscillation amplitude, (3) the oscillation frequency, and (4) the type of solid material.

8.5.5.1.1 Base Case 2-D Oscillatory Flow

For the oscillatory-flow simulations, the base case forcing function, at 27.98 Hz, is:

$$\text{Mass flux} = 6.17215^*\cos(2^*\pi^*27.98^*t + 1.56556) \ (\text{kg/m}^2\text{-s}) \qquad (8.30)$$

This function is applied at the west (left) fluid boundary. Figure 8.59 shows the variation of the mass flux with the crank angle. All the following 2-D

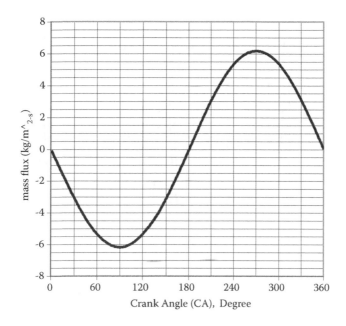

FIGURE 8.59
Mass-flux forcing function as a function of crank angle, *CA*, degrees, for the base-oscillatory-flow case.

oscillatory-flow cases were run until cycle-to-cycle convergence was obtained, and only after that were the data extracted.

For characterizing the oscillatory flow, the friction factor is compared with the experimental correlation obtained by Gedeon, Equation 8.28, for the involute-foil. The friction factor plotted versus the crank angle is shown in Figure 8.60. The values obtained by the present simulation fall below the Gedeon correlation. This was expected because the correlation was obtained from experimental results on an actual involute-foil regenerator while the present 2-D simulation represents an idealized case, with flow through a foil-channel that does not flow around foils in adjacent layers (i.e., there are no obstacles in the flow path).

In order to characterize the heat transfer that takes place during an oscillatory-flow run, the mean Nusselt number is plotted with respect to the crank angle in Figure 8.61. For comparison, the experimental mean Nusselt correlation also obtained by Gedeon, Equation (8.29), is used. However, Gedeon's correlation represents a mean over the length of a stack of 42 layers tested in the oscillatory-flow rig at Sunpower Inc.

The present work focuses only on the region from the middle of layer 3 to the middle of layer 4. So the length over which the Nusselt number is averaged in the present work is equal to the thickness of one layer only. That is done in order to stay away from the ends where entrance effects can distort the results. On the other hand, calculating a mean Nusselt number over the length of six layers would not have been representative of the actual geometry.

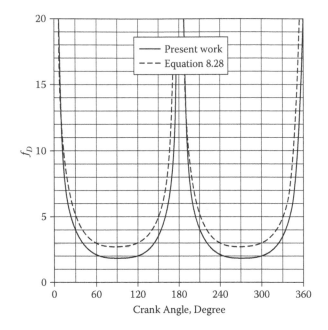

FIGURE 8.60
Darcy friction factors, f_D, as functions of crank angle, CA, degrees. Comparisons of values calculated from two-dimensional computational fluid dynamics base-oscillatory-flow case (50 × 20 grids/segment) with Gedeon involute-foil correlation of Equation (8.28).

Furthermore, experimental testing done at UMN looked at the mean Nusselt number calculated between two layers similar to what has been done in the present work. The shape of the mean-Nusselt-number, or Nu_m, plot versus the crank angle is different from the Gedeon correlation, and this difference arises from how the Nusselt number is averaged. The experiments done at UMN show a Nusselt number curve similar to the present work. However, the comparison with the Gedeon correlation is useful for the maximum Reynolds number regions that are located around 90° and 270° crank angle, where flow rates are maximum in the two directions. At these locations, the 2-D analysis lies slightly below the correlation.

By integrating the fluid enthalpy crossing a plane section in the middle of layer 3 over the whole cycle, one can calculate the net enthalpy loss over the cycle. If one integrates the solid-conduction heat transfer at the middle of layer 3 over the whole cycle, a net-conduction loss over the cycle can be obtained. Because both losses are crossing the plane at the middle of layer 3, they can be added together to obtain a total axial heat loss over the cycle. For the base-oscillatory-flow case, Table 8.20 shows these 2-D CFD axial heat-loss results.

8.5.5.1.2 Effect of Changing the Thermal Contact Resistance

In order to study the effect of TCR between the layers, a change was made from the zero TCR of the base case to an infinite TCR at the interfaces

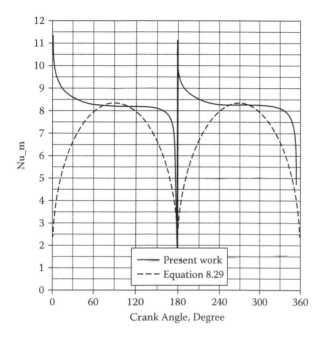

FIGURE 8.61

Mean Nusselt numbers, Nu_m, as functions of crank angle, CA, degrees. Comparison of values calculated from two-dimensional computational fluid dynamics (CFD) base-oscillatory-flow case (50×20 grids/segment) with Gedeon involute-foil correlation of Equation (8.29). (CFD assumes perfect thermal contact between layers, or zero thermal contact resistance, TCR.)

between the layers (or from perfect thermal contact to perfect thermal insulation, between solid layers). The effect of thermal contact resistance is important because in reality the contact between the layers is not perfect. This added contact resistance impedes the solid conduction from layer to layer. This contact resistance between the layers causes discontinuities in the solid-wall temperature profile between the hot and cold sides of the interface. The changed wall-temperature profile should affect the heat transfer between the wall and the fluid which, in turn, should change the plot of the Nusselt number. However, the friction factor is not affected (as expected). Figure 8.62 shows a comparison of the Nusselt number behavior between the base case, zero-TCR, and the infinite TCR.

TABLE 8.20

Two-Dimensional (2-D) Computational Fluid Dynamics (CFD) Base Oscillatory Flow Case Enthalpy, Conduction and Total Axial Heat Losses

	Enthalpy Loss, W	Conduction Loss, W	Total Loss, W
Base case	1.722	1.174	2.896

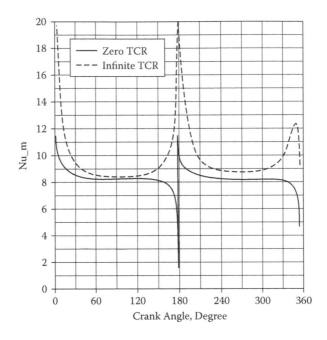

FIGURE 8.62
Mean Nusselt numbers, Nu_m, as functions of crank angle, CA, degrees. Comparison of values calculated from two-dimensional computational fluid dynamics oscillatory-flow cases (50 × 20 grids/segment) for (1) zero thermal contact resistance (TCR) (perfect thermal contact, base case) and (2) infinite TCR (adiabatic contact or perfect insulation).

The infinite TCR (adiabatic contact) has caused the Nusselt number to rise, especially in the regions of low Reynolds numbers close to where the flow reverses (near 180° and 360° crank angle). This was expected. When no TCR is present, the solid wall temperature on each side of the contact between the solid layers is the same. When the infinite TCR is introduced, a temperature difference takes place in the solid wall between the two sides of the contact area. The fluid flowing in the channel past the contact between the two solid layers sees the discontinuity in the wall temperature profile. The increased delta-T between the wall and the fluid causes an increase in the heat transfer and that is reflected in the higher Nusselt number. At lower Reynolds numbers, the fluid has more time to absorb the heat, and the effect of the infinite thermal contact resistance is more pronounced. However, when the flow stops, as it does when it switches direction, the temperature between the fluid and the solid equalizes, and delta-T becomes very small, tending to zero. That introduces a discontinuity in the calculation of the Nusselt number, and Figure 8.62 shows that discontinuity at 180° and 360° of crank angle. The other change that has happened is related to the difference between the cooling and the heating parts of the cycle. The cooling half happens from zero to 180° of crank angle when the flow goes from the cold side to the

TABLE 8.21

Heat Loss Comparison of Zero-Thermal Contact Resistance (TCR) Base Case to Infinite TCR Case

	Enthalpy Loss, W	Change	Conduction Loss, W	Change	Total Loss, W	Change
Base case	1.722		1.174		2.896	
Infinite TCR	1.960	13.8%	0.531	−54.7%	2.491	−14.0%

hot side, and the heating half happens from 180° to 360°. When zero TCR was present (base case), the mean Nusselt number curves for the two halves looked the same. However, with infinite TCR introduced, the heating half of the cycle shows a higher mean Nusselt number.

In terms of heat loss, Table 8.21 shows the results of integrating the enthalpy loss and the conduction loss over the whole cycle. In the case of the solid conduction, increasing the TCR from zero (base case) to infinite has resulted in a reduction of 54.7%. The enthalpy loss has increased by 13.8%. However, the total loss has decreased by 14%. This suggests that it is a good idea to increase the thermal contact resistance in order to reduce the heat loss. (This is approximated for the case of the 3-D involute-foil disks by alternating orientation of the involute-foil segments and offsetting separating ring locations, in adjacent disks, thus greatly reducing the contact area and increasing thermal resistance between disks.)

8.5.5.1.3 Effect of Changing the Oscillation Amplitude

The increase of the oscillation amplitude results in an increase of the maximum Reynolds numbers at 90° and 270° CA. The friction factor has dropped, and that is consistent with the steady-state simulations. When the Reynolds number is increased, the friction factor becomes smaller. The results show that Nusselt number stays much the same.

In terms of heat loss, Table 8.22 shows the results of integrating the enthalpy loss and the conduction loss over the whole cycle, for the base case and the increased oscillation amplitude case.

A reduction in conduction loss occurred after the oscillation amplitude was increased by a factor of three. However, due to the higher flow, the

TABLE 8.22

Heat Loss Comparison of Base Case ($Re_{max} = 50$) and Increased Oscillation Amplitude Case ($Re_{max} = 150$)

	Enthalpy Loss, W	Change	Conduction Loss, W	Change	Total Loss, W	Change
Base case	1.722		1.174		2.896	
$Re_{max}=150$	18.249	10.6×	0.956	−18.6%	19.204	6.6×

TABLE 8.23

Heat Loss Comparison of Base Case ($Re_{max} = 50$, $Re_\omega = 0.229$), the Increased
Oscillation Amplitude Case ($Re_{max} = 150$, $Re_\omega = 0.229$), and the Increased Oscillation
Frequency Case ($Re_{max} = 150$, $Re_\omega = 0.687$)

	Enthalpy Loss, W	Change	Conduction Loss, W	Change	Total Loss, W	Change
Base case $Re_{max} = 50$, $Re_\omega = 0.229$	1.722		1.174		2.896	
Increased-oscillation amplitude $Re_{max} = 150$, $Re_\omega = 0.229$	18.249	10.6×	0.956	−18.6%	19.204	6.6×
Increased-oscillation frequency $Re_{max} = 150$, $Re_\omega = 0.687$	13.466	7.8×	1.009	−14.1%	14.474	5.0×

enthalpy loss increased by a factor of 10.6. The effect on the total heat loss is an increase by a factor of 6.6. The reduction in the axial solid conduction is attributed to the higher radial heat flow between gas and metal due to the higher instantaneous gas mass flow.

8.5.5.1.3 Effect of Changing the Oscillation Frequency

The effect of increasing the oscillation frequency was also studied. A frequency that is three times the base-case frequency was chosen for testing this effect. Therefore, frequency increased from 27.98 Hz to 84 Hz while the Valensi number, Re_ω or Va, increased from 0.229 to 0.687; this also increased Re_{max} by a factor of 3. A change in frequency would alter both the fluid flow and the heat transfer behavior.

In terms of heat loss, Table 8.23 shows the results of integrating the enthalpy loss and the conduction loss over the whole cycle for the base case ($Re_{max} = 50$, $Re_\omega = 0.229$), the increased-oscillation-amplitude case ($Re_{max} = 150$, $Re_\omega = 0.229$), and the increased oscillation-frequency case ($Re_{max} = 150$, $Re_\omega = 0.687$).

For the increased oscillation-frequency case, a 14% reduction in the axial conduction took place while an increase (by a factor of 7.8) in the enthalpy flux occurred when compared to the base case. The net effect is about a factor of 5 increase in the total axial heat loss. Again, the reduction in axial solid conduction is attributed to the higher radial heat flow between gas and metal due to the higher instantaneous gas mass flow.

8.5.5.1.4 Effect of Changing the Solid Material

The impact of solid-material properties on performance is of interest. Pure metals are known to have higher conductivity than alloys. Nickel was

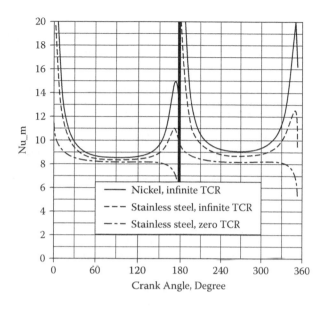

FIGURE 8.63
Mean Nusselt number comparison between nickel and stainless steel, both with infinite thermal contact resistance (TCR) (adiabatic contact) between layers, and the base case (stainless steel, zero TCR, or perfect contact).

chosen for fabrication of the prototype test regenerator, due to limitations of cost and time—because a nickel regenerator could be fabricated via LiGA alone. However, nickel has a higher conductivity than stainless steel (the preferred material) by about 5.5 times. For this comparison of nickel and stainless steel materials, the contact between layers has been set to infinite thermal contact resistance (TCR); thus, the reference stainless-steel case for this material study was not the base case (which had zero TCR). (Infinite TCR is closer to the real 3-D geometry, due to the reduced contact area between the disks.) As expected, no change in the friction factor was detected upon changing the material from stainless steel to nickel. Figure 8.63 shows the behavior of the mean Nusselt number upon changing the material. Because of the higher conductivity of the nickel, the wall temperature profile was flatter than that for stainless steel for the length of one layer, between two interfaces of infinite thermal contact resistance. That should cause a change in the heat transfer. Note that Figure 8.63 compares the cases of nickel with infinite TCR (adiabatic contact), stainless steel with infinite TCR (adiabatic contact), and stainless steel with zero TCR (perfect contact, the base case).

The results show that for infinite TCR between sections, the mean Nusselt number has increased overall for nickel compared to that for stainless steel, especially at low Reynolds numbers (i.e., near flow reversals at ~180° and 360°).

TABLE 8.24

Heat Loss Comparison between Stainless Steel (SS) and Nickel Materials, Both with Infinite Thermal Contact Resistance (TCR) (or Adiabatic Contact) at Interfaces between Layers (or Disks)

	Enthalpy Loss, W	Change	Conduction Loss, W	Change	Total Loss, W	Change
Infinite TCR and SS	1.960		0.531		2.491	
Infinite TCR and nickel	1.862	–5.1%	0.724	36.3%	2.586	3.8%

Again, this is explained by the fact that within a given layer, the nickel temperature profile is flatter, resulting in higher temperature differences between the wall and the fluid. The lower-conductivity stainless-steel material can maintain a steeper temperature profile within one layer. This steeper profile is closer to the bulk-temperature profile of the fluid, leading to smaller temperature differences between fluid and solid. A comparison of the axial heat losses for nickel and stainless steel is given in Table 8.24 (both with infinite TCR or adiabatic contact at the interfaces between layers).

In this case, a 36.3% increase in the axial conduction took place while a decrease (by 5.1%) in the enthalpy flux occurred when nickel is compared with stainless steel, both cases with infinite TCR between layers. The net effect is about a 3.8% increase in the regenerator axial heat loss using nickel. The increase in the axial conduction is attributed to the higher thermal conductivity of nickel (compared to stainless steel). The infinite TCR approximates the effect of the limited-surface-area contact between disks and greatly reduces the impact of nickel's 5.5 times greater, than stainless steel, thermal conductivity.

8.5.5.2 Results of Model (2), 3-D CFD Calculations for Straight-Channel Layers Approximation of Involute-Foil Layers

As discussed earlier, the grid for the involute-foil problem is quite dense, and it was not feasible for use in oscillatory-flow simulations. The 3-D straight-channel grid was subjected to oscillatory flow boundary conditions and the results follow (see geometry in Figure 8.54).

8.5.5.2.1 Steady-Flow Simulation (Model 2, 3-D, Straight-Channel Layers)

A 3-D steady-state simulation at a Reynolds number of 50 was performed in order to compare with the literature and with the 2-D results. This comparison is based on the Darcy friction factor plotted against the dimensionless hydrodynamic axial coordinate, $x+$. The same correlation from Shah and London (1978), discussed earlier, is used. In Figure 8.64, the Shah and

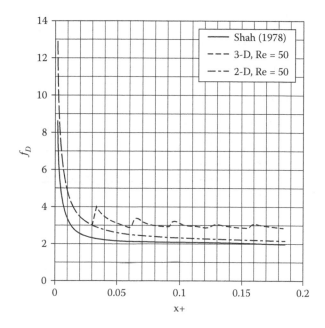

FIGURE 8.64
3-D straight-channel layers, steady-flow, friction-factor comparisons with 2-D results and Shah (1978) literature correlation—all at Reynolds number, $Re = 50$.

London correlation is below the 2-D CFD results in the entry section and then matches well for larger values of $x+$.

Also seen in Figure 8.64, the 3-D results agree with the 2-D results for the first layer. But upon entering the second layer, the flow encounters some resistance from the perpendicularly oriented second layer. That is where the friction factor goes up and departs from the agreement with the 2-D results. Then the flow tries to settle again until it encounters another geometry change upon entry into the third layer. As it moves through the stack of layers, the behavior of the fluid flow settles into periodicity, with small increases in friction factor upon entering each layer and with an average value above the 2-D prediction. This behavior was expected, and the simulation provided an answer regarding the magnitude of the friction factor increase.

In order to characterize the heat transfer, the mean Nusselt number is plotted with respect to the dimensionless thermal axial coordinate, x^*, in Figure 8.65 and is compared to the results from the 2-D simulations and the correlation developed by Stephan (1959). As discussed earlier, the alternating orientation of the layers is expected to improve the heat transfer, relative to 2-D and uniform-channel flow. That should result in higher 3-D Nusselt number values at each flow-channel discontinuity; this can be observed in Figure 8.65.

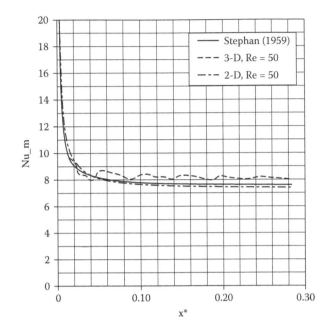

FIGURE 8.65
3-D straight-channel layers, steady-flow, mean-Nusselt-number comparisons with 2-D results and Stephan (1959) literature correlation—all at Reynolds number, $Re = 50$.

8.5.5.2.2 Oscillatory-Flow Simulation (3-D Straight-Channel Layers)

8.5.5.2.2.1 Base Case 3-D Oscillatory Flow For the 3-D straight-channel oscillatory-flow simulations, the same forcing-function for the mass flux as in the 2-D simulations was used (see Equation 8.30). As in the 2-D simulations, it took about 10 cycles to establish cycle-to-cycle convergence. The expectation again was that both the friction factor and the Nusselt number from the 3-D results would be higher.

As for the 2-D simulations, the friction factor is plotted (see Figure 8.66) against the crank angle in degrees and compared to the 2-D case and the experimental correlation for involute-foil by Gedeon (Equation 8.28).

As expected, the 3-D friction-factor values are higher than the 2-D results at all crank angles and are more in line with the experimental involute-foil correlation values from Gedeon (Equation 8.28). The mean-Nusselt-number results determined from the 3-D simulation are compared (see Figure 8.67) with the 2-D results and the experimental involute-foil correlation from Gedeon (Equation 8.29). As mentioned earlier, the correlation from Gedeon averages the Nusselt number over the length of a stack of layers while the present work uses only the length of one layer to obtain the mean Nusselt number. As expected, the Nusselt number values are higher than the values

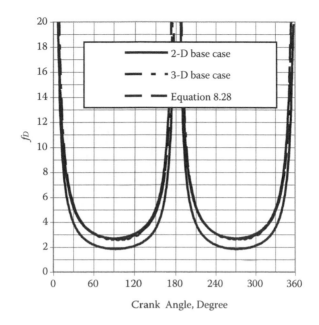

FIGURE 8.66
3-D straight-channel layers, oscillatory-flow, friction-factor comparison to 2-D base-case oscillatory-flow results and the Gedeon (Equation 8.28) experimental involute-foil correlation.

for the 2-D parallel-plate simulation and also are higher than the maximum value generated by the Gedeon correlation of Equation (8.29).

8.5.5.2.2.2 Effect of Changing the Thermal Contact Resistance (3-D Straight-Channel-Layers) As in the 2-D work, in order to study the effect of thermal contact resistance between the layers, an infinite thermal-contact-resistance (TCR) condition was imposed at the interfaces between the layers (replacing the zero-TCR of the base case). The expectation is that the friction factor will not change when compared with the perfect contact case. However, the mean Nusselt number should behave similar to how it behaved when the same condition was imposed in the 2-D study. That is, the Nusselt number should increase overall, especially at the low Reynolds numbers that are encountered when the flow switches direction (near 180° and 360°). Figure 8.68 shows a comparison of the mean-Nusselt-number behavior among the 3-D zero-TCR (perfect contact) and the infinite-TCR (adiabatic contact) cases and the infinite-TCR (adiabatic contact) 2-D case. When compared to the 3-D zero-TCR case (i.e., perfect contact), the infinite-TCR (adiabatic contact) case has a higher mean Nusselt number, especially in the regions of lower Reynolds numbers close to where the flow reverses. This was expected and is similar

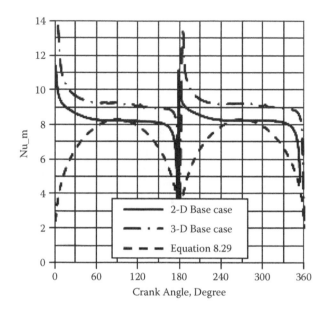

FIGURE 8.67
3-D straight-channel layers, oscillatory-flow, mean-Nusselt-number comparison to 2-D base-case oscillatory-flow results and the Gedeon (Equation 8.29) experimental involute-foil correlation.

to what happened in 2-D when the infinite thermal contact resistance condition was imposed.

8.5.5.3 Results, Model (1), 3-D Steady-Flow Simulation of Involute-Foil Layers

In this section, the results for the 3-D involute-foil layer simulations (see Figure 8.53) are presented and discussed. This 3-D computational domain resembles more closely the actual microfabricated design in terms of the shape of the channel. However, the attempt to capture the geometry of the channel better resulted in a dense grid. Even after reducing the length of the stack to only two layers, the cell count was close to 2.7 million. As explained earlier, one can use a two-layer long unit repeatedly by taking the velocity and temperature profile from the outlet and applying it back to the inlet in order to simulate a longer stack. The drawback of this technique is that oscillatory-flow simulation is not possible. Instead of one oscillatory-flow run, simulations have been performed using steady-state conditions at several Reynolds numbers.

The 3-D involute-foil grid actually consists of two types of grids, half and full length, with a repeating unit consisting of a one-half layer entry,

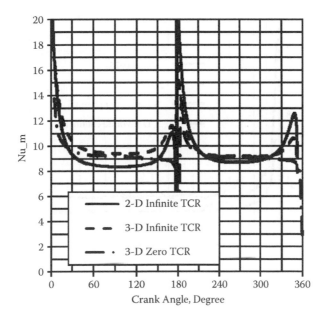

FIGURE 8.68
Mean-Nusselt-number comparison for 3-D straight-channel layers, oscillatory-flow cases, with zero TCR (perfect contact) and infinite TCR (adiabatic contact) between layers—and for the 2-D oscillatory-flow infinite-TCR (adiabatic contact) case.

a full layer, and a one-half layer exit (see Figure 8.53). Cutting the first layer in half allows for the two grids to line up properly when passing the boundary conditions (profile) from one grid to the other. By the same token, two repeating units will line up properly so that the boundary profile can be passed, where the geometry is continuous. As shown in Figure 8.53, the repeating-grid unit captures one full involute-foil channel in the middle. The other channels are also simulated as full because of the periodic boundary conditions applied to the sides. In fact, when periodicity is considered, this grid simulates a full ring of channels. The contact between the layers is perfect (i.e., the thermal contact resistance is zero).

Meshing this geometry presented a considerable challenge. Even though the channels were ensured to line up properly, in order not to get discontinuities when passing the boundary profile, one has to ensure that the cell coordinates of the exit face match the cell coordinates of the inlet face. The ends of the channel with their round shape did not allow for a structured mesh, and an unstructured mesh had to be employed. However, the generation of the unstructured mesh is harder to control when it comes to matching cell coordinates between inlet and outlet faces. This required linking the inlet and the outlet faces, which is a tedious process.

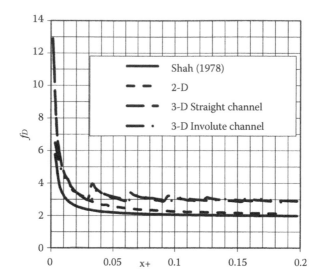

FIGURE 8.69
Steady-state friction-factor comparison at Reynolds number, $Re = 50$ as a function of dimensionless length. Calculated via the 3-D involute-foil and straight-channel layer simulations, the 2-D parallel-plate simulation—and compared with the Shah (1978) correlation.

8.5.5.3.1 Steady State, Re = 50, Comparisons of Friction Factor and Mean Nusselt Number

As for the 2-D parallel-plate and the 3-D straight-channel layers, a 3-D involute-foil layers simulation at Reynolds number 50 was performed. In order to compare with the 2-D results and with the literature, the Darcy friction factor is plotted (see Figure 8.69) against the dimensionless hydrodynamic axial coordinate, $x+$. The same literature correlation, Shah (1978), is used.

The 3-D involute-foil layers simulation shows a variation in friction factor (the saw shape) similar to the 3-D straight-channel layers, as expected. One thing to keep in mind is that the length of the involute-foil layer is 15 microns longer in the flow direction than the 3-D straight-channel layer. While work was in progress on this project, it was learned that the actual fabricated layers were shorter than originally intended. The 3-D straight channel was adapted to the shorter length, and simulations were performed that way. However, the 3-D involute-foil layer length simulated was kept at the original length. The above comparison captures this difference graphically by showing that the layer-to-layer rise in friction factor for the 3-D involute-foil layers happens after the rise shown by the 3-D straight-channel layers.

In order to characterize the heat transfer, the 3-D involute-foil mean Nusselt number is plotted (see Figure 8.70) with respect to the thermal axial coordinate and is compared to the results from the 2-D simulation, 3-D straight-channel

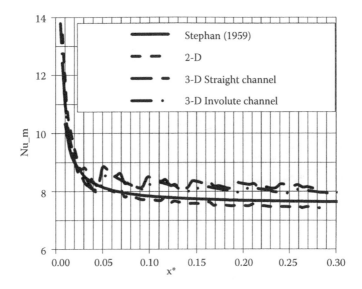

FIGURE 8.70

Steady-state mean-Nusselt-number comparisons at Reynolds number, $Re = 50$, as a function of dimensionless length. Calculated via the 3-D involute-foil and straight-channel layer simulations, the 2-D parallel-plate simulation—and compared with the Stephan (1959) correlation.

simulation, and the correlation developed by Stephan (1959). The variation in mean Nusselt number, Nu_m, is similar to that encountered for the 3-D straight-channel grid. The out-of-step layer-to-layer transition is again explained by the previously described difference in the lengths of the layers.

8.5.5.3.2 Summary: Steady 3-D Involute-Foil Friction-Factor Results for All Reynolds Numbers

Simulations at other values of Reynolds number were performed in order to determine the variation of the friction factor with the Reynolds number. A total of six Reynolds numbers were used: $Re = 50, 94, 183, 449, 1005,$ and 2213.

Figure 8.71 compares the variation of the friction factor at Reynolds numbers of 50, 94, and 183. The friction factor is lower at higher Reynolds numbers, as expected. As in previous simulations, the regular friction factor increases observed at the transition from layer to layer are present at the other Reynolds numbers. The fact that the transitions do not line up is an artifact of the plotting versus the dimensionless axial coordinate, $x+$, which includes the Reynolds number in the denominator.

In Figure 8.72, the results obtained from the 3-D involute-foil simulations have been compared with experiments done by UMN on 8- and 10-layer large-scale stacks. The same graph shows results from a correlation developed by Kays and London (1964) for staggered plate heat exchangers and a correlation developed

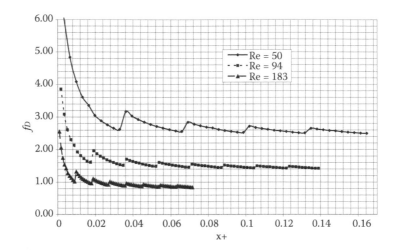

FIGURE 8.71
Steady 3-D involute-foil simulation friction-factor at different Reynolds numbers, Re = 50, 94, and 183—as a function of the dimensionless length, x^+.

by Gedeon (Equation 8.28) based on involute-foil experiments performed in the NASA/Sunpower oscillating-flow test rig. The results from the current 3-D involute-foil simulations match the UMN experiments. Both the CFD and UMN data match Gedeon's correlation, Equation (8.28), at the low end of the Reynolds number range (~100 to 200). However, Equation (8.28) provides higher friction-factor values at the high end of the Reynolds number range (~1000). This may be due to the roughness associated with the manufacturing process.

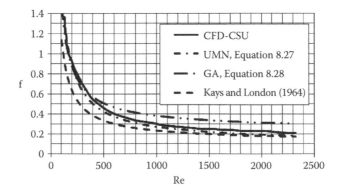

FIGURE 8.72
3-D involute-foil-simulation friction factor (CFD-CSU-8) compared with University of Minnesota (UMN) large-scale measurements (8 and 10 layers), Gedeon involute-foil experimental correlation (GA-42-1, original stacking), and Kays and London (1964) staggered-plate correlation.

8.5.6 Summary and Conclusions

8.5.6.1 Conclusions, Model (3), 2-D Parallel-Plate Simulations (Steady and Oscillating Flow) (See Figure 8.55)

The 2-D parallel-plate steady-state simulations showed agreement with the correlations from the literature (Shah, 1978; Stephan, 1959). A baseline case was chosen for this study using helium as the working fluid at pressure = 2.5×10^6 Pa (25 atm), and stainless steel solid material with zero thermal contact resistance (TCR) between the six layers. The oscillation frequency and amplitude chosen resulted in maximum Reynolds number, $Re_{max} = 50$, and Valensi number, $Re_\omega = 0.229$. For this base case, the enthalpy loss computed throughout the cycle equals 1.722 W, while the axial conduction loss equals 1.174 W with a total axial heat-transfer loss of 2.896 W. This total value, plus viscous-flow losses, must be minimized for optimum regenerator design. Table 8.25 shows this case as well as four other cases examined for Model (3).

Changing the solid thermal contact resistance from zero to infinity between the layers (Case 2) resulted (compared to Case 1) in 54.7% reduction in the axial conduction, 13.8% increase in the enthalpy loss, with a 14% reduction in the regenerator axial heat loss. Increasing the velocity amplitude of the oscillation (by a factor of 3), Case 3, resulted (compared to Case 1) in an 18.6% reduction in the axial conduction, an increase (by a factor of 10.6) in the enthalpy loss, with a factor of 6.6 increase in the regenerator axial heat loss. Increasing the frequency of the oscillation (by a factor of 3), Case 4, resulted (compared to Case 1) in a 14% reduction in the axial conduction, an increase by a factor

TABLE 8.25

Summary for All Cases Examined for Model (3)

Case	Description	Enthalpy Loss W (Change from Baseline, Except Case 5)	Conduction Loss W (Change from Baseline Except Case 5)	Total W (Change from Baseline Except Case 5)
1	*Baseline case*: $Re_{max} = 50$, $Re_\omega = 0.229$, thermal contact resistance (TCR) = 0, stainless steel (SS)	1.722	1.174	2.896
2	$Re_{max} = 50$, $Re_\omega = 0.229$, TCR = inf., SS	1.96 (13.8%)	0.531 (−54.7%)	2.491 (−14%)
3	$Re_{max} = 150$, $Re_\omega = 0.229$, TCR = 0, SS	18.249 (10.6×)	0.956 (−18.6%)	19.204 (6.6×)
4	$Re_{max} = 150$, $Re_\omega = 0.687$, TCR = 0, SS	13.466 (7.8×)	1.009 (−14.1%)	14.474 (5×)
5	$Re_{max} = 50$, $Re_\omega = 0.229$, TCR = inf., nickel (compared with Case 2)	1.862 (−5.1%)	0.724 (36.3%)	2.586 (3.8%)

of 7.8 in the enthalpy loss, with a net effect about a factor of 5 increase in the regenerator heat loss. Increasing the thermal conductivity of the solid material (by changing the solid material from stainless steel to nickel), Case 5, resulted (compared to Case 2, with both cases have infinite TCR) in a 36.3% increase in the axial conduction, a 5% decrease in the enthalpy loss, with a net effect about 3.8% increase in the regenerator heat loss.

It can be seen from the above cases (1 and 2) that changing the TCR from zero (lower bound) to infinity (upper bound) has the net effect of decreasing the regenerator heat loss. In the case of stainless steel (Case 2), the axial conduction reduction (–54.7%) offsets the increase in the enthalpy loss (13.8%) by net decrease in the heat loss (–14%). However, this is not the case for nickel (Case 5), where the thermal conductivity (5.5 times greater than stainless steel) resulted in net increase in heat loss of 3.8% (see Cases 2 and 5). Both changing the velocity amplitude of the oscillation or the frequency resulted in an increase by a big factor (about 9×) in the enthalpy loss.

8.5.6.2 Conclusions, Model (2), 3-D Straight-Channel Layer Simulations (Steady and Oscillating Flow) (See Figure 8.54)

For the 3-D straight-channel layer steady-state simulation, both the friction factor and the mean Nusselt numbers are in agreement with the 2-D simulation values in the first layer. Upon entering the second layer, the 3-D effects become obvious, and they persist as the axial coordinate advances. At the entrance of every layer, the forced reorientation of the flow results in small rises of both the friction factor and the mean Nusselt number with subsequent decreases as the flow settles into each new layer. Overall, the plots of the 3-D friction factor and the mean Nusselt number tend to flatten out as the flow reaches a fully developed condition, but at values above the 2-D simulation results.

For the oscillatory-flow simulations of the 3-D straight-channel layers, the friction factor shows an overall increase compared to the 2-D oscillatory-flow simulation and is in agreement with the experimental involute-foil correlation from Gedeon (see Ibrahim et al., 2007). The mean Nusselt number also shows an overall increase compared to the results from the 2-D simulation. It shows a higher value when compared to the correlation from Gedeon (see Ibrahim et al., 2007) at the maximum Reynolds number. The shapes of both the friction factor and mean Nusselt number curves are similar to the shapes observed in the 2-D simulations. So, the effect of going from 2-D to 3-D resulted in shifts upward of both friction factor and mean Nusselt number curves, as might be expected because the 2-D simulations do not include the flow perturbations resulting from flow around the ends of the foil layers.

Changing the thermal contact resistance from zero to infinite between the solid layers in the 3-D straight-channel layers with oscillating flow had no effect on the friction factor, as expected. However, the mean Nusselt number increased overall but with a more pronounced increase at the crank angles near where the flow switches direction. Although the actual values for the mean Nusselt number are higher in the 3-D simulations, the more expedient 2-D simulations capture well the behavior of the mean Nusselt number as the thermal contact resistance is changed. It should be noted that while the experimental correlation for N_a was obtained for 42 layers, the calculation made in Model (2) was only done for one layer.

8.5.6.3 Conclusions, Model (1), 3-D Involute-Foil Layer Simulations (Steady Flow Only) (See Figure 8.53)

As with the 3-D straight-channel layers, the simulations for the 3-D involute-foil layers show increases in both friction factor and mean Nusselt number at the geometric transitions between the layers. Furthermore, at Reynolds number 50, these increases are similar for the 3-D straight-channel and the 3-D involute-foil simulations. From this similarity, one can infer that using a simpler grid such as that used for the 3-D straight-channel layers can capture to a good extent the steady-state 3-D effects of the 3-D involute-foil layers for low Reynolds numbers. This is important in terms of the practicality of doing CFD simulations. Although more computing power is available now than ever, the researcher still must compromise between computing time and accuracy.

The overall friction factor matched the experimental results (obtained from Ibrahim et al., 2007). That lends credence to the simulations performed for the 3-D involute-foil layers. The detailed work that went into constructing the grid, running the CFD simulations, and postprocessing the data provided meaningful results, validated by experiment. By their nature, the simulations can in a more expedient way go beyond what the experiments can do and provide further predictions for optimization or comparison with other designs.

The technique of analyzing a repeating unit recursively has also been validated. That allows for steady-state simulations of a stack consisting of a large number of layers by using a repeating unit that is only two layers thick. That, in turn, not only saves on computation time and resources but also makes the simulation of a large stack feasible.

8.6 Structural Analysis of Involute Foil Regenerator

8.6.1 Plan for Structural Analysis of the Phase I Involute-Foil Regenerator Concept

The involute-foil regenerator consists of a regular array of repetitive elements laid out in an axisymmetric pattern.

At the lowest level, the regenerator features consist of curved "foil-like" elements. These were to be investigated for stiffness, resonant frequency,

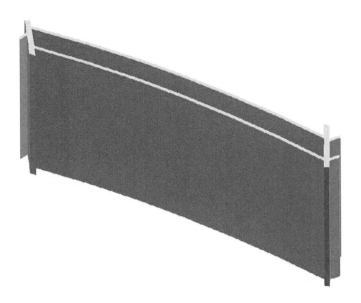

FIGURE 8.73
Basic structural unit cell for the Phase I involute-foil regenerator.

and the response to various loadings. Of particular interest was the change in curvature resulting from a thermal-expansion growth with constrained endpoints. Because these low-level elements are essentially curved beams, analysis may be done using published analytical solutions rather than finite-element analysis.

At a slightly higher level are the individual, closed cells comprising two foil elements and the segments of the circular walls to which they are attached, as illustrated in Figure 8.73. At the highest level are arrays of closed cells ranging from a few to the entire regenerator disk similar to the array shown in Figure 8.74 (and stacks of these disks).

Note that Figures 8.73 and 8.74 correspond to the Phase I involute-foil concept, in which the direction (slope) of the foils alternate from one ring of foil elements to the next. For the final involute-foil concept, analyzed later, the direction of the foils in any given disk was in the same general direction from ring to ring, though the average slope varied. Also, in this final concept, the rings in alternate disks were staggered, so that the rings in alternate disks did not line up; this permitted flow through the regenerator to redistribute radially, if needed, to provide more uniform flow in the primary-axial-flow direction.

Of interest at the highest level (Figure 8.74, for example, showing several rings of the Phase I concept) are the overall structural stiffness and how the structure responds to various loadings. Of particular importance are how it responds to (1) clamping forces that might be used to hold the regenerator in place within the engine and to (2) thermal stresses imposed by nonuniform heating or heating with constrained outer boundaries.

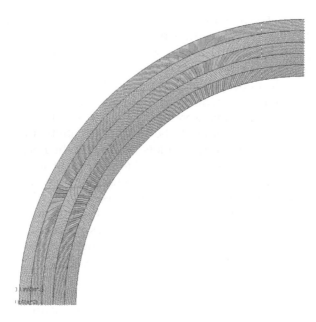

FIGURE 8.74
Top view of regenerator model showing involutes and annular rings.

8.6.1.1 Finite Element Analysis (FEA) Model of Phase I Involute-Foil Concept and Results of FEA

An FEA of the microfabricated regenerator was performed to examine the effects of axial compression on the regenerator. Axial compression and a slight interference were the mechanism used at the Stirling Technology Co. (now Infinia Corporation) to capture the regenerators within the heater head. Fixing the regenerator within the heater head is extremely important as any relative movement between the regenerator and the heater head can quickly lead to regenerator failure.

A quarter-model, two-dimensional CAD drawing was created with inner and outer diameters representative of an existing regenerator (see Figure 8.74). The annular rings were equally spaced, similar to the regenerator proposed for testing in the oscillating flow rig. Appendix G, by Gedeon Associates, presents the involute math used to generate the involute patterns; Appendix F presents other considerations used in defining the geometry of the segmented-involute foils. In order to simplify the modeling, an equal number of regenerator sections were used in each of the rings. The CAD drawing was imported into the Algor finite element package, and a three-dimensional model with plate elements was created. The thickness of the annular rings and the regenerator walls were taken directly from the proposed oscillating flow rig regenerator.

Boundary conditions were applied to the model. The first load case to be investigated was the axial insertion and compression of the regenerator into the heater head region. The bottom face surface was constrained from translation in the z-axis. A displacement of 59 μm (0.0015 inches) in the negative z direction (compression) was applied to the top surface. This displacement represents about 10% of a nominal regenerator compression. Algor did not then have the capability to directly model periodicity and could only model quarter- or half-model symmetry. Quarter-model symmetry was chosen and the applicable *x* and *y* translation constraints were applied to the side faces. Ibrahim et al. (2004d) shows figures illustrating (1) the 3-D model with boundary condition definitions, and 3-D colored contour plots illustrating the results of (2) z-axis displacements, (3) *x*-axis displacements, and (4) equivalent stresses.

The results indicate an extremely stiff structure that may not be conducive to the nominal compression levels (previously used by STC Co./Infinia Corp. for random-fiber regenerators). Obtaining only 10% of the nominal compression requires nearly 2000 lb$_f$ and leads to stresses beyond yield for the 316 stainless. This leads to the conclusion that the regenerator should be installed with a known force instead of a known displacement. It is important to note that the radial growth is extremely small and indicates that the regenerator will not deform significantly, changing channel size, during installation.

During installation, all attempts should be made to ensure that the regenerator is installed on axis and that no side loading is placed on the regenerator. A second FEA case was run to investigate the effects of side loads on this regenerator configuration. The boundary conditions are similar to the previous case except that a 10 radial point load was substituted for the axial displacements. Figure 8.75 illustrates the displacement from the force with

FIGURE 8.75
Displaced regenerator due to 10 pound side load.

one outline representing the undisplaced regenerator and the other outline representing the displaced regenerator with the side load applied.

The model displaced approximately 0.787 mm (0.020 inches) near the point load and expanded a similar amount in the adjacent quadrant. During heater head installation, the regenerator would be constrained by the heater head, which is not modeled in this load case. This regenerator is much less stiff in the radial direction. Any installation processes will need to ensure that the regenerator is not loaded in this way.

Refinements to this Algor model would have been difficult, as node and element counts were already very large. ANSYS is capable of modeling periodicity and would most likely be used for any later modeling. Any refinements, however, would not change the stiffness significantly or lead to other conclusions about the installation of a microfabricated involute regenerator.

The preliminary analysis showed that the involute regenerator is of sound design, structurally. The extremely high axial stiffness ensures that at appropriate axial compression force, the regenerator can be held firmly in place with negligible regenerator deformation. Special care is needed for regenerator installation to prevent potential lateral deformation due to misalignment, as the radial stiffness is relatively low.

8.6.2 Structural Analysis of Revised Involute-Foil Regenerator Structure

8.6.2.1 Summary

These results also indicate that the revised regenerator structure has high axial stiffness, and the stress level is sensitive to a radial side disturbance. Potential further work is also discussed. It should be noted that in these structural analyses, stainless steel was used (the originally planned microfabrication material), while the thermal test data (friction factor and Nusselt number) were taken using a nickel regenerator.

8.6.2.2 Introduction

The regenerators of interest are constructed of high-porosity material that readily conducts heat radially (i.e., between gas and matrix) and has a high surface area. Most current space-power regenerators are made from random fibers, which are somewhat difficult to manufacture in a precisely repeatable manner and are susceptible to deformation; these problems can lead to performance losses. The CSU NRA regenerator microfabrication contract team proposed a microfabricated involute-foil regenerator to potentially replace random-fiber and wire-screen regenerators. Figures 8.76 and 8.77 illustrate the geometry of the annular rings and the involute sections of the final regenerator design. Early analysis showed the potential for significant gains in performance efficiency and reductions of manufacturability variability while improving structural integrity (Qiu and Augenblick, 2005).

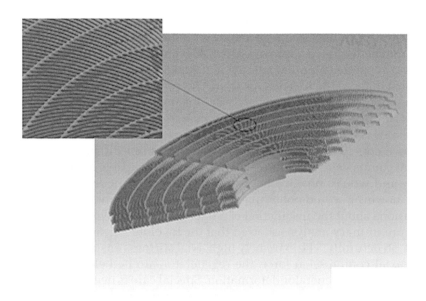

FIGURE 8.76
Partial of the solid model.

To ensure that the stiffness and the stress levels meet the design criteria, linear stress analysis was carried out on this proposed new regenerator. This section presents the results of a FEA on the microfabricated, involute-foil regenerator under 44 N (10 lbs) axial force and 4.4 N (1 lb) side disturbance force.

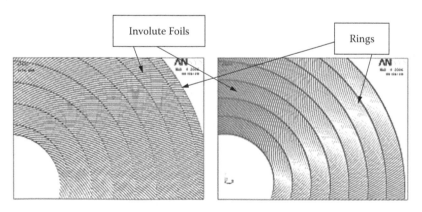

FIGURE 8.77
Geometry of layers.

TABLE 8.26

Material Properties of Stainless Steel 316L Used
in Finite Element Analysis

Item	Value
Young's modulus	1.9×10^{11} N/m² (2.796×10^7 psi);
Poisson's ratio	0.3
Tensile strength	4.97×10^8 N/m² (7.21×10^4 psi)
Yield strength	1.8×10^8 N/m² (2.61×10^4 psi)

8.6.2.3 FEA (Revised Involute-Foil Structure)

A FEA model was created to represent the geometric characteristics and three load cases were applied to the finite-element model to examine the stiffness and the stress levels.

Without using symmetry or periodic-symmetry conditions, all 360° of the geometry was included in the finite-element model. The thicknesses of the annular rings and the involute segments were much smaller than the other two dimensions. To allow the model to be handled by the computer power available, the FEA model was simplified as surfaces in 3-D space with four layers in axial direction. ANSYS shell element shell63 was used in the FEA to reduce the size of the model. The regenerator was made from stainless steel 316L, and the assumption that the material properties were not sensitive to temperature change was used in the analysis.

8.6.2.3.1 Geometric Model

Thicknesses of the involute sections, inside annular rings, and the outside annular ring are 12.7 m (0.0005 inch), 25.4 m (0.001 inch), and 127 m (0.005 inch), respectively. Total box volume was about 270 mm³ (0.0165 in³); the mass volume was about 44.5 mm³ (0.0027163 in³). The porosity was about 84%. Individual disk thickness (axial direction) was 250 μm. The FEA was based on the preferred stainless steel material, even though nickel was chosen for convenience in early testing of the performance of the involute-foil geometry (see the stainless-steel properties used in Table 8.26).

8.6.2.3.2 FEA Model

ANSYS 3-D shell element "shell63" with six degrees of freedom at each node was used. The total number of elements was 136,422. The total number of nodes was 170,220. Four layers were modeled. Ibrahim et al. (2007) shows an illustration with the four layers modeled.

8.6.2.3.3 Boundary and Loading Conditions

Case 1 (Axial Compression):

It is extremely important that the regenerator be properly fixed within the heater head, as any relative movement between the regenerator and the heater head can lead to regenerator structural oscillation and failure. Axial compression and a slight press-fit mechanism are currently used by Infinia to stabilize their regenerators within the heater head. A 44 N (10 lb) axial force was uniformly distributed on the top surface to simulate the axial fit, and the bottom-face surface was constrained from translation in the axial direction. In order to avoid rigid-body motion in the FEA, the minimum constraint condition $U_x = R_x = R_y = R_z = 0$ and $U_y = R_x = R_y = R_z = 0$ were used on two nodes of the inside circle. Boundary conditions for Cases 1, 2, and 3 are illustrated in Ibrahim et al. (2007).

Case 2 (Radial Side Load):

During installation, the regenerator should be installed so that it is under axial compression with no side loading. To simulate the disturbance from a side load, a 4.45 N (1 lb) side force acting on about 0.047% of the top layer outside annular ring was added to Case 1 to investigate the side load effect. The bottom-face surface was constrained in the axial direction and the inside circle of the bottom face was fixed in rotation and translation directions to avoid the rigid-body motion.

Case 3 (Distributed Radial Side Load):

This load case was created to investigate the stress sensitivity with respect to the side-load acting area. The boundary conditions and the load conditions were similar to those of Case 2 except that a 4.45 N (1 lb) radial side load acted on 10% of the top layer outside annular ring.

8.6.2.3.4 FEA Results for Cases 1, 2, and 3

The FEA results for Cases 1, 2, and 3 are summarized in Table 8.27. The results are illustrated via 15 color contour plots in Ibrahim et al. (2007).

TABLE 8.27

Maximum Displacement and Von Mises Stress

Load Case	Displacent U_x (in)	Displacement U_y (in)	Displacement U_z (in)	Total Displacement (in)	Von Mises Stress (psi)
Case 1	0.734e-6	0.736e-6	0.646e-6	0.804e-6	1732
Case 2	0.111e-3	0.148e-3	0.289e-5	0.148e-3	40624
Case 3	0.462e-4	0.304e-4	0.735e-6	0.462e-4	6374

8.6.2.3.5 Structural Analysis Summary and Conclusions

FEA of the microfabricated involute-foil regenerator shows that the regenerator had very high average axial direction stiffness (3.75e7 lb/in). Without any radial side disturbance, the stress level was much lower than the material yielding strength. If the radial side disturbance such as misalignment was localized in a small area, as in Case 2, Von Mises stress was beyond the material yielding strength, and permanent deformation could occur in that area, which may decrease the Stirling efficiency. In order to prevent local permanent deformation, the radial side load must be small or the disturbance area must be large, as in Case 3.

In summary, the proposed microfabricated involute-foil regenerator has high axial stiffness. The stress level is sensitive to the radial side disturbance, which therefore requires special caution and appropriate processing during installation to prevent lateral permanent deformation.

8.7 Stirling Engine Regenerator Results

The fabrication process described earlier was used to fabricate the Stirling engine regenerator for Phase III of this project. Part of a typical part is shown in Figure 8.30. The nickel webs are approximately 15 μm in width and are arranged in an involute pattern similar to the first regenerator (Ibrahim et al., 2007). The thickness of each disk is approximately 475 μm.

8.7.1 Regenerator Inspection and Installation

Figure 8.78 shows the regenerator in its shipping fixture soon after arrival at Sunpower. Some of us were expecting that, because of very tight manufacturing tolerances, the exposed surface of the regenerator stack would look like a smooth cylinder, with the divisions between individual disks barely visible. That is not quite the way it appears, as the close-up view (Figure 8.79) shows.

Some disks are seen to be much thinner than others, and there are thickness variations within individual disks resulting in visible gaps in several places. The local disk thickness even drops to zero in some cases. (See the upper center of the photo.)

8.7.1.1 Outer Diameter Measurements

Measurements of the assembled regenerator outer diameter (OD) on the Sunpower optical comparator show that the regenerator is slightly slimmer than the nominal OD by 0.009 or 0.016 mm, depending on who was taking

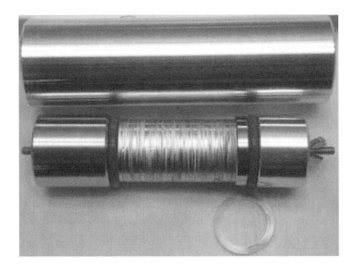

FIGURE 8.78
The regenerator in its fixture after arrival at Sunpower Inc. (Athens, Ohio).

the measurement. Figure 8.80 shows the regenerator surface as it appears on Sunpower's optical comparator. One end of the regenerator appeared to be slightly bigger than the other.

Based on the measured dimensions, the mean diametric gap between regenerator and heater head could be anywhere from 19 to 51 microns (based on the largest regenerator measurements combined with the smallest head, or vice versa). Assuming concentric location, the worst-case radial gap would be about 25 microns, which is still small compared to the 85 micron involute-foil channel gap.

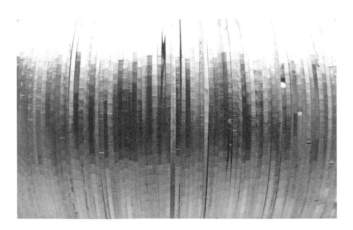

FIGURE 8.79
Close-up view of the regenerator.

FIGURE 8.80
The regenerator surface as it appears on Sunpower Inc.'s (Athens, Ohio) optical comparator.

8.7.1.2 Overall Length

The regenerator stack length, as measured, varied by 0.14 mm, depending on how tightly the clamping wing-nut at the end of the holding fixture was screwed down. The stack was elastically flexible, probably as a result of the many little gaps produced by disk thickness variations.

8.7.1.3 Disk Thickness Variation

We also measured the individual disk thickness by using the optical comparator to measure the distance between steps in the projected profile (see Figure 8.80). Plotted in Figure 8.81 are the results of two linear traverses along the regenerator with the regenerator rotated 120° between the two. The average disk thickness decreases toward one end of the regenerator at the same time as the disk-to-disk scatter increases. The mean disk thickness is 0.465 mm, and the standard deviation is 0.045 mm, according to calculations via Excel.

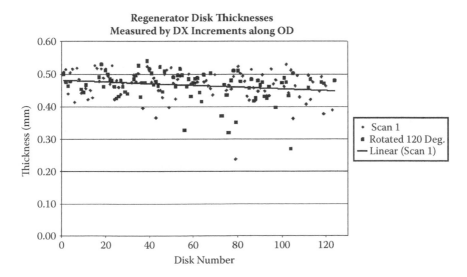

FIGURE 8.81
Disk thickness measurements.

8.7.1.4 Optical Comparator Methodology

The optical comparator places the regenerator assembly on a table where its somewhat fuzzy shadow-image is displayed on a screen with cross-hairs, as shown in Figure 8.80. Two dials move the table in X and Y directions, and a digital readout displays the table position to an accuracy of 2 microns.

Regenerator diameter measurements required positioning the horizontal cross-hair at a height representative of the local regenerator surface. We ignored several "bumps" where certain disks protruded from the surface by as much as 100 microns. We found that such bumps were localized (disappeared upon rotating the regenerator by 10°, or so) and easily pushed back into place. We are assuming they will continue to be easily pushed into place during final regenerator assembly.

Measuring individual disk thicknesses involved positioning the vertical cross-hair at the step transition between successive disks, which was sometimes clear but often a rounded fuzzy bump, difficult to discern, as Figure 8.80 shows. We scanned most of the regenerator length this way, and then rotated the regenerator by about 120° and scanned it again. It was difficult to scan the first few and last few disks because of visual interference from the holding fixture at the two ends of the regenerator.

8.7.2 FTB Test Results versus Sage Predictions (*Sunpower Inc. and Gedeon Associates*)

Some additional information about the final design of the regenerator as tested in the engine is given in Appendix J.

8.7.2.1 Summary of Results

Testing of the microfabricated regenerator in the FTB convertor produced about the same efficiency as testing with the original random-fiber regenerator. But the high thermal conductivity of nickel was responsible for significant performance degradation with the microfabricated regenerator. Had the microfabricated regenerator been made from a low-conductivity material, the efficiency would have been higher by a factor of ≈ 1.04. Had the FTB engine been completely designed to take full advantage of the microfabricated regenerator's low flow resistance, the efficiency would likely have been higher still. In any event, there was agreement between Sage computer modeling and the test data, validating the use of Sage to design and optimize future microfabricated regenerators.

Comparing test measurements to Sage model predictions is not as easy as it might seem. The primary test measurements are electrical output power delivered to a load and gross thermal input to the heating elements surrounding the engine head. The primary Sage outputs are pressure-volume (PV) power delivered to the piston and net thermal input through the engine boundaries. One is forced to either convert test measurements to Sage outputs or vice versa. The approach taken here is the first one. Electrical power

TABLE 8.28A

Comparison between Frequency-Test Bed (FTB) Engine Test Data and Sage Model Data

Test Data	Random-Fiber Regenerator			Microfabricated Regenerator	
Date	7/12/04	9/20/04	9/20/04	1/7/08	1/7/08
Test point	4	3	11	2	3
Pressure charge (bar)	32.94	36.39	36.39	31.22	31.15
Frequency (Hz)	106.7	105.4	105.4	104.9	103
T head (C)	650	650	649.5	650	649.5
T rejection (C)	35	30	30	30.1	30.1
Piston amplitude (mm)	4.5	4.6	4.55	4.5	4.5
Electrical power output (W)	88.90	85.75	85.75	82.6	89.1
Alternator current phase (deg)	122.00	87.1	86.6	125.7	103.1
Alternator electrical efficiency	*0.88*	*0.91*	*0.91*	*0.87*	*0.91*
Estimated PV power output (W)	*101.0*	*93.8*	*93.8*	*95.0*	*98.0*
Heat input gross (W)	303.9	303	303.6	310.1	325.1
Heat leak insulation (W)	−56.3	−56.3	−56.3	−71.7	−71.7
Heat input net (W)	*247.6*	*246.7*	*247.3*	*238.4*	*253.4*
Electrical efficiency	0.3590	0.3476	0.3467	0.3465	0.3516
PV efficiency	*0.4078*	*0.3804*	*0.3795*	*0.3984*	*0.3867*

output is converted to estimated PV power output and gross heat input to net heat input. More about how these things are done will be presented later.

First, here are the results: Table 8.28 compares FTB engine test data against Sage model predictions for the original random-fiber regenerator and the microfabricated regenerator. The random-fiber data points selected for comparison are those with piston and displacer amplitudes and phase close to the microfabricated data points.

TABLE 8.28B

Comparison between Frequency-Test Bed (FTB) Engine Test Data and Sage Model, Sage Predictions

Sage Comparison	Random-Fiber Regenerator			Microfabricated Regenerator	
Date	7/12/04	9/20/04	9/20/04	1/7/08	1/7/08
PV power output (W)	107.47	113.59	112.72	105.45	110.45
Heat input net (W)	242.3	254	251.7	222.2	234.1
PV efficiency	0.4435	0.4472	0.4478	0.4746	0.4718
Sage/Test Ratios					
PV power output ratio	1.06	1.21	1.20	1.11	1.13
Heat input net ratio	0.98	1.03	1.02	0.93	0.92
PV efficiency ratio	1.09	1.18	1.18	1.19	1.22

TABLE 8.29

Solid Conduction Losses for 250- and 500-Micron
Disk Thicknesses

Disk thickness (micron)	250	500
Average solid conduction (W)	3.8	11.8

The numbers in italic font ("alternator electrical efficiency," "estimated PV power output," "heat input net," and "PV efficiency" are indirectly derived, see Appendix A. Sage consistently overpredicts PV power by about 10% to 20% but does much better predicting heat input— within a few percent in all cases once nickel regenerator conduction losses are factored in.

8.7.2.2 Correcting for Nickel Regenerator Conduction

During Phase II, we estimated the thermal conduction losses in the nickel part of the microfabricated regenerator as installed in the FTB engine. The estimated loss was a function of disk thickness with the values calculated as shown in Table 8.29.

The 3.8 W loss for the 250-micron disk case was already built into the Sage simulation as a result of the heat-transfer correlations used for modeling the microfabricated regenerator being derived from a test sample with that disk thickness. The additional 8 W estimated conduction loss for the 500 micron thick case—the thickness actually used in the FTB regenerator—was *not* included in the simulation. One can argue that it is reasonable to add 8 W to the Sage net heat input values, which would bring them significantly closer to the test values. With the 8 W addition, the last part of Table 8.30 would be as presented.

As a result, Sage comes within about 3% to 4% of the net heat input calculated for the actual tests, suggesting that the added nickel thermal conduction is real.

Alternately, one can ask what the tested efficiency might have been had the microfabricated regenerator been made of a low-conductivity material. In that case, the evidence suggests that the test heat input would have been about 12 W lower and the electrical efficiency higher by a factor of about 1.04. So the tested electrical efficiency for data point 3 on January 7, 2008 (see Table 8.30), might have been 36.53%, instead of 35.16%.

8.7.2.3 Regenerator Flow Friction and Enthalpy Loss Trades

We have understood all along that the FTB engine was not optimal for demonstrating the microfabricated regenerator. It did not permit taking full advantage of the low flow resistance offered by the involute foil structure. The FTB engine requires a certain amount of pressure-drop power dissipation across the heat-exchanger plus regenerator flow path in order to balance the power produced by the displacer drive rod. We decided not to modify that rod for the microfabricated regenerator redesign. As a result, the microfabricated

TABLE 8.30

Comparison between Frequency-Test Bed (FTB) Engine Test Data and Sage Model Prediction (Including 8 W Conduction Losses)

Test Data

Date	7/12/04	9/20/04	9/20/04	1/7/08	1/7/08
Test point	4	3	11	2	3
Estimated PV power output (W)	*101.0*	93.8	93.8	95.0	98.0
Heat input gross (W)	303.9	303	303.6	310.1	325.1
Heat leak insulation (W)	–56.3	–56.3	–56.3	–71.7	–71.7
Heat input net (W)	*247.6*	*246.7*	*247.3*	*238.4*	*253.4*
Electrical efficiency	0.3590	0.3476	0.3467	0.3465	0.3516
PV efficiency	*0.4078*	*0.3804*	*0.3795*	*0.3984*	*0.3867*
Sage Comparison					
PV power output (W)	107.47	113.59	112.72	105.45	110.45
Heat input net (W)	242.3	254	251.7	**230.2**	**242.1**
PV efficiency	0.4435	0.4472	0.4478	0.4581	0.4562
Sage/Test Ratios					
PV power output ratio	1.06	1.21	1.20	1.11	1.13
Heat input net ratio	0.98	1.03	1.02	**0.97**	**0.96**
PV efficiency ratio	1.09	1.18	1.18	**1.15**	**1.18**

Note: Changed values are shown in bold italic font.

regenerator pressure drop is higher than it might have been, and there are also higher pressure drops in other components.

Table 8.31 compares the main losses for the two FTB regenerators, as simulated by Sage. Flow-resistance losses are tabulated as available-energy losses, which are the actual pumping losses multiplied by the appropriate temperature ratio, $T_{ambient}/T_{hx}$ (ambient/heat exchanger temperature), in effect assuming that some of the pumping loss suffered at high temperatures is recoverable. Enthalpy flow losses are actual thermal energy flows that add directly to net heat input.

The italic "Wdis" entries (shown in Table 8.31) are the simulated powers delivered to the displacer at its observed amplitude and phase angle after all flow-friction dissipations have been accounted for. In the case of the random-fiber regenerator, there is an extra 1.95 W drive power left over, suggesting additional flow resistance in the actual engine—probably in the regenerator. In the case of the microfabricated regenerator, there seems to be slightly less overall flow dissipation than modeled.

According to the way Sage saw things during the design process, it had to maintain about 5.9 W of pumping-dissipation losses in the microfabricated regenerator design, which it did by distributing the losses as indicated in Table 8.31. It managed to do this while at the same time reducing enthalpy flow losses by about 6 W (40%) compared to the random-fiber

TABLE 8.31

Sage Model Prediction, Enthalpy Losses Referenced to
Frequency-Test Bed (FTB) Engine Test Data

	Random Fiber	Microfab
Test Reference		
Date	7/2/2004	1/7/2008
Test point	2	3
Sage AEfric, Available Energy-Friction, Losses (W)		
Rejector	0.32	0.41
Jet diffuser C	NA	0.61
Regenerator	3.34	4.7
Jet diffuser H	NA	0.34
Acceptor	0.16	0.14
Wdis	*1.95*	*–0.35*
Total	5.8	5.9
Sage Enthalpy Flow (W)		
Regenerator	16.5	10.4

enthalpy loss. Sage would have done better if it had not had to maintain the 5.9 W pumping dissipation losses, although by exactly how much is not clear. But every watt saved in pumping dissipation is another watt added to PV power output.

The 6 W reduction of regenerator enthalpy loss in Table 8.31 suggests that the microfabricated regenerator should have produced slightly higher engine efficiency than the random-fiber regenerator. But, because of the increased nickel regenerator conduction, this was not the case. Engine efficiency was about the same. If future Sage models correctly account for regenerator solid conduction, then it appears that Sage will come very close to predicting the performance of a microfabricated regenerator.

8.7.2.4 Estimated Alternator Efficiency

In Table 8.28a, the alternator efficiency is not measured directly but rather estimated from a simple alternator loss model calibrated to the data. The estimated PV power output is then the measured electrical output divided by the estimated alternator efficiency. Appendix K shows more details in estimating the alternator efficiency.

8.7.2.5 Net Heat Input

The net heat input in Tables 8.28 and 8.30 is derived from the total electrical input to the heater elements, less insulation heat loss estimated from separate testing and data analysis. The insulation heat loss is a relatively large number. It was 56.3 W or about 19% of the gross heat input for the random-fiber regenerator tests and 71.7 W or about 23% of the gross heat input for the microfabricated regenerator tests. The two are different because the method of heating the head was different for the two cases. For the random-fiber tests, there were heating elements directly attached to the head—on the end dome and acceptor walls. For the microfabricated tests, the heating elements were attached to a nickel block bolted to the head.

In both cases, the insulation loss is measured with the test setup brought to operating temperature with the engine not running. The total electrical heat input is measured, and the thermal losses down the engine structures are calculated. The difference is attributed to insulation loss. The principle sources of error are the calculated losses down the various engine and structural components.

In the case of the random-fiber tests, the engine structure during heat-leak testing includes the pressure wall, regenerator, displacer cylinder, displacer, and heater support structure.

In the case of the microfabricated tests, the heat-leak testing was done two ways. First, the test was run, as described above, with the full engine in place. Second, a test was done with a "dummy" engine, consisting of only a pressure wall stuffed with fibrous ceramic insulation. The second method of testing is more accurate because there are fewer components, besides the heater insulation, down which heat is flowing. It is the basis for the 71.7 W reported in the table.

The 71.7 W insulation loss for the microfabricated regenerator tests is probably more accurate than the 56.3 W for the random-fiber regenerator tests, although the error bands are unknown.

8.8 Overall Involute-Foil Conclusions and Recommendations for Future Work

The microfabricated regenerator was assembled into the Sunpower frequency-test-bed (FTB) convertor, and the convertor was tested. The test results showed a PV power output of 98 W and an electrical efficiency of 35.16%. The Sage model results came within 3% to 4% of the test data for the net heat input (see Table 8.30).

Testing in the FTB convertor produced about the same efficiency as testing with the original random-fiber regenerator. But the high thermal conductivity of nickel, the material used for the microfabricated regenerator, was responsible for significant performance degradation. Had the microfabricated regenerator been made from a low-conductivity material, the efficiency would have been higher by a factor of 1.04. Had the FTB engine been completely redesigned to take full advantage of the microfabricated regenerator's low flow resistance, the efficiency would likely have been higher still. In any event, there was agreement between Sage computer modeling and the test data, validating the use of Sage to design and optimize future microfabricated regenerators.

It should be noted that the above work was selected as a NASA Tech Brief entitled, "Microfacbricated Segmented-Involute-Foil Regenerator for Stirling Engines," LEW-18431-1.

Beyond this Phase III effort, the microfabrication process needs to be further developed to permit microfabrication of higher-temperature materials than nickel. NASA and Sunpower are currently developing an 850°C engine for space-power applications. And a potential power/cooling system for Venus applications could need regenerator materials capable of temperatures as high as 1200°C. Early Mezzo attempts to EDM stainless steel, using a LiGA-developed EDM tool, involved a burn time (dependent on EDM machine setting) that was much too large to be practical. Some possible options for further development of a microfabrication process for high-temperature involute-foils are:

1. Optimization of an EDM process for high-temperature materials that cannot be processed by LiGA only. Burn times can be greatly reduced by higher-power, EDM-machine settings than originally used, in Phase I, by Mezzo, but "overburn" (i.e., the gaps between the EDM tool and the resulting involute-foil channels) increases with higher powers.

2. Development of a LiGA-only process for some high-temperature alloy or pure metal that would be appropriate for the regenerator application. Pure platinum would work but has very high conductivity, which would tend to cause larger axial regenerator losses and is very expensive.

3. Microfabrication of an appropriate ceramic material for high-temperature regenerators. Structural properties of ceramics, which tend to be brittle, would be a concern. Matching of ceramic-regenerator and metal-regenerator-container coefficients-of-thermal-expansion would also likely be a problem area.

9

Segmented-Involute-Foil Regenerator— Large-Scale (Experiments, Analysis, and Computational Fluid Dynamics)

9.1 Introduction

To support the development of actual scale involute-foils, a large-scale involute-foil test rig was designed. Design, fabrication, and testing of the large-scale involute-foils were carried out at the University of Minnesota, with support from the rest of the National Aeronautics and Space Administration (NASA) Research Award (NRA) team. Gedeon Associates (Athens, Ohio) provided system simulation support, and Cleveland State University provided detailed computational fluid dynamics (CFD) support. The results of the large-scale tests are discussed in detail in this chapter.

The main objective is to determine the performance characteristics of the involute-foil regenerator. This was partially done in the NASA/Sunpower (Athens, Ohio) oscillatory test rig under conditions that approached those of the engine, with the engine regenerator feature size and at engine oscillation frequency. However, it was also desired to investigate details of the flow and heat transfer that cannot be measured in an engine-size regenerator. To do so, we scaled the microfabricated regenerator up in size (by a factor of 30 as will be discussed later) while maintaining the ratios of effects important to the flow and heat-transfer processes. Such scaled-up systems are "dynamically similar" systems, and the methodology is labeled "dynamic similitude." Similitude is frequently used in testing and design. Examples can come from the turbomachinery field where, for instance, the actual scale is too large for direct testing (hydroturbines) or the engine scale is too small and speeds too high for high-resolution testing (aircraft gas turbines). The dynamic similitude was applied to the microfabricated regenerator in order to measure local flow features and local heat-transfer rates that cannot be captured with certainty using engine-size test measurements, or computation. Such flow features include the growth within the regenerator matrix of discrete jets that emerge from the heater or cooler channels; the redistribution of flow in response to any nonuniformities of entrance flow or temperature that may be present; thermal transport in the near-wall region where the axial temperature gradients in the wall and the matrix may be different, leading to near-wall radial thermal gradients, and radial thermal transport; unsteady heat transfer from the matrix solid material to the nearby fluid to investigate flow oscillation effects on the local and instantaneous heat transfer rates.

Some other measurements are complementary to the Sunpower test-rig data, including measurement of porous medium parameters, such as the permeability and the inertial coefficient used for the porous medium model.

Computational work to simulate such measurements was conducted in parallel. When the computational simulation results are confirmed by comparison with test data, the code can be used for wider applications, such as for parametric studies to support the engine and regenerator design.

9.2 Dynamic Similitude

The first step in applying dynamic similitude is to identify the important effects in the flow that must be properly modeled in the mock-up model. Appendix B presents dynamic similitude in more detail and lists many parameters that might be applied. The parameters of importance to this simulation are discussed below.

For dynamically similar terms in the momentum equation, we have the Reynolds number based upon the cycle maximum local velocity, u_{max}, and the hydraulic diameter of the flow passage within the porous medium, d_h, and the Valensi number, a dimensionless frequency of oscillation that presents the ratio of the time for diffusion of momentum by viscosity across a distance equal, $d_h/2$, to a cycle characteristic time, $1/\omega$, in seconds per radian. The Valensi number is also called the "kinetic Reynolds number." Simon and Seume (1988) gave ranges of representative values of these two dimensionless parameters for Stirling engine heat exchangers: heaters, coolers, and regenerators (see Chapter 3).

For similarity in heat conduction within the solid matrix under oscillatory-flow conditions, we would match the Fourier number, $Fo = \frac{\alpha\tau}{L_c^2}$, where α is the thermal diffusivity of the regenerator solid material, τ is the oscillation period, and L_c is the thermal diffusion depth. The Fourier number represents the ratio of the cycle time to the time required for thermal penetration to a distance L_c.

For similarity in convective heat transfer, we consider the dimensionless heat transfer coefficient, the Nusselt number:

$$Nu = \frac{h d_h}{k_f}$$

where

$$h = \frac{\dot{q}''}{(T_{solid,surface} - T_{fluid,bulk})}$$

In oscillatory-flow, the Nusselt number, with its heat transfer coefficient, h, based on the instantaneous heat flux and instantaneous solid surface to bulk mean fluid temperature difference is not always well behaved (Niu et al., 2003b). Nevertheless, it can be used in a quasi-steady way or during the portion of the cycle when the advection is strong (avoiding the times of flow reversal).

The Valensi and Reynolds numbers were applied for the flow setup, and the Fourier number was applied for the matrix material choice and the matrix feature sizing to establish a dynamically similar mock-up test section that could be used in our oscillatory-flow facility.

9.3 Large-Scale Mock-Up Design

For engine operation data on which to base our large-scale mock-up design, we requested NASA to consider a "pattern engine" that is representative of modern Stirling engines being developed for space power. The pattern engine would be one designed for operation with a random wire regenerator. We found that a scale factor of 30 from the pattern engine was suitable for our test facility. If it had been much larger, our flow velocities would be too low to accurately measure, and if it had been much smaller, our features sizes would be too small for accurate measurements and the fabrication cost would be excessive. Then we selected the stroke, piston diameter, and frequency for use in our large-scale experiment that matched the dimensionless numbers. When the stroke is 252 mm, the piston diameter is 216 mm, the frequency is 0.2 Hz, the Reynolds number is 77.2, and the Valensi number is 0.53 in our large-scale experiment, a reasonable match to the dimensionless numbers for the microfabricated regenerator. Assuming that a slice of the microfabricated regenerator would be approximately 250 μm thick, we chose a nominal thickness of 30×250 μm = 7.5 mm. So that we could use standard 5/16 inch stock, we modified this thickness to be 7.9 mm.

The regenerator was installed in an oscillatory-flow drive that provides pure sinusoidal oscillatory-flow via a piston cylinder and a scotch yoke drive. The first few meshes of one end of the regenerator are heated by passing alternating current (AC) through resistance wires, and the other end is attached to a shell and tube heat exchanger to provide cool working fluid for the portion of the cycle during which fluid enters the regenerator from the cooler. We computed Fourier numbers of the matrix material of the pattern engine to learn that they were high enough to indicate that the temperature was always spatially uniform within the regenerator matrix, in spite of the rapid temporal changes in temperature throughout the cycle. We applied the criterion that the mock-up must also have spatially uniform temperatures to learn that any metal was suitable, but ceramics or plastics were not. We chose aluminum.

The next step involved choices for fabrication. The focus of the large-scale experiment is on measurements of flow, heat transfer, and return to uniformity from a maldistribution of entrance mass flux or temperature. The microfabricated regenerator is of an annular design (see Section 8.2.6.4) that cannot be scaled up in its entirety to a factor of 30 and still be operational in our oscillatory-flow facility. Thus, we chose to model only a portion of it.

Three manufacturing processes were considered for fabricating the large-scale mock-up regenerator matrix: three-dimensional (3-D) printing, fused deposition modeling (FDM), and wire electric discharge machining (EDM) (Sun et al., 2004). The wire EDM method, which could be done within our shop, met the requirements and seemed to be the most reasonable method. The choice of aluminum was influenced by the EDM processing choice, for one can cut more rapidly if the material is aluminum, versus stainless steel, another reasonable choice. Several large-scale regenerator samples were fabricated by wire EDM. The process and results were satisfactory.

9.3.1 Large-Scale Mock-Up (LSMU) Final Design

The microfabricated regenerator is of an annular design that cannot be scaled up in its entirety to a factor of 30 and still be operational in our oscillatory-flow facility. Thus, only a 30° sector of it was chosen for modeling. Two geometries are shown in Figure 9.1a and 9.1b. The second pattern

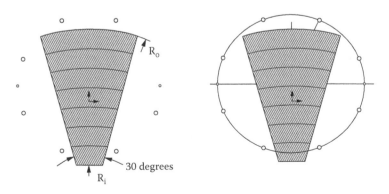

FIGURE 9.1
(a) First pattern geometry (six ribs). (b) Second pattern geometry (seven ribs).

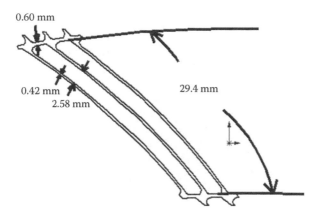

FIGURE 9.2
Geometry of typical channels.

geometry is achieved by shifting and flipping the first geometry. Figure 9.2 shows the dimensions of typical LSMU channels. The channel width is 2.58 mm and the fin thickness is 0.42 mm. The channel length changes from 54.56 mm to 32.72 mm as one passes from the inner radius to the outer radius of the first pattern layer. For the channels of the inner and outer edge of the second pattern layer, the lengths are 22.43 mm and 14.32 mm, respectively. For the channels of the center area, the lengths change from 49.73 mm to 34.65 mm. The layers were stacked to make the LSMU assembly. Figure 9.3 shows the mesh of the first pattern layer stacked on top of the second pattern layer.

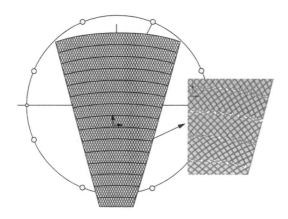

FIGURE 9.3
The mesh of the first pattern layer stacked on top of the second pattern layer.

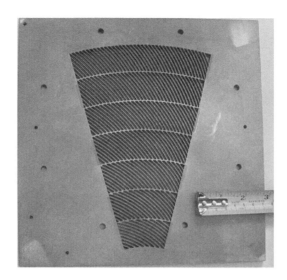

FIGURE 9.4
The large-scale mock-up (LSMU) layer fabricated by wire electric discharge machining (EDM).

The LSMU regenerator layers were fabricated by wire EDM. Figure 9.4 shows a photo of a LSMU layer, which is a 30° sector of the first pattern design. By inspection, the surface appears smooth with a matte finish. Literature indicates the roughness obtained by wire EDM is 0.05 μm.

9.3.2 The Operating Conditions

For engine operational data on which to base the LSMU design, a "pattern engine" was selected. It is representative of modern Stirling engines being developed for NASA space-power applications. The pattern engine was designed for operation with a random wire regenerator. Operational data for this pattern engine and the hydraulic diameter of the microfabricated regenerator were used to calculate the Reynolds number and Valensi number for our pattern engine with a microfabricated regenerator installed. The computed Reynolds numbers varied from 19.7 to 75.7 from the hot end to the cold end of the regenerator, and the Valensi number varied from 0.12 to 0.6. Then the stroke, piston diameter, and frequency of the oscillatory-flow test facility were selected for use in the large-scale experiment that matched these dimensionless numbers. A Scotch-yoke mechanism is employed to produce precise sinusoidal movement of the piston with a zero-mean velocity in the oscillatory-flow test facility. Figure 9.5 shows a schematic of the oscillatory-flow generator, the details of which were given in a NASA report (Seume et al., 1992).

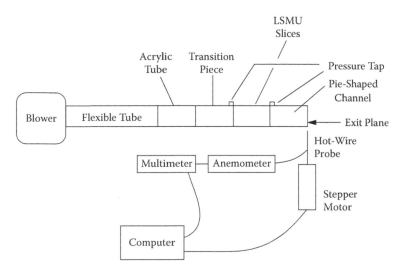

FIGURE 9.5
The experimental setup of unidirectional flow test.

Figure 9.1a shows the geometry of the first pattern regenerator layer (with six ribs), where the outer radius, R_o, is 284.25 mm and the inner radius, R_i, is 77.25 mm. The second pattern regenerator layer (seven ribs) has the same outer and inner radii as the first pattern. The scale factor is 30 times the actual size. For the first pattern regenerator layer, scaled-up channel lengths change from 54.56 mm for the slots near the inner radius to 32.72 mm for the slots near the outer radius. Scaled-up channel width is 2.58 mm (101 mils). The hydraulic diameter of the channel, d_h, is 4.87 mm (192 mils). The wall thickness is 0.42 mm (17 mils). The plate thickness is 7.95 mm (312 mils or 5/16 inches). The porosity is 86%. The area of the 30° sector of an annulus is $\pi \times (R_o^2 - R_i^2) \times 30/360 = 0.01959$ m².

The stroke, piston diameter, and operating frequency are selected to match the Valensi number and Reynolds number of the pattern engine. For air, at ambient pressure and temperature, viscosity, $v = 15.9 \times 10^{-6}$ m²/s. The stroke and piston diameter are set to the most suitable values of the options available in the present rig: stroke = 178 mm (7 inch); piston diameter, $D_p = 216$ mm (8.5 inch). Assuming incompressible fluid and balancing the volumetric flows in the regenerator and the driving piston/cylinder zone, we obtain $Re_{max} = 74.5$ and $Va = 0.47$ which is a reasonable match to the dimensionless numbers for the pattern engine. The local maximum velocity, U_{max}, is 0.24 m/s. Adolfson (2003) found that quasi-steady velocity measurements with hot-wire anemometry could be made with less than 10% uncertainty when the velocity exceeds 0.12 m/sec.

TABLE 9.1

Comparison of Microfabricated and Large-Scale Mock-Up (LSMU)
Involute-Foil Regenerators

Item	Microfabricated Regenerator	LSMU Regenerator
Geometry		
Channel width (mm)	0.086	2.58
Channel wall thickness (mm)	0.014	0.42
Regenerator layer thickness (mm)	0.25	7.9
Hydraulic diameter, d_h (mm)	0.162	4.87
Working Conditions		
Working medium	Helium	Air
Operating frequency (Hz)	83	0.2
Pressure (MPa)	2.59	0.101
Temperature of the hot end (K)	923	313
Temperature of the cold end (K)	353	303
U_{max} of the hot end (m/s)	3.7	0.24
U_{max} of the cold end (m/s)	2.85	0.24
Kinematic viscosity of the hot end (m^2/s)	32.3×10^{-6}	15.9×10^{-6}
Kinematic viscosity of the cold end (m^2/s)	6.48×10^{-6}	15.9×10^{-6}
Reynolds number, Re_{max}, of the hot end	19.7	74.5
Reynolds number, Re_{max}, of the cold end	75.7	74.5
Valensi number, Va, of the hot end	0.12	0.47
Valensi number, Va, of the cold end	0.6	0.47

Table 9.1 shows a comparison of microfabricated regenerator and LSMU
regenerator geometry and working conditions.

9.4 The LSMU Experiments under Unidirectional Flow

The Darcy friction factor, the permeability, and the inertial coefficient of the
LSMU layers were measured under unidirectional flow.

9.4.1 The LSMU Experimental Setup

Figure 9.5 shows the setup of the experiments under unidirectional flow.
At one side of the LSMU slices, the transition piece is connected with a fan
by a 11.08 m (12 feet) long flexible tube and a 0.54 m (21 inches) long acrylic
tube. At the other side, the transition piece connects the LSMU plates with
a sector-of-an-annulus shaped opening. The transition piece is for transitioning from a round to a sector-of-annulus cross section. It consists of nine

FIGURE 9.6
Two transition sections (without screens).

layers with the sector-of-annulus shaped opening and one layer with the round opening, which are shown in Figure 9.6. The thickness of one layer is 12.7 mm (0.5 inch). Screen material (not shown in Figure 9.6) is sandwiched between every two layers to help with the flow diffusion. The pressure drop across the LSMU layers is measured by a micro manometer. The hot-wire anemometer is used to measure the velocity of the outlet flow. The voltage readings of the anemometer are input to the multimeter and then stored on the computer.

9.4.2 Traverse of the Hot-Wire Probe over the Area of a Sector of an Annulus

A program was written in the C programming language to traverse the hot-wire probe over the area of a sector of an annulus. The measurement grid is shown in Figure 9.7. Dots show the locations at which velocity measurements were taken. In the radial direction, the increment is 9 mm. In the upper area, including areas 1a, 1b, 2a, and 2b, the angular increment is $\pi/60$. In the lower area, the angular increment is $\pi/120$. There are 348 grid points in total. The order in which the probe visited the various areas is 1a, 2a, 2b, 1b, 3, to 10. The probe visited the centroid, shown by a cross in Figure 9.7, before and after each area was visited. This allowed a check for time variations.

9.4.3 Friction Factor

9.4.3.1 LSMU Layers

The Darcy velocity in the LSMU layers (i.e., the approach velocity to the LSMU) was measured with eight plates or layers in the LSMU for this friction-factor measurement. The local velocity inside the channels, V, of the LSMU test section is calculated by dividing the Darcy velocity by the porosity. The Reynolds number is based on local velocity inside the channels and the hydraulic diameter of the channels, which is 4.87 mm. Also, the pressure drop, $\Box p$, across the

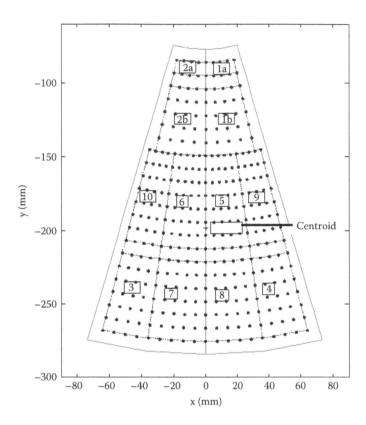

FIGURE 9.7
The measurement grid.

LSMU test section was measured, and the friction factor f was obtained. The result will be compared with the friction factors of continuous and staggered channels of parallel plates, a random-fiber matrix, and a woven-screen matrix in Figure 9.8. The following paragraphs describe the origins of the friction-factor data to be compared with the measured LSMU friction factor.

9.4.3.2 Friction Factor for Different Geometries

First, the equivalent continuous channel geometry to the LSMU regenerator is described. The LSMU channel length varies from 54.5 mm to 32.7 mm for the six-rib plate. The average length of 43.6 mm is chosen to calculate the aspect ratio for the continuous channel comparison case. Thus, the aspect ratio is $43.6/2.58 = 17$. The hydraulic diameter of the equivalent continuous channels is 4.87 mm. For fully developed flow in a continuous channel, $f*Re = 89.9$ when the aspect ratio is 20; $f*Re = 96$ for an infinite aspect ratio (Munson et al., 1994). The friction factor of fully developed flow of a 43.6 mm long by 2.58 mm wide continuous channel is calculated by interpolation as $f = 88.78/Re$.

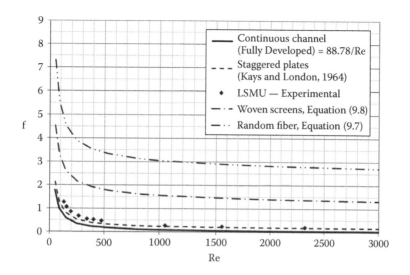

FIGURE 9.8
Darcy-Weisbach friction factor versus Reynolds number for different geometries. The Reynolds number is based on the local velocity and hydraulic diameter.

Figure 9.8 shows the friction coefficient versus *Re* for different geometries. These geometries are (1) continuous channel (β = 0.86); (2) staggered plates (β = 0.86), Kays and London (1964); (3) LSMU (β = 0.86); (4) woven screen (β = 0.9), Equation (3.36) (Gedeon, 1999); and (5) random fiber (β = 0.9), Equation (3.35) (Gedeon, 1999). In Equations (3.35) and (3.36), the local bulk mean velocity (not the Darcy velocity) is used to calculate the Reynolds number.

9.4.4 Permeability and Inertial Coefficient

The Darcy-Forchheimer equation (a steady-flow form of the one-dimensional [1-D] momentum equation) can be written in terms of the empirical coefficients permeability, K, and inertial coefficient, C_f as:

$$\frac{\Delta p}{L} = \frac{\mu}{K} U + \frac{C_f}{\sqrt{K}} \rho U^2 \qquad (9.1)$$

Equation (9.1) could be rewritten as:

$$\frac{\Delta p}{LU} = \frac{\mu}{K} + \frac{C_f}{\sqrt{K}} \rho U \qquad (9.2)$$

The Darcy velocity U can be calculated by $U = \beta \cdot V$, where β is the porosity.
From the plot of $\frac{\Delta p}{LU}$ versus U, the intercept is taken to be $\frac{\mu}{K}$. Permeability is calculated by dividing the dynamic viscosity by the intercept value.

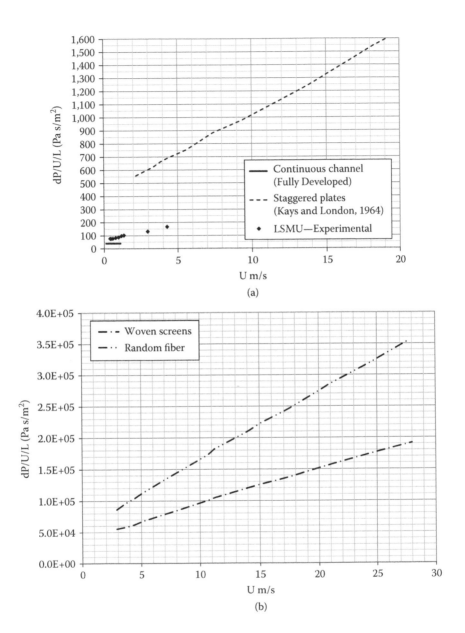

FIGURE 9.9
(a) and (b) $\Delta p/L/U$ versus U for different geometries.

Figure 9.9a,b shows the $\frac{\Delta p}{L \cdot U}$ versus U plots for continuous channels, staggered plates, LSMU layers, woven screens, and random fibers, respectively. For continuous channels, there is no inertial effect (because the flow path contains no "obstacles"). The plot is a horizontal line, and $\frac{\mu}{K}$ is taken to be 40.7. Table 9.2 shows the permeability for different geometries.

TABLE 9.2

Comparison of Permeability for Different Geometries

	D_h (mm)	K (m²)	$K/D_h{}^2$
Continuous channels	4.872	4.54E-07	1/52.3
Staggered plates[a]	1.539	4.26E-08	1/55.6
Large-scale mock-up (LSMU) layers	4.872	2.40E-07	1/98.98
Woven-screen matrix[a]	0.2	5.07E-10	1/78.9
Random fibers[a]	0.2	3.47E-10	1/115.2

[a] Extrapolated into Darcy region.

The choice of the hydraulic diameter of the woven-screen matrix and random fibers will not affect the value of $K/D_h{}^2$, which is proven by the following. The friction pressure drop relationship can be rewritten as

$$\frac{\Delta p}{V \cdot L} = \frac{f \cdot \rho \cdot V}{2D_h} \tag{9.3}$$

Because $Re = VD_h/\upsilon$ and $U = V\beta$, Equation (9.3) can be written as

$$\frac{p}{U \cdot L} = \frac{f \cdot \rho \cdot Re \cdot \upsilon}{2\beta \cdot D_h^2} \tag{9.4}$$

As shown by Equation (9.2), when the flow velocity is small, $\frac{\Delta p}{L \cdot U}$ is dominated by the viscous term, $\frac{\mu}{K}$. Thus,

$$\frac{p}{U \cdot L} = \frac{f \cdot \rho \cdot Re \cdot \upsilon}{2\beta \cdot D_h^2} = \frac{\mu}{K} \tag{9.5}$$

$$\frac{K}{D_h^2} = \frac{2\beta}{f \cdot Re} \tag{9.6}$$

For random-fiber matrix, from Equation (3.36),

$$f Re = 192 + 4.53 Re^{0.933} \tag{9.7}$$

For a woven-screen matrix, from Equation (3.35),

$$f Re = 129 + 2.91 Re^{0.897} \tag{9.8}$$

Therefore, for small Reynolds numbers, the value of $f Re$ is a constant, 192 and 129 for random-fiber and woven-screen matrix, respectively. Thus, $K/D_h{}^2$ does not change with the selection of the hydraulic diameter of the matrix, provided that the porosity remains fixed.

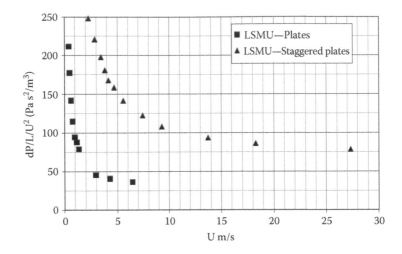

FIGURE 9.10
$p/L/U^2$ versus U for large-scale mock-up (LSMU) plates.

Equation (9.2) could be rewritten as:

$$\frac{\Delta p}{LU^2} = \frac{\mu}{KU} + \frac{C_f}{\sqrt{K}}\rho \qquad (9.9)$$

The inertial coefficient, C_f, can be evaluated accurately as the velocity goes toward infinity. However, due to the limitation of the equipment available, 6.5 m/s is the maximum mean velocity that was reached. Figure 9.10 shows the $\frac{\Delta p}{LU^2}$ versus U plot for LSMU plates and staggered plates, respectively. For LSMU plates, $\frac{\mu}{K6.473} + \frac{C_f}{\sqrt{K}}\rho$ is taken to be 34.6. For staggered plates, $\frac{\mu}{K27.37} + \frac{C_f}{\sqrt{K}}\rho$, is taken to be 77.5. The inertial coefficient values are 0.00939 for LSMU plates and 0.01075 for staggered plates.

9.4.5 Friction Factor of Various LSMU-Plate Configurations

Figure 9.11 shows the comparison of Darcy friction factor of various LSMU plate configurations, including 8 LSMU plates, 10 LSMU plates, 5 aligned LSMU plates, and 8 LSMU plates (double thickness). For the "aligned" test, five 6-rib LSMU plates are stacked together and tested under the unidirectional flow. Figure 9.12 shows a picture of five aligned LSMU plates. The fins are aligned throughout the entire area. The total thickness of the five plates is 39.7 mm. The hydraulic diameter, D_h, of the flow channel is 4.87 mm. The ratio of the length to the hydraulic diameter is 8.15. For laminar flow in a continuous channel, the ratio of the entrance length to the hydraulic diameter is $0.06*Re$. Under current test conditions, the Reynolds number varied from 207 to 1618. Thus, the entrance length changes from 12.4 D_h to 97.1 D_h and the flow of the five aligned plates is

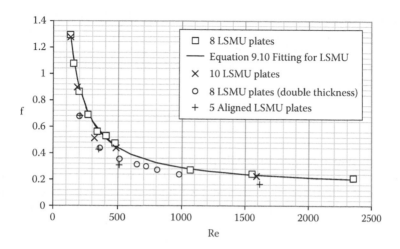

FIGURE 9.11
Comparison of Darcy friction factors of various LSMU plate configurations.

in the developing regime. Figure 9.11 shows the aligned plates have lower friction factor values than those for the standard LSMU plate arrangement. This is because the flow through the aligned plates is continuous and has minimal flow separation, whereas the LSMU plates under the standard arrangement would have wakes from trailing edges and separation on leading edges.

A comparison can be made between the case where 10 LSMU plates are stacked under the standard configuration and the case where 8 LSMU plates

FIGURE 9.12
A picture of five aligned LSMU plates.

are stacked similarly. The two cases compare very closely. The shorter assembly has only slightly larger friction factor values. This is an indication that the flow develops rapidly within the assembly, perhaps in the first three or four plates. The fitting equation for friction factor versus Reynolds number for the eight LSMU plates is given in Equation (9.10) below:

$$f = \frac{153.1}{R_e} + 0.127 R_e^{-0.01} \qquad (9.10)$$

This equation also fits the 10 LSMU plate data.

To determine what the friction factor would be with the LSMU geometry but with plates that are twice as thick, we stacked two, 6-rib plates together and two 7-rib plates together, and then repeated, giving four groups with the eight LSMU plates (or the equivalent of four double-thickness plates). Figure 9.11 shows that the friction factor is reduced from that with normal stacking with this new stacking order. The reason is that there are fewer flow redistributions from 6-rib geometry to 7-rib geometry, or the reverse, with this stacking order.

9.5 The Jet Penetration Study

9.5.1 The Jet Penetration Study for the Round Jet Generator

The geometry of the round jet generator is shown in Figure 9.13. The fabricated round jet generator is shown in Figure 9.14 with round holes are arranged in an equilateral triangle pattern. The hole diameter is 20 mm and the center-to-center spacing is 40 mm. The jet generator is 30.5 cm (12 inches) long giving a hole L/D ratio of 15.2.

9.5.1.1 The Experimental Setup

Because the diameter of the hot-wire support tube is 4.57 mm, a spacer for hot-wire insertion was fabricated that is thicker than 4.57 mm, smaller than, but comparable in size to, the thickness of an LSMU plate (7.95 mm or 5/16 inch). Insertion of the hot-wire probe is thus limited to the plenum between the jet generator, and the matrix and velocity data are not taken between LSMU layers. Temperature data are used to document the thermal field in the matrix, as affected by the penetrating jet. For this measurement, a thermocouple wire, which is much thinner than the velocity probe, is passed through a much thinner spacer inserted between two adjacent LSMU layers in the test. The thermal effect of the jet flow within the regenerator is documented by temperature profiles from the thermocouple traversed in the radial direction (of the test section which is a sector of an

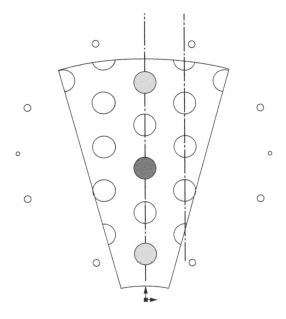

FIGURE 9.13
The round jet generator and the plenum shape, a sector of an annulus. The center shaded jet is the primary jet.

annulus—see Figure 9.13) across the primary jet of the test section and its two neighboring jets on that radius (also shown in Figure 9.13). Traverses are taken at several axial locations to document the widening thermal effects of the jets as one moves away from the jet generator. The thermal field is due to the low temperature of the jet being dispersed within the matrix,

FIGURE 9.14
The round jet generator.

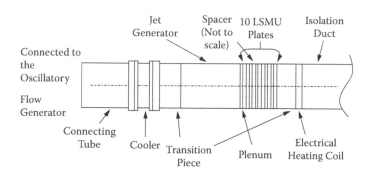

FIGURE 9.15
The oscillatory flow test facility.

hydrodynamically and thermally, and, possibly, being turned radially away from its center by the matrix material.

Figure 9.15 shows a schematic of the large-scale oscillatory-flow test facility. The main components of the facility are the oscillatory-flow generator, a cooler, two transition pieces (one on each end of the regenerator), a jet generator, 10 LSMU plates, an electrical heating coil, and an isolation duct. The cooler is a compact heat exchanger used to heat the passenger compartment of a car (called a heater core). One transition piece is put between the regenerator and the heating coil; the other transition piece is located between the jet generator and the cooler. The isolation duct is a long, open tube. It has active mixing to isolate the experiment from the room conditions. For the oscillatory-flow generator, the stroke is 178 mm, the piston diameter is 216 mm, and the frequency is 0.2 Hz, selected to match the Reynolds number and the Valensi number of the microfabricated regenerator in the pattern engine (79 and 0.53, respectively). There is a plenum between the jet generator and the LSMU plates which allows the hot-wire traverse for documenting the periodicity of the jets along a radius through the centerlines of the holes highlighted in Figure 9.13. The nominal thickness of the plenum, δ, is determined such that the axial-flow area, $\pi d_c^2/4$, equates to the radial-flow area, $\pi d_c \delta$. The diameter of the jet channel, d_c, is 20 mm; thus, the nominal thickness for the spacer, δ, is 5 mm. This plenum allows a single hot-wire probe, with a support tube diameter of 4.57 mm, to be inserted into the plenum for taking velocity measurements.

A spacer consisting of two 0.76 mm (0.030 inch) thick stainless-steel sheets (right and left of Figure 9.16) was inserted between two adjacent LSMU plates to allow the thermocouple wire used to take temperature profiles to pass through the test matrix. The opening of the spacer, which is the gap between two stainless steel sheets, is 0.51 mm (0.020 inch).

Thermocouples of type E with a diameter of 76 μm (3 mils) are used for unsteady temperature measurements within the LSMU plate test section. The time constant of the thermocouple is 0.05 sec, which means the

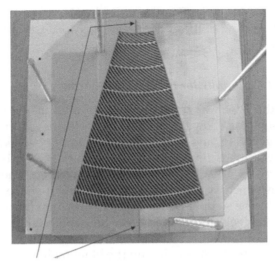

Slots for Thermocouples

FIGURE 9.16
The spacer on the LSMU plates.

thermocouple can sufficiently quickly respond to changes in flow temperature. The thermocouple measuring the temperature within the LSMU matrix is mounted on a stepper motor-driven rail so that it can be traversed inside the spacer and between two LSMU plates. The spacer can be moved to other axial locations within the LSMU matrix, allowing temperature documentation at various axial locations. The temperatures within the LSMU matrix, $T(x, r, \theta)$, are presented as functions of x, the distance in the axial direction; r, the distance along the centerline radius; and θ, the crank angle. One stationary thermocouple is located at one end of the jet generator and adjacent to the plenum. It is for measuring the cold end temperature, $T_c(\theta)$. Another stationary thermocouple is located at the end of the transition piece which is adjacent to the LSMU plates. It is for measuring the hot end temperature, $T_h(\theta)$. At each location, these three temperatures are taken at a sampling frequency of 500 Hz for 50 cycles. To eliminate temperature drift, a dimensionless temperature is calculated as:

$$\phi(x, r, \theta) = \frac{T(x, r, \theta) - T_c}{T_h - T_c} \qquad (9.11)$$

The cold end temperature, $T_c(\theta)$, is averaged over the portion of one cycle when the flow is passing from the jet generator to the LSMU regenerator plates. This gives an average temperature, T_c, for each cycle. The hot end temperature, $T_h(\theta)$, is averaged over the portion of one cycle when the flow is passing from the heater to the LSMU regenerator plates. This gives an average temperature, T_h, for one cycle. The dimensionless temperature $\phi(x,r,\theta)$ is calculated for each reading of the cycle. Averages of $\phi(x,r,\theta)$ are taken over an ensemble of 50 cycles.

9.5.1.2 Jet-to-Jet Uniformity for the Round Jet Array

To verify that the flow was uniformly distributed in the round jet generator under oscillatory flow conditions, the velocities within the plenum between the jet generator and the LSMU regenerator plates were measured. Results are given versus time within an oscillation cycle based upon an ensemble average of 50 cycles. Figure 9.13 shows the round jet generator and the plenum which is a sector of an annulus in shape. The hot-wire probe is driven by the stepper motor to move horizontally along a line that passes through the centerlines of the three holes highlighted in Figure 9.13. Each velocity measurement is taken at a sampling frequency of 500 Hz for 50 cycles. Velocity profiles taken during the blowing half of the cycle, when the flow is passing from the jet generator to the LSMU plates, are shown in Figure 9.17. The origin of the horizontal axis is the center of the center jet. Velocity profiles during the drawing half of the cycle, when the flow is passing from the LSMU plates back to the jet generator, are shown in Figure 9.18. Velocity profiles show that the jets from the three round channels shown are similar to one another. This confirms that when the center jet is interrogated, the data are representative of data for flow from all interior jets.

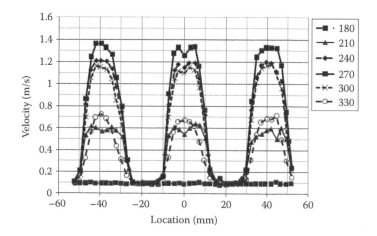

FIGURE 9.17
Velocity profiles during the blowing half of the cycle, when the flow is passing from the jet generator to the LSMU plates. The origin of the horizontal axis is the center of the center jet.

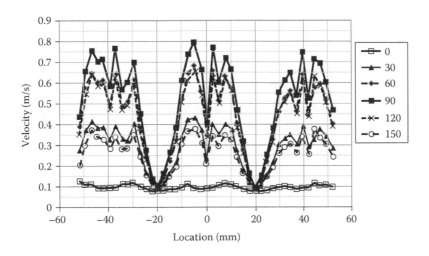

FIGURE 9.18
Velocity profiles during the drawing half of the cycle, when the flow is passing from the LSMU plates back to the jet generator. The origin of the horizontal axis is the center of the center jet.

When the crank angle is 270°, the average velocity of the jet is 0.988 m/s. From the mass conservation equation for incompressible flow, the area mean flow velocity within the round jet can be calculated:

$$U_c A_c = \frac{\pi D_p^2}{4} U_p \tag{9.12}$$

where A_c is the open area of the jet generator, U_c is the average velocity over the round jet generator, D_p is piston diameter, and U_p is the piston velocity. In the oscillatory-flow generator, the piston moves in a sinusoidal fashion. The displacement, X_p, can be calculated from the stroke and the frequency:

$$X_p = \frac{Stroke}{2} \cos(2\pi ft) \tag{9.13}$$

The piston velocity can be obtained by taking the first derivative of the piston displacement:

$$U_p = \dot{X}_p = \pi f \ Stroke \ \sin(2\pi f t) \tag{9.14}$$

The opening area of the jet generator is 4308 mm². The piston velocity is:

$$U_p = 0.112 \sin(2\pi f t) \ m/s$$

The average velocity over the round jet generator is:

$$U_c = 0.95 \sin(2\pi f t) \ m/s$$

which is close to 0.988 m/s, which is the average velocity measured by the hot wire when the crank angle is 270°.

9.5.1.3 Some Important Parameters in the Round Jet Generator and the LSMU Plates

The displacement of the fluid particle within the round jet also can be calculated from the mass conservation equation for incompressible flow:

$$X_c A_c = \frac{\pi D_p^2}{4} X_p \tag{9.15}$$

So, $X_c = 757 \cos(2\pi f t)$ mm.

The amplitude ratio, A_R, is the fluid displacement during half a cycle divided by the tube length. For the round jet generator, which is 305 mm (12 inches) long, the amplitude ratio is 2.48. The maximum Reynolds number in the round jet generator is:

$$Re_{max} = \frac{Uc_{max}d}{v} = \frac{0.95 \times 0.02}{15.9 \times 10^{-6}} = 1195$$

The Valensi number in the round jet generator is:

$$Va = \frac{d^2 \varpi}{4v} = \frac{0.02^2 \times 2\pi \times 0.2}{4 \times 15.9 \times 10^{-6}} = 7.9$$

The flow velocity within the LSMU plates can be calculated from:

$$U_r A_r = \frac{\pi D_p^2}{4} U_p \tag{9.16}$$

where A_r is the open area of the LSMU plates. The flow velocity within the LSMU plates is:

$$U_r = 0.243 \sin(2\pi f t) \ m/s$$

The displacement of the fluid particle within the LSMU plates can be calculated as:

$$X_r A_r = \frac{\pi D_p^2}{4} X_p$$

where $X_r = 194 \cos(2\pi f t)$ mm.

For the LSMU of 10 plates, which is 79.4 mm long, the amplitude ratio is 2.44.

9.5.1.4 Jet Penetration of the Round Jet Generator

Dimensionless temperatures at six axial locations have been measured. The six locations are between the plenum on the jet generator side and the first

LSMU plate, between plates 2 and 3, between plates 3 and 4, between plates 5 and 6, between plates 8 and 9, and after plate 10. Color contour plots indicating dimensionless temperature as functions of crank angle and radial location, at six different horizontal (axial) locations, are given in Ibrahim et al. (2007).

During the blowing half of the cycle, when the flow is passing from the jet generator to the LSMU plates, the crank angle changes from 180° to 360°. Three cold jets, which are distinguished from the rest of the region, can be seen in several of the color contour plots in Ibrahim et al. (2007). One of these plots shows that, downstream, the jet edges are nearly imperceptible. The jet penetration depth is about the thickness of eight LSMU plates, which is 63.5 mm. The hydraulic diameter of the LSMU plates is 4.872 mm, so the jet penetration depth is about 13 times the hydraulic diameter. A movie of the jet penetration was generated and can be obtained from T. Simon at UMN (contact: tsimon@me.umn. edu.). Results of the round-jet study are summarized in Section 9.5.1.5.

9.5.1.5 Jet Growth and the Fraction of Inactive Matrix Material

The center jet's edge, as the jet expands along the axial flow direction, is defined by assuming the edge occurs at that point at which the dimensionless temperature is the average of the maximum and the minimum dimensionless temperatures found in traversing across the jet—at a certain axial location and a certain crank angle. One way of representing this assumption about the jet width, b, is via the temperature expression:

$$\phi(x, b/2, \theta) = \frac{1}{2}(\phi_{max}(x, \theta) + \phi_{min}(x, \theta)) \tag{9.17}$$

where $\pm b/2$ represent the two edges of the jet along the radial direction, with the center of the jet at the origin. Throughout the blowing half cycle, this jet diameter remains almost invariant with crank angle. The dimensionless temperature at 270° crank angle is chosen to evaluate the jet growth. Figure 9.19 shows the jet growth along the axial direction. (Note that the jet edges are difficult to identify between plates 5 and 6, and the jet width there is extrapolated.)

Figure 9.20 shows the jet penetration in the matrix and the jet penetration depth x_p. The fraction of inactive matrix material is the fraction of matrix material that is not participating fully in thermal exchange with the working medium over the jet penetration depth, x_p. The fitting equation of Figure 9.19 can be used to get jet diameter, $b(x)$, over $0 < x/d_h < 13$ for the LSMU plates. For one jet, the corresponding matrix area is a hexagon with side length of 23.1 mm, which is shown in Figure 9.21. The area of the hexagon is A_j. The fraction of inactive matrix material is calculated by:

$$F = \frac{1}{x_p} \int_0^{x_p} \frac{\left[A_j - \frac{\pi b(x)^2}{4}\right]}{A_j} dx \tag{9.18}$$

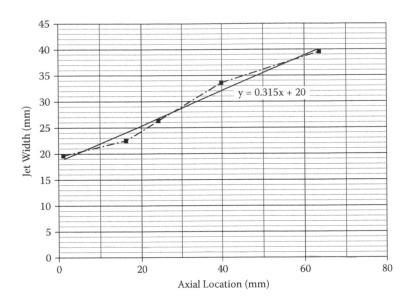

FIGURE 9.19
Jet width of the center jet at crank angle 270°.

A value of 47% is found for the LSMU plates with the round jet generator.

9.5.2 Jet Penetration Study for the Slot Jet Generator

Figure 9.22 shows the slot jet generator and the plenum which is a sector of an annulus in shape. The fabricated slot jet generator is shown in Figure 9.23. The slot channels are separated by the fins in the slot jet generator. The

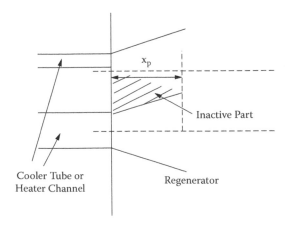

FIGURE 9.20
Jet penetration.

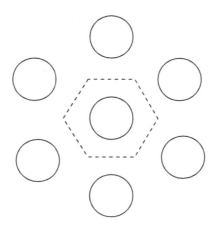

FIGURE 9.21
Area assigned to each jet.

channel width is 8.5 mm and the fin thickness is 23 mm. The jet generator is 30.5 cm (12 inches) long. The experimental setup is shown in Figure 9.15.

9.5.2.1 Jet-to-Jet Uniformity of the Slot Jet Array

To verify that the flow is uniformly distributed in the slot jet generator under oscillatory-flow conditions, the velocities within the plenum between the jet generator and the LSMU regenerator plates are measured. Results are given versus time within an oscillation cycle based upon an ensemble average of 50 cycles. The hot-wire probe is driven by the stepper motor to move

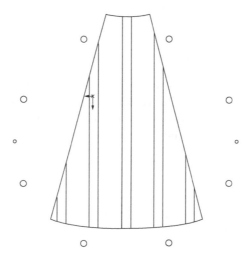

FIGURE 9.22
The slot jet generator and the plenum shape, a sector of an annulus.

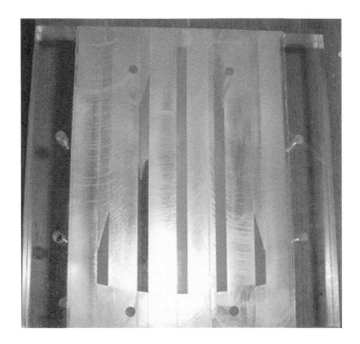

FIGURE 9.23
The slot jet generator.

horizontally along a line that passes through the center of the jet generator, normal to the slots.

The velocity measurement is taken at a sampling frequency of 500 Hz for 50 cycles. Velocity profiles taken during the blowing half of the cycle, when the flow is passing from the jet generator to the LSMU plates, are shown in Figure 9.24. The origin of the horizontal axis is the center of the center jet. Velocity profiles show that the jets from the three slot channels shown are similar to one another. Velocity profiles during the drawing half of the cycle, when the flow is passing from the LSMU plates back to the jet generator, are shown in Figure 9.25. These velocity profiles show that the velocities in the center are slightly lower than the velocities that are distant from the center. Figure 9.26 shows the geometry of the six-rib LSMU plate and the traversing route of the hot-wire probe. The flow in the center is very close to the rib of the six-rib LSMU plate, which decreases the flow velocity. Figure 9.27 shows the geometry of the seven-rib LSMU plate and the traversing route of the hot-wire probe.

When the crank angle is 270°, the average velocity of the jet is 0.783 m/s. From the mass conservation equation for incompressible flow, the flow velocity within the slot jet can be calculated:

$$U_h A_h = \frac{\pi D_p^2}{4} U_p \tag{9.19}$$

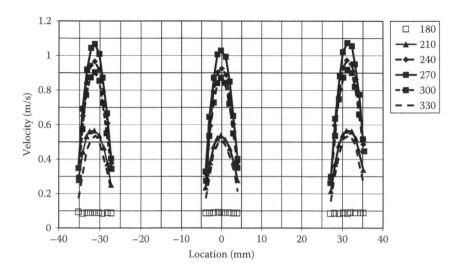

FIGURE 9.24
Velocity profiles during the blowing half of the cycle, when the flow is passing from the jet generator to the LSMU plates. The origin of the horizontal axis is the center of the center jet.

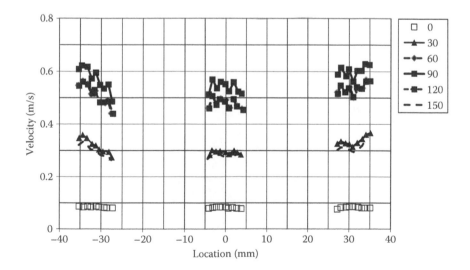

FIGURE 9.25
Velocity profiles during the drawing half of the cycle, when the flow is passing from the LSMU plates back to the jet generator. The origin of the horizontal axis is the center of the center jet.

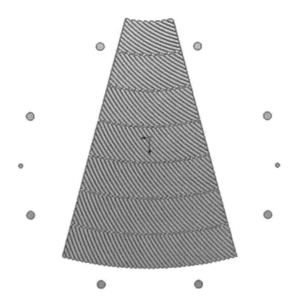

FIGURE 9.26
Geometry of the six-rib LSMU plate.

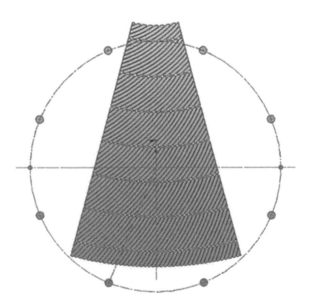

FIGURE 9.27
Geometry of the seven-rib LSMU plate.

where A_h is the open area of the slot jet generator, U_h is the average velocity over the slot jet generator, D_p is piston diameter, and U_p is the piston velocity.

In the oscillatory-flow generator, the piston moves in a sinusoidal fashion. The displacement, X_p, can be calculated from the stroke and the frequency:

$$X_p = \frac{Stroke}{2} \cos(2\pi ft) \qquad (9.20)$$

The piston velocity can be obtained by taking the first derivative of the piston displacement:

$$U_p = X_p = \pi f \, Stroke \, [\sin(2\pi ft)] \qquad (9.21)$$

The open area of the slot jet generator is 5278 mm^2. The piston velocity is:

$$U_p = 0.112 \sin(2\pi ft) \text{ m/s}$$

The average velocity over the slot jet generator is:

$$U_c = 0.776 \sin(2\pi ft) \text{ m/s}$$

which corresponds very well with 0.783 m/s, which is the average velocity measured by hot wire when the crank angle is 270°.

9.5.2.2 Some Important Parameters in the Slot Jet Generator

The displacement of the fluid particle within the slot jet also can be calculated from the mass conservation equation for incompressible flow:

$$X_h A_h = \frac{\pi D_p^2}{4} X_p \qquad (9.22)$$

So, $X_h = 618 \cos(2\pi ft)$ mm.

The amplitude ratio, A_R, is the fluid displacement during half a cycle divided by the tube length. For the slot jet generator, which is 304.8 mm (12 inches) long, the amplitude ratio is 2.03.

The maximum Reynolds number in the slot jet generator is:

$$Re_{max} = \frac{U_{h,max}d}{v} = \frac{0.776 \times 0.017}{15.9 \times 10^{-6}} = 830$$

The Valensi number in the slot jet generator is:

$$Va = \frac{d^2\omega}{4v} = \frac{0.017 \times 2\pi \times 0.2}{4 \times 15.9 \times 10^{-6}} = 5.7$$

9.5.2.3 Jet Penetration of the Slot Jet Generator

Dimensionless temperatures at five axial locations were measured. The five locations were between the plenum on the jet generator side and the first LSMU plate, between plates 3 and 4, between plates 5 and 6, between plates 6 and 7, and between plates 8 and 9. Color contour plots indicating dimensionless temperature as functions of crank angle and radial location at five different locations within the matrix are shown in Ibrahim et al. (2007). A movie of the jet penetration was generated and can be obtained from T. Simon (contact tsimon@me.umn.edu).

During the blowing half of the cycle, when the flow is passing from the jet generator to the LSMU plates, the crank angle changes from 180° to 360°. Three cold jets, which are distinguished from the rest of the region, can be seen in between the plenum and the first LSMU plate, via one of the color contour plots in Ibrahim et al. (2007). Downstream the jet edges are nearly imperceptible (i.e., between plates 6 and 7). The jet penetration depth is about the thickness of six LSMU plates, which is 47.6 mm. The hydraulic diameter of the LSMU plates is 4.872 mm, so the jet penetration depth is about 10 times the hydraulic diameter. The results of the slot jet generator study are summarized in Section 9.5.2.4.

9.5.2.4 Jet Growth and the Fraction of Inactive Matrix Material

Throughout the blowing half cycle, the slot jet width remains almost invariant with crank angle. The dimensionless temperature at 270° crank angle is chosen to evaluate the jet growth. Figure 9.28 shows the jet growth. The jet

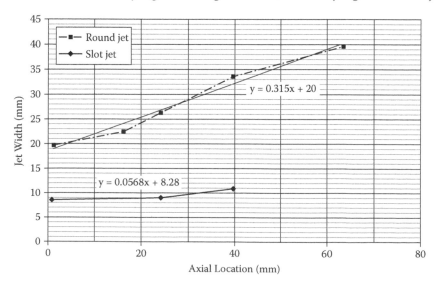

FIGURE 9.28
Jet width of slot jet and round jet at crank angle 270°.

TABLE 9.3

Results of the Jet Penetration Study—Large-Scale Mock-Up (LSMU)

Jet Geometry	Penetration Depth, x_p/d_h (Multiples of the Matrix Hydraulic Diameter)	"Inactive" Fraction of the Matrix Volume, F (between the Edge of the Matrix and the Penetration Depth)	Dimensionless Total Volume of "Inactive" Matrix, $F\,x_p/d_h$ (Normalized by the Volume $A_j d_h$)
Round jet	13	0.47	6.1
Slot jet	10	0.69	6.9

edges are difficult to identify between plates 6 and 7, and the jet width there is found by extrapolation.

The fraction of inactive matrix material is calculated by:

$$F = \frac{1}{x_p} \int_0^{x_p} \frac{[S-b(x)]L}{SL} dx \qquad (9.23)$$

where $b(x)$ is the jet width, x_p is the jet penetration depth, S is the jet center-to-center spacing (31.5 mm), and L is the jet length. A value of 69% is found for the LSMU plates with the slot jet generator. This compares to 47% for the round jets entering the LSMU regenerator and 55% for the round jets entering the 90% porous screen regenerator studied by Niu et al. (2003a, 2003b).

Two fundamental parameters were extracted from this study, both at the maximum-velocity location within the cycle when the jets are immerging into the matrix: (1) the depth into the matrix at which the thermal signature of an individual jet can no longer be distinguished (the "jet penetration depth") and (2) a measure of the matrix volume fraction that resides outside of the jets over the matrix volume that extends from the end of the matrix to the penetration depth (the "fraction of inactive material," F). Because the jets diffuse while they penetrate, F is not to be taken too literally. In fact, considerable heat transfer between the matrix and the jets occurs beyond the "edge" of the emerging jets. These two values are given in Table 9.3.

9.6 Unsteady Heat-Transfer Measurements

Unsteady heat transfer with the LSMU plates was investigated. The LSMU dynamically simulates the microfabricated regenerator plates of the segmented-involute-foil regenerator.

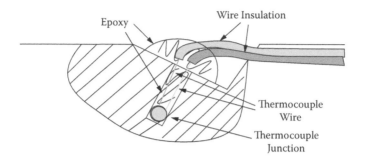

FIGURE 9.29
Embedded thermocouple.

9.6.1 Embedded Thermocouple

Figure 9.29 shows a sketch of one embedded thermocouple. The drilled hole is 0.30 mm (0.012 inch) in diameter and 2 mm (0.080 inch) deep. Figure 9.30 shows a picture of one of the embedded thermocouple installations.

9.6.2 Experimental Procedure

There are three thermocouples mounted in the six-rib plate and three thermocouples mounted in the seven-rib plate. The three thermocouple locations are labeled 1 to 3 along the radial direction pointing to the center of the arc, as shown in Figure 9.31. For the case presented in this report, 10 plates are stacked in the design order. The six-rib plate with

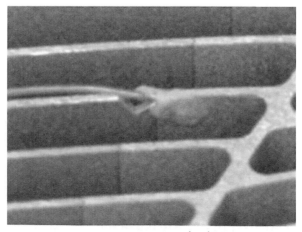

FIGURE 9.30
One embedded thermocouple.

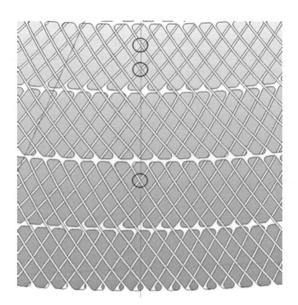

FIGURE 9.31
Locations of the embedded thermocouples.

thermocouples is plate 5 and the seven-rib plate with thermocouples is plate 6. One thermocouple is traversed between plates 5 and 6 to measure the air temperature, shown in Figure 9.32. The thermocouple junctions in plate 5 are near the traversing thermocouple, and the thermocouple junctions in plate 6 are more distant from the traversing thermocouple. For

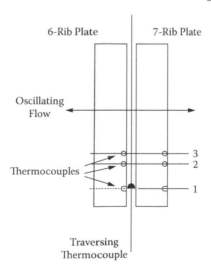

FIGURE 9.32
Temperature measurement locations for the case presented.

every embedded thermocouple location, air temperatures are measured at five locations on the same radial line as that on which the embedded thermocouples reside: –2 mm, –1 mm, 0, 1 mm, and 2 mm away from the embedded thermocouple radial location, labeled "1" through "5," respectively. Two runs with different warm-up times, both sufficiently large, were conducted. Table 9.4 shows the case names of the two runs. Case B13 is chosen to show the data processing.

TABLE 9.4

Case Names of the Two Runs

	Case	Date	Number of Plates	Tra. t.c. between Plate	Embedded t.c. loc.	Tra. t.c. loc.
1	A11	8/11	10	5 and 6	1	1
2	A12	8/11	10	5 and 6	1	2
3	A13	8/11	10	5 and 6	1	3
4	A14	8/11	10	5 and 6	1	4
5	A15	8/11	10	5 and 6	1	5
6	A21	8/11	10	5 and 6	2	1
7	A22	8/11	10	5 and 6	2	2
8	A23	8/11	10	5 and 6	2	3
9	A24	8/11	10	5 and 6	2	4
10	A25	8/11	10	5 and 6	2	5
11	A31	8/11	10	5 and 6	3	1
12	A32	8/11	10	5 and 6	3	2
13	A33	8/11	10	5 and 6	3	3
14	A34	8/11	10	5 and 6	3	4
15	A35	8/11	10	5 and 6	3	5
16	B11	8/16	10	5 and 6	1	1
17	B12	8/16	10	5 and 6	1	2
18	B13	8/16	10	5 and 6	1	3
19	B14	8/16	10	5 and 6	1	4
20	B15	8/16	10	5 and 6	1	5
21	B21	8/16	10	5 and 6	2	1
22	B22	8/16	10	5 and 6	2	2
23	B23	8/16	10	5 and 6	2	3
24	B24	8/16	10	5 and 6	2	4
25	B25	8/16	10	5 and 6	2	5
26	B31	8/16	10	5 and 6	3	1
27	B32	8/16	10	5 and 6	3	2
28	B33	8/16	10	5 and 6	3	3
29	B34	8/16	10	5 and 6	3	4
30	B35	8/16	10	5 and 6	3	5

Note: Tra., Transversing; t.c., thermocouple; loc., location.

After at least 5 hours of warm-up time, data collection begins:

Step 1: Hot end and cold end plenum temperatures of the LSMU test setup are collected for 20 cycles.

Step 2: Solid temperatures at location 1 in the six-rib plate and the seven-rib plate, and air temperature around location 1 are collected simultaneously. Data are measured over 50 cycles.

Step 3: Hot end and cold end plenum temperatures of the LSMU test setup are collected for 20 cycles.

Step 4: Solid temperatures at location 2 in the six-rib plate and the seven-rib plate, and air temperature around location 2 are collected simultaneously. Data are measured over 50 cycles.

Step 5: Hot end and cold end plenum temperatures of the LSMU test setup are collected for 20 cycles.

Step 6: Solid temperatures at location 3 in the six-rib plate and the seven-rib plate, and air temperature around location 3 are collected simultaneously. Data are measured over 50 cycles.

Step 7: Hot end and cold end plenum temperatures of the LSMU test setup are collected for 20 cycles.

9.6.3 LSMU Unsteady Heat-Transfer Measurement Results

Assuming the axial temperature distribution of the fin is linear, one can perform an energy balance of the plate fin in the vicinity of the embedded thermocouple:

$$h(x,r,t)A_s(T_f(x,r,t) - T_s(x,r,t)) = mC\frac{\partial T_s(x,r,t)}{\partial t} \qquad (9.24)$$

where h is the convective heat transfer coefficient, A_s is the surface area of the plate, T_f is the air temperature, T_s is the temperature of the plate, m is the mass of the plate, and C is the specific heat of the plate material. Equation 9.24 becomes:

$$h(x,r,t)(T_f(x,r,t) - T_s(x,r,t)) = \rho C\frac{s}{2}\frac{\partial T_s(x,r,t)}{\partial t} \qquad (9.25)$$

where s is the plate thickness and ρ is the density of plate material. The convective heat transfer coefficient can be calculated from the measured temperature differences between the air and the plate and temporal gradients of the plate temperature. The Nusselt number can be obtained from $Nu = \frac{hD_h}{k}$,

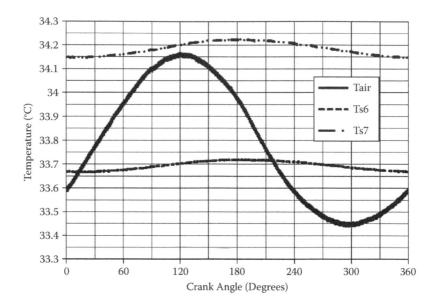

FIGURE 9.33
Air temperature at location 3 (T_{air}), solid temperatures of the six-rib plate (T_{s6}), and the seven-rib plate (T_{s7}), at location 1 in case B13.

where D_h is the hydraulic diameter of the channel (4.87 mm), and k is the thermal conductivity of the air.

Figure 9.33 shows the air temperature at location 3 and the solid temperatures of the six-rib plate and the seven-rib plate at location 1 in case B13. Consider the axial difference between the traversing thermocouple junction and the embedded thermocouple junction in the nearest plate (the six-rib plate). Because there is an axial gradient in the system, the air temperature at the location of the air thermocouple is not the air temperature at the axial location of the plate thermocouple junction. A small correction is made. It is assumed that the axial temperature gradient of the air can be obtained from the temperature difference between the six-rib and seven-rib plates. This temperature gradient is used to shift the air temperature from the air thermocouple location 1.3 mm (0.05 inch) to the embedded thermocouple location (six-rib plate). Figure 9.34 shows the temperature difference between the air and solid after the shift. This temperature difference profile is not balanced, which means the value of the peak does not match the valley. Figure 9.35 shows the comparison of the Nusselt number of current experiment with the correlation from the NASA/Sunpower oscillating-flow test rig. Recall that similar measurements by Niu et al. (2003a, 2003b, 2003c), but in a wire-screen matrix, showed a similar plot of Nusselt number versus cycle position. The following features were noted: The heat flux computed from

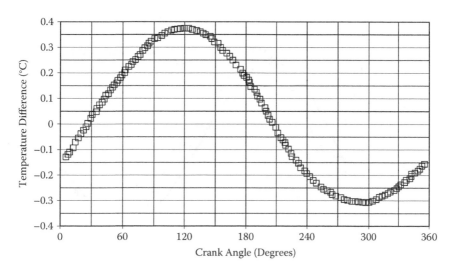

FIGURE 9.34
The temperature difference between the air and solid.

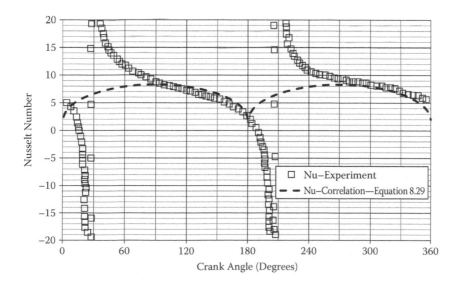

FIGURE 9.35
Comparison of the Nusselt number of the current LSMU experiment with the correlation from the NASA/Sunpower Inc. oscillating-flow test rig.

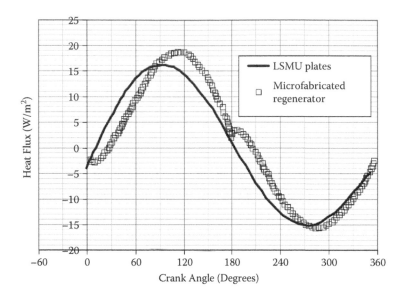

FIGURE 9.36
Comparison of the heat flux of the current LSMU experiment with that given by the correlation from the NASA/Sunpower Inc. test rig and the measured temperature difference.

the solid temporal gradient is zero when the temperature difference is not, creating a zero Nusselt number. The temperature difference becomes zero when the heat flux is not, creating an infinite Nusselt number. When Niu et al. (2003b) compared the measured results to correlation results computed by assuming quasi-steady behavior, the comparison was close only when the fluid velocity was near the peak value or during the deceleration part of the cycle. We expected similar behavior here.

Figure 9.36 shows the comparison of the heat flux from this LSMU experiment with the correlation from the NASA/Sunpower oscillating-flow test rig. The lack of symmetry of Nusselt number and heat flux of the current experiment result from the lack of symmetry of the temperature difference profile. A balanced temperature difference profile can be generated by a small shift, moving the curve of Figure 9.34 vertically 0.04°C. This is within the uncertainty in measuring a temperature within this system. This gives the symmetric curve we expect (because the measurement is in the axial center of the LSMU plates). Figure 9.37 shows the temperature difference between the air and solid based after this shift. Figures 9.38 and 9.39 show the Nusselt number comparison and heat flux comparison after this shift is made.

In the following, only the original temperature shift is applied (to effectively move the air thermocouple to the axial location of the metal thermocouple).

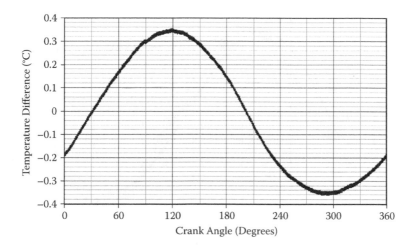

FIGURE 9.37
The temperature difference between the air and solid, with the shifted air temperature.

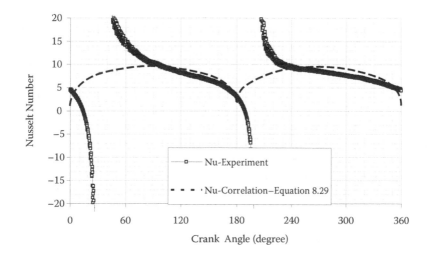

FIGURE 9.38
Comparison of the Nusselt number of current LSMU experiment (with the shifted air temperature) with the correlation from the NASA/Sunpower Inc. oscillating-flow test rig (Equation 8.29).

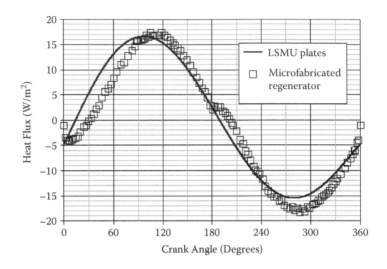

FIGURE 9.39
Comparison of the heat flux of the current LSMU experiment with the correlation from the NASA/Sunpower Inc. oscillating-flow test rig and the measured temperature difference with the shifted air temperature.

The Nusselt numbers of the six cases for which the traversing thermocouple is at location 3 (closest to the embedded thermocouple) are calculated for cycle positions 120° and 300°. The results are shown in Table 9.5. The average value is 7.14 and the root-mean-square (rms) is 1.07.

For the following, the second temperature shift is applied (to make the temperature difference plot symmetric, essentially to get the mean air temperature and the mean metal temperature equal to one another). The Nusselt numbers of the six cases for which the traversing thermocouple is at location 3 (closest to the embedded thermocouple), are calculated at cycle positions 120° and 300°. The results are shown in Table 9.5. The average value is 7.07 and the rms is 0.86. We note that at a cycle position of 300°, the local velocity in our test is 0.21 m/sec. With this, the correlation of Equation (8.29)

TABLE 9.5

Nusselt Number for Different Cases

Case	Crank Angle 120°	Crank Angle 300°
A13	7.7302	7.5696
A23	8.2965	6.545
A33	6.5366	5.3525
B13	7.3799	8.6731
B23	7.2222	7.9134
B33	5.1717	7.2729

TABLE 9.6

Nusselt Number for Different Cases
Based on the New Shift

Case	Crank Angle 120°	Crank Angle 300°
A13	7.7302	7.5696
A23	7.3766	7.2917
A33	6.0522	5.7459
B13	8.2363	7.6841
B23	7.7836	7.3048
B33	6.1025	5.9403

(Gedeon, 1999), gives a Nusselt number of 9.1. The average value in Table 9.6 is 7.07, which is about 22% lower than the Gedeon value. The difference is 2.4 standard deviations, so we expect it to be significant. We note that the microfabricated regenerator had roughness at the entrance of each channel due to debris, whereas our LSMU did not. Roughness would tend to enhance heat transfer.

10

Mesh Sheets and Other Regenerator Matrices

10.1 Introduction

At the National Defense Academy of Japan, Noboru Kagawa and his staff and students have developed a "mesh-sheet" regenerator that has some similarities to wire screen but with the meshes much more regular than traditional wire screen (Furutani et al., 2006; Kitahama et al., 2003; Matsuguchi et al., 2005, 2008; Takeuchi et al., 2004; Takizawa et al., 2002). Wire screen is formed from wires woven to form an approximately square grid of wires perpendicular to which the fluid in a Stirling regenerator flows. In contrast, the mesh sheets, which also have approximately square openings for fluid flow, are formed by chemical etching of a metal sheet. The chemical etching used to fabricate the mesh sheets allows freedom in choice of the various detailed dimensions of the meshes not available to fabricators of traditional wire-screen mesh. Attempts have been made to optimize the detailed dimensions of the mesh sheets for a particular Stirling engine, the ~3 kW NS03T (Kagawa, 1988, 2002). Several mesh-sheet combinations (two types of mesh sheets used alternately in the regenerator stack) have also been tested in the NS03T and in a relatively new double-acting ~3-kW engine, the SERENUM05 (Kagawa et al., 2007; Matsuguchi et al., 2009).

Matt Mitchell developed several types of etched foil regenerators (Mitchell et al., 2005). One of these was tested in the NASA/Sunpower oscillating-flow test rig. This work is summarized below.

At Sandia Laboratories a flat-plate regenerator was designed and fabricated for use in a thermoacoustic Stirling engine (Backhaus and Swift, 2000, 2001). This work, reported in a U.S. Department of Energy (DOE)/Sandia National Laboratories paper, will be summarized. Segmented-involute-foils are intended to achieve some of the benefits of flat-plate regenerators, while avoiding some of their difficulties.

10.2 Mesh-Sheet Regenerators

Noboru Kagawa and his staff and students at the National Defense Academy of Japan have developed mesh-sheet regenerators (Furutani et al., 2006; Kitahama et al., 2003; Matsuguchi et al., 2005, 2008; Takeuchi et al., 2004; Takizawa et al., 2002). These regenerators are somewhat similar to wire-screen regenerators. However, mesh-sheet fabrication via chemical etching has allowed freedom of design in choosing solid and fluid-flow dimensions in working toward the goal of optimization of the dimensions to maximize regenerator and engine performance. Several generations of these mesh sheets have all been tested in the approximately 3 kW NS03T engine (Kagawa (1988, 2002), which was originally designed for use as a residential heat pump. More recently, mesh-sheet-combination regenerators have been tested in the NS03T engine and the double-acting SERENUM05 engine (Kagawa et al., 2007; Matsuguchi et al., 2009).

10.2.1 NS03T 3 kW Engine in Which Mesh Sheets Were Tested

A schematic of the NS03T engine is shown in Figure 10.1. Tables 10.1 and 10.2 give the engine design parameters and specifications.

10.2.2 Regenerator Canister of 3 kW NS03T Engine

The canister-type regenerator is 60 mm in length. Stacked wire screen was originally used as a matrix. Each screen had a 3 mm diameter hole in its center, and a rod passed through the hole to form the cylindrical matrix. The 70 mm matrix diameter was the same as the piston diameter. Three wheel-shaped plates with 1 mm thickness were inserted in the matrix, and two were put at both ends to decrease side-leakage loss.

10.2.3 Design of Grooved Mesh Sheets

Figures 10.2 and 10.3 show schematic views of a grooved mesh sheet. As shown in Figure 10.3, holes and grooves are arranged on a thin disk. These small square holes and shallow grooves were etched on one of the circular surfaces of a metal disk. The disk has a flat edge around the etched area. The edge is 0.1 mm in width and is 0.04 mm in thickness. The precisely cut edge reduces the side leakage of the working gas. For the shaft of the matrix holder (with the wheel-shaped rims), each sheet has a hole, 3 mm in diameter, in its center.

The holes, grooves, and shape of the mesh sheets are designed and manufactured using an advanced etching technology. Table 10.3 shows the geometric parameters of the original wire-screen matrices and those of mesh sheets 1, 2, 3, and 4; all of these (and some other mesh sheets) have been tested in the

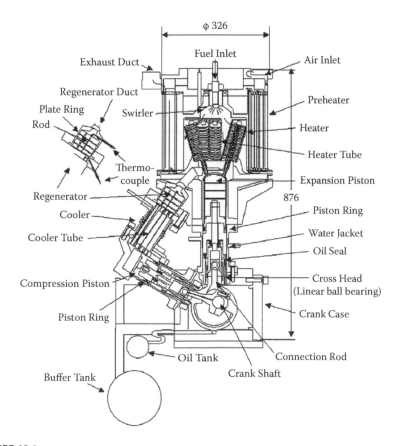

FIGURE 10.1
NS03T 3 kW engine (dimensions in mm). (From Takizawa, H. et al., 2002, Performance of New Matrix for Stirling Engine Regenerator, Paper 20057, *Proceedings of 37th Intersociety Energy Conversion Engineering Conference*, Washington, DC, July 29–31. With permission.)

~3 kW NS03T Stirling engine. As shown in Table 10.3, the four mesh sheets have the same pitch between holes, p_x and p_y, but other parameters such as opening width, l, thickness, t, depth of the grooves, d_g, and so forth, vary among the mesh sheets. To compare the mesh sheets with the wire screens, it is best to use the geometric characteristic parameters of open area ratio, β, porosity, ϕ, and specific surface area, σ, which are defined in Table 10.4.

Mesh sheet 1 has the smallest porosity of the matrices shown in Table 10.3. The #200 wire screen has twice the specific surface area as the #100 screen, so the #200 screen can store more heat to reduce reheat losses. Mesh sheet 3 was designed to be 20% thicker than the other mesh sheets to provide a large heat capacity. Mesh sheet 4 was made of nickel to realize a larger heat capacity than the other mesh sheets, which are made of stainless steel 304. Table 10.5 compares the properties of nickel and stainless steel 304.

TABLE 10.1

Design Parameters of the 3 kW NS03T Engine

Items	Parameter
Main fuel	Natural gas
Working fluid	Helium
Mean pressure	3–6 MPs
Maximum expansion space temperature	975 + 50°K
Compression space temperature	<323°K (water cooling)
Engine speed	–1500 rpm
Maximum output power	>3 kW
Maximum thermal efficiency	32%
Mass	<75 kg

Source: From Takizawa, H. et al., Performance of New Matrix for Stirling Engine Regenerator, Paper 20057, 37th Intersociety Energy Conversion Engineering Conference (IECEC), Washington, DC, July 29–31, 2002. © 2002 IEEE.

TABLE 10.2

3 kW NS03T Engine Specifications

Items	Data
1. Engine	
Type	Two piston
Swept Volume	
Expansion	192 cm^3
Compression	173 cm^3
Volume phase angle	100°
2. Piston	
Bone × Stroke	
Expansion	70 mm × 50 mm
Compression	70 mm × 45 mm
3. Regenerator	
Type	Canned
Number	1
Dead volume	150 cm^3
Matrix outer diameter × length	70 mm × 54 mm

Source: From Takizawa, H. et al., Performance of New Matrix for Stirling Engine Regenerator, Paper 20057, 37th Intersociety Energy Conversion Engineering Conference (IECEC), Washington, DC, July 29–31, 2002. © 2002 IEEE.

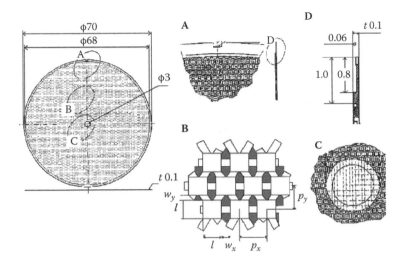

FIGURE 10.2
Plane sketch of grooved mesh sheet. (From Takeuchi, T. et al., 2004, Performance of New Mesh Sheet for Stirling Engine Regenerator, Paper AIAA 2004-5648, A Collection of Technical Papers, 2nd International Energy Conversion Engineering Conference, Providence, RI, August 16–19. With permission.)

The regenerated heat increases as the heat capacity of the regenerator solid becomes larger. Table 10.5 shows that nickel has a 3% larger heat capacity than stainless steel. Table 10.3 shows that the dimensions of the openings, frame width, and pitch of openings are the same for sheet 3 (stainless steel) and sheet 4 (nickel); only the dimensions of the groove are different for these

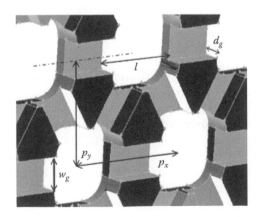

FIGURE 10.3
Three-dimensional (3-D) sketch of grooved mesh sheet. (From Takeuchi, T. et al., 2004, Performance of New Mesh Sheet for Stirling Engine Regenerator, Paper AIAA 2004-5648, A Collection of Technical Papers, 2nd International Energy Conversion Engineering Conference, Providence, RI, August 16–19. With permission.)

TABLE 10.3

Geometric Parameters of Wire Screens and Grooved Mesh Sheets Tested in the NS03T Stirling Engine

Dimensions	Wire Screens (Holes/Inch) #100	#150	#200	Mesh Sheet 1 (Sheet 1)	Mesh Sheet 2 (Sheet 2)	Mesh Sheet 3 (Sheet 3)	Mesh Sheet 4 (Sheet 4)
Opening (mm)	0.132	0.113	0.066	0.242	0.280	0.300	0.300
Wire diameter d_m (mm)	0.122	0.056	0.061	—	—	—	—
Width of frame (mm)	—	—	—	w_x: 0.199 w_y: 0.140	w_x: 0.161 w_y: 0.102	w_x: 0141 w_y: 0.082	w_x: 0.141 w_y: 0.082
Pitch p (mm)	0.254	0.169	0.127	p_x: 0.441 p_y: 0.382	p_x: 0.441 p_y: 0.382	p_x: 0.441 p_y: 0.382	p_x: 0.441 p_y: 0.382
Width of groove w_g (mm)	—	—	—	0.120	0.120	0.150	0.140
Depth of groove d_g (mm)	—	—	—	0.060	0.060	0.072	0.060
Thickness t (mm)	0.244	0.112	0.122	0.100	0.100	0.120	0.100
Open area ratio β	0.270	0.447	0.270	0.315	0.433	0.487	0.502
Porosity ϕ	0.582	0.726	0.582	0.543	0.606	0.667	0.668
Specific surface area σ (mm)	10.751	16.503	21.502	21.058	17.990	14.487	16.141

Source: From Takeuchi, T., et al., 2004, Performance of New Mesh Sheet for Stirling Engine Regenerator, Paper AIAA 2004-5648, A Collection of Technical Papers, 2nd International Energy Conversion Engineering Conference, Providence, RI, August 16–19. Reprinted with permission of the American Institute of Aeronautics and Astronautics.

two sheets; this leads to a slight difference in porosity and differences in open area ratio and specific surface area.

The reheating capacity of the regenerator, H_R, is 2.6 times the total heat input, Q_{in}. Therefore, in the case of output, W_{out} = 3 kW and efficiency, η_{gross} = 27%, it is estimated that Q_{in} and H_R become 11.2 kW and 28.9 kW, respectively. (These numbers imply waste heat is 11.2 − 3 = 8.2 kW. It is interesting to use these numbers to show the importance of the regenerator in achieving good engine performance. Suppose the regenerator were removed. Then, to maintain the same 3 kW output power, Q_{in} would have to increase to also provide the previous regenerator reheat of 28.9 kW—that is, now Q_{in} = 11.2 + 28.9 = 40.1 kW. Therefore, engine efficiency would drop from 27% to 100*(3/40.1) = 7.5%, a significant decrease. And waste heat would increase from 8.2 kW to 40.1 − 3 = 37.1 kW. And, to actually achieve this 3 kW output with no regenerator, it would be necessary, if possible, to redesign both heater and cooler

TABLE 10.4

Definitions of Open Area Ratio, Porosity, and Specific Surface Area in Terms of the Matrix Dimensions Given in Table 10.3 and Figures 10.2 and 10.3

Matrix Type	Equations
1. Opening Area Ratio β	
Wire screen	$\beta = \left\lfloor \dfrac{l}{P} \right\rfloor^2$
Mesh sheet	$\beta = \left\lfloor \dfrac{l^2}{P_x P_y} \right\rfloor$
2. Porosity φ	
Wire screen	$\phi = 1 - \dfrac{\pi d_m \sqrt{p^2 + d_{m^2}}}{4p^2}$
Mesh sheet	$\phi = \dfrac{l^2 t + w_g d_g \left\lfloor w_x + 4/w_y \sqrt{3} \right\rfloor}{p_x p_y t}$
3. Specific Surface Area σ	
Wire screen	$\sigma = \dfrac{\pi \left\lfloor \sqrt[2]{p^2 + d_m^2 - d_m} \right\rfloor}{2p^2}$
Mesh sheet	$\sigma = \dfrac{2\left\lfloor p_x p_y - l^2 \right\rfloor + \dfrac{4}{t} + d_g \left\lfloor w_x + \dfrac{8}{\sqrt{3}} w_y - \left\lfloor 2 + \dfrac{8}{\sqrt{3}} \right\rfloor w_g + p_x - \right.}{p_x p_y^t}$

Source: From Takeuchi, T. et al., 2004, Performance of New Mesh Sheet for Stirling Engine Regenerator, Paper AIAA 2004-5648, A Collection of Technical Papers, 2nd International Energy Conversion Engineering Conference, Providence, RI, August 16–19. Reprinted with permission of the American Institute of Aeronautics and Astronautics.

TABLE 10.5

Properties of Nickel and Stainless Steel 304

	Density (kg/m³)	Specific Heat (kJ/kg-°K)	Thermal Capacitiance (kJ/m³-°K)	Thermal Conductivity (kJ/m³-°K)
Nickel	8906	0.46	4090	64
Stainless steel 304	7930	0.50	3960	16.3

Source: From Takeuchi, T. et al., 2004, Performance of New Mesh Sheet for Stirling Engine Regenerator, Paper AIAA 2004-5648, A Collection of Technical Papers, 2nd International Energy Conversion Engineering Conference, Providence, RI, August 16–19. Reprinted with permission of the American Institute of Aeronautics and Astronautics.

to achieve the new, much larger values, of heat input and rejection. Such a low efficiency, and likely heavier, engine—would be impractical. Thus, the regenerator is a critical component in helping Stirling engines achieve efficiencies large enough to compete with other heat engines.)

In the case of mesh sheet 3 whose heat penetration depth is thought to be insufficient, H_R increases by 3% by using sheet 4 made of nickel, and the regenerated heat capacity increases about 430 W. The opening of sheet 4 was designed to be the same as sheet 3, because the pressure loss of sheet 3 is slightly low compared with conventional wire screens.

10.2.4 Design of New Mesh Sheets without Grooves, and with a Different Type of Groove

The grooves used in the earlier mesh sheets reduce flow resistance. However, they also tend to make the mesh sheets thicker, increase porosity, and therefore increase the dead volume of the engine. Test results, to be discussed later, indicated that although the grooved mesh sheets improved the regenerator performance, the performance of the engine did not change significantly due to the increased dead volume (Matsuguchi et al., 2005).

Dimensions of the mesh sheet were reoptimized using the general purpose finite-element method fluid analysis program, FIDAP, to ensure manufacturable dimensions (Matsuguchi et al., 2005). Three layers of matrix material were simulated with periodic boundary conditions applied at all sides of the matrix material. About 60,000 tetrahedral elements were used. Helium flow was assumed laminar and steady, with uniform inflow. The temperature difference between fluid and matrix material was 1°K, and the matrix solid temperature was uniform. Helium inflow was at 4 MPa and 900°K. Governing equations modeled were continuity, the Navier-Stokes equations, and the energy equation.

Figure 10.4 shows schematic views of the earlier grooved mesh sheets developed by Kagawa and Takizawa (also see Figure 10.3), the mesh sheets developed by Matsuguchi et al. (2005), and a mesh sheet proposed in Furutani et al. (2006). Table 10.6 compares mesh sheet dimensions, so far as possible, for grooved mesh sheet 3, and mesh sheets 5, 6, and 7. Mesh sheet 5 has no grooves. Mesh sheets 6 and 7 have a different type of groove than mesh sheet 3, located around the square opening.

10.2.5 Definition of NS03T Engine Performance Parameters

Engine performance was measured at the National Defense Academy (Kagawa, 2002). Thermocouples, pressure transducers, and crank-angle pickup sensors were installed in the engine. The output power was measured by a dynamometer. A data logger and a digitizer converted the analog signals received from the transducers and sensors to digital signals. The data

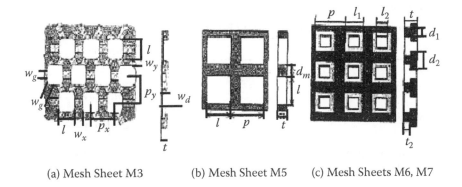

(a) Mesh Sheet M3   (b) Mesh Sheet M5   (c) Mesh Sheets M6, M7

FIGURE 10.4
Schematic comparisons of mesh sheets 3 (grooved), 5, 6, and 7. (From Furutani, S., Matsuguchi, A., and Kagawa, N., 2006, Design and Development of New Matrix with Square-Arranged Hole for Stirling Engine Regenerator, Paper AIAA 2006-4017, *Proceedings of 4th International Energy Conversion Engineering Conference*, San Diego, CA, June 26–29. With permission.)

were acquired by a personal computer (PC). A fuel (natural gas) flow meter provided data to the PC. Data for airflow-rate and cooling water flow rates to the cooler and the expansion and compression cylinders were input to the PC manually. Calculations of the engine performance, a display of the results and the engine operating conditions, and acquisition and saving of the data were automatically carried out by software written with a graphical programming language (HP Vee).

Table 10.7 gives the definitions of the powers, heats, losses, and efficiencies. The work loss due to pressure drop, ΔW_p, means the difference between the actual indicated work, W_{ind}, and a calculated indicated work, based on an assumption of no pressure losses. Effective heat input, Q_{eff}, and regenerator loss, Q_{rloss}, are derived from heat fluxes in the engine (Takizawa et al., 2002). The typical operating conditions are shown in Table 10.8.

TABLE 10.6

Mesh Sheet Dimensions and Geometric Characteristic Parameters, for Mesh Sheets 3, 5, and 6 (Units in mm, Except for Dimensionless Ratios, β and ϕ)

Matrix	l	w_x	w_y	p_x	p_y	t	w_g	w_d	β	ϕ
M3	0.3	0.141	0.082	0.441	0.382	0.12	0.15	0.072	0.487	0.667

Matrix	$l1$	$l2$	d_m	$d1$	$d2$	p	t	β	ϕ
M5	0.65		0.15			0.8	0.1	0.66	0.66
M6	0.68	0.58		0.1	0.2	0.78	0.1	0.65	0.65
M7	0.68	0.66		0.1	0.12	0.78	0.1	0.738	0.738

Source: From Matsuguchi et al., 2008, *Proceedings of the Japan Society of Mechanical Stirling*, 11. Reprinted with permission of the JSME.

TABLE 10.7

Definitions of Powers, Heat Flows, and Efficiencies

Items	Equations
1. Power	
Expansion work (power)	$W_e = n \oint P_e dV_e$
Compression work (power)	$W_c = n \oint P_c dV_c$
Indicated work (power)	$W_{ind} = W_e - W_c$
Pressure loss	$\Delta WP = (\oint P_e dV_e + \oint P_c dV_c - W_{ind})/2$
2. Quantity of Heat	
Heat Input	
Cooling heat in cooler (quantity)	Q_c
Cooling heat in exp. cyl. water jacket (quantity)	Q_{ec}
Cooling heat in comp. cyl. water jacket (quantity)	Q_{cc}
Regenerator loss	$Q_{rloss} = Q_c + Q_{ec} - W_c$
Effective heat input	$Q_{eff} = Q_{rloss} + W_e$
3. Efficiency	
Indicated efficiency	$\eta_{ind} = W_{ind}/Q_{eff}$
Internal conversion efficiency	$\eta_{int} = W_{ind}/W_e$

Source: From Takeuchi, T. et al., 2004, Performance of New Mesh Sheet for Stirling Engine Regenerator, Paper AIAA 2004-5648, A Collection of Technical Papers, 2nd International Energy Conversion Engineering Conference, Providence, RI, August 16–19. Reprinted with permission of the American Institute of Aeronautics and Astronautics.

TABLE 10.8

N503T Engine Operating Conditions Used for Regenerator Testing

Items	Parameter
Mean pressure	2–5 MPa
Engine speed	650–1200 rpm
Heater temperature	$993 \pm 10°K$
Expansion space temperature	883–908°K
Compression space temperature	293–303°K (water cooling)

Source: From Takeuchi, T. et al., 2004, Performance of New Mesh Sheet for Stirling Engine Regenerator, Paper AIAA 2004-5648, A Collection of Technical Papers, 2nd International Energy Conversion Engineering Conference, Providence, RI, August 16–19. Reprinted with permission of the American Institute of Aeronautics and Astronautics.

10.2.6 NS03T Stirling Engine Performance with Various Matrices

Kitahama et al. (2003) compares engine performance with wire screens #100, #150, and #200 with that of mesh sheets 1, 2, and 3. Mesh sheet 3 (stainless steel) and 4 (nickel) performance is discussed in Takeuchi et al. (2004). Matsuguchi et al. (2005) introduces a nongrooved mesh sheet and compares engine performance of nongrooved mesh sheet 5 with that of grooved mesh sheet 3. Furutani et al. (2006) introduces a different type of groove to reduce mesh sheet contact area, and attempts to optimize the design, yielding mesh sheet 6; also, NS03T engine performance is compared for regenerators constructed with mesh sheets 5 and 6.

10.2.6.1 Comparisons of NS03T Engine Performance for Wire-Screen and Grooved Mesh Sheets 1, 2, 3, and 4 (Kitahama et al., 2003; Takeuchi et al., 2004)

10.2.6.1.1 Power Comparisons: Wire Screens #100, #150, #200 and Mesh Sheets 1, 2, 3

Figures 10.5 and 10.6 represent the behaviors of specific indicated work, $W_{ind}/(P_{e,mean} \cdot v_e \cdot n)$, for each matrix. In the figures, the data series are fitted with curves to clarify the behaviors. As shown in the figures, specific indicated work for the various matrices is plotted with changing helium

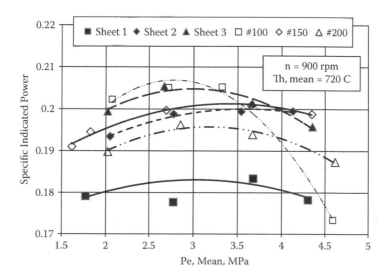

FIGURE 10.5
Specific indicated power as a function of pressure. (From Kitahama, D. et al., 2003, Performance of New Mesh Sheet for Stirling Engine Regenerator, Paper AIAA 2003-6015, *Proceedings of 1st International Energy Conversion Engineering Conference*, Portsmouth, VA, August 17–21. With permission.)

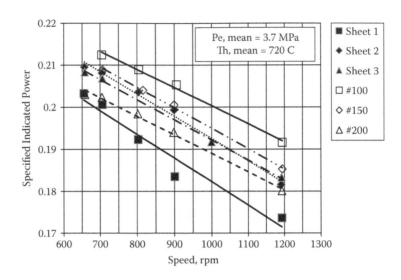

FIGURE 10.6
Specific indicated power as a function of engine speed. (From Kitahama, D. et al., 2003, Performance of New Mesh Sheet for Stirling Engine Regenerator, Paper AIAA 2003-6015, *Proceedings of 1st International Energy Conversion Engineering Conference*, Portsmouth, VA, August 17–21. With permission.)

pressure in the expansion space at 900 rpm engine speed, and versus the engine speed at 3.7 MPa, respectively. Figure 10.5 shows that specific indicated work increases with increasing expansion space pressure. Sheet 2, sheet 3, and wire screens #100 and #150 appear to yield similar peak specific indicated works, and #200 and sheet 1 yield less. The peaks occur at different mean pressures. Under other engine operating conditions, sheets 2 and 3 yielded 20% higher power and efficiency than the conventional stacked screen wire. As shown in Figure 10.6, specific indicated work decreases with increasing speed.

10.2.6.1.2 Efficiency Comparisons: Wire Screens #100,
#150, #200 and Mesh Sheets 1, 2, 3

Figures 10.7 and 10.8 show the experimental results for indicated efficiency, η_{ind}. In Figure 10.7, η_{ind} is nearly constant with increasing $P_{e,mean}$ except for wire screen #100. Efficiency of #100 decreases significantly with increasing $P_{e,mean}$. As shown in Figure 10.8, η_{ind} decreases approximately linearly with increasing speed for each of the matrices.

10.2.6.1.3 Pressure Loss Comparisons: Wire Screens #100,
#150, #200 and Mesh Sheets 1, 2, 3

The specific pressure losses, $\Delta W_P/(P_{e,mean} \cdot \nu_e \cdot n)$, are shown in Figures 10.9 and 10.10. These losses are caused by frictional resistance in the heat exchangers. The loss reduces indicated work directly. In Figure 10.9, specific pressure

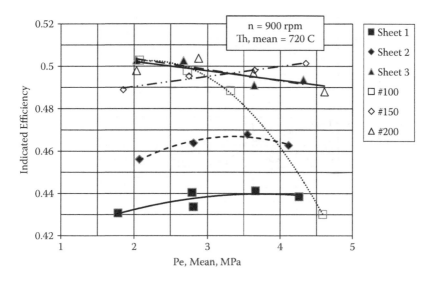

FIGURE 10.7
Indicated efficiencies as functions of expansion space pressure. (From Kitahama, D. et al., 2003, Performance of New Mesh Sheet for Stirling Engine Regenerator, Paper AIAA 2003-6015, *Proceedings of 1st International Energy Conversion Engineering Conference*, Portsmouth, VA, August 17–21. With permission.)

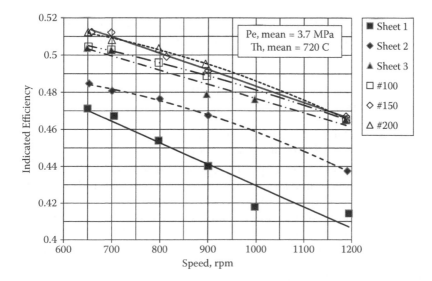

FIGURE 10.8
Indicated efficiencies as functions of engine speed. (From Kitahama, D. et al., 2003, Performance of New Mesh Sheet for Stirling Engine Regenerator, Paper AIAA 2003-6015, *Proceedings of 1st International Energy Conversion Engineering Conference*, Portsmouth, VA, August 17–21. With permission.)

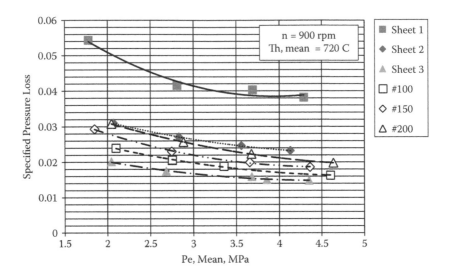

FIGURE 10.9
Specific pressure loss as functions of expansion space pressure. (From Kitahama, D. et al., 2003, Performance of New Mesh Sheet for Stirling Engine Regenerator, Paper AIAA 2003-6015, *Proceedings of 1st International Energy Conversion Engineering Conference*, Portsmouth, VA, August 17–21. With permission.)

loss of each matrix decreases slightly with increasing expansion space pressure. Sheet 3 has the least loss, and sheet 1 has the largest loss among all the evaluated wire-screen and mesh-sheet matrices. The actual difference in loss between sheets 3 and 1 ranges from ~150 W to ~300 W. Figure 10.10 shows that the specific pressure losses increase rapidly with increasing engine speed. Sheet 1 has the largest pressure loss values and the largest rate of increase with speed.

10.2.6.1.4 Regenerator Loss Comparisons: Wire Screens #100, #150, #200 and Mesh Sheets 1, 2, 3

Figures 10.11 and 10.12 show specific regenerator loss, $Q_{rloss}/(P_{e,mean} \cdot V_e \cdot n)$, of the matrices. Regenerator loss, which includes the reheat and thermal conduction losses, is calculated from rejected heat from the cooler, cooling heat at the water jacket, and the compression power. Therefore, the derived data have some uncertainty and scatter in the figures. Taking this uncertainty and scatter into consideration, it can only be stated that the specific regenerator losses of the various matrices are about equal. With increases in pressure, specific regenerator loss decreases slightly as shown in Figure 10.11. As shown in Figure 10.12, specific regenerator loss gradually increases with the engine speed. Sheet 1 shows rapid increase and has the largest specific regenerator loss among the compared matrices at 1200 rpm. The reason that the regenerator loss of sheet 1 increases rapidly is due to the working fluid velocity.

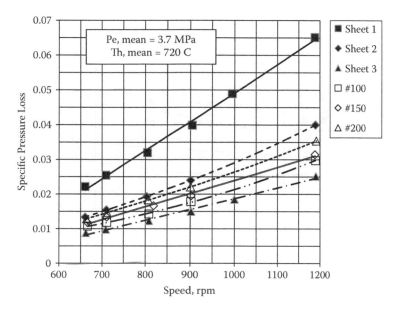

FIGURE 10.10
Specific pressure loss as functions of engine speed. (From Kitahama, D. et al., 2003, Performance of New Mesh Sheet for Stirling Engine Regenerator, Paper AIAA 2003-6015, *Proceedings of 1st International Energy Conversion Engineering Conference*, Portsmouth, VA, August 17–21. With permission.)

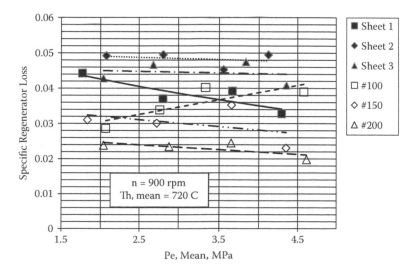

FIGURE 10.11
Specific regenerator loss as a function of pressure. (From Kitahama, D. et al., 2003, Performance of New Mesh Sheet for Stirling Engine Regenerator, Paper AIAA 2003-6015, *Proceedings of 1st International Energy Conversion Engineering Conference*, Portsmouth, VA, August 17–21. With permission.)

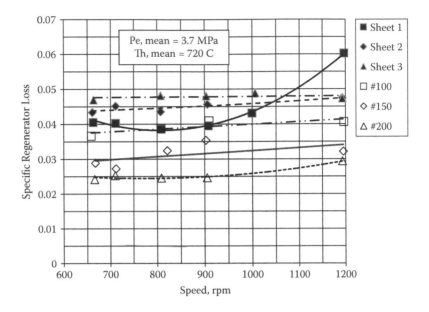

FIGURE 10.12
Specific regenerator loss as a function of engine speed. (From Kitahama, D. et al., 2003, Performance of New Mesh Sheet for Stirling Engine Regenerator, Paper AIAA 2003-6015, *Proceedings of 1st International Energy Conversion Engineering Conference*, Portsmouth, VA, August 17–21. With permission.)

This velocity is decreased by rapidly increasing specific pressure loss, and decrease of heat transfer in the matrix with increasing engine speed.

10.2.6.1.5 Efficiency Comparisons: Mesh Sheets 3 (Stainless Steel) and 4 (Nickel)

Mesh sheet 3 and 4 data comparisons, discussed in this and the next four paragraphs, are taken from Takeuchi et al. (2004). Figure 10.13 shows the experimental results for indicated efficiency, η_{ind}. Sheets 3 and 4 have about the same peak magnitudes, but sheet 4 peaks at a higher mean expansion space pressure. Thus, sheet 4 is more suitable for higher load operations than sheet 3.

10.2.6.1.6 Power Comparisons: Mesh Sheets 3 (Stainless Steel) and 4 (Nickel)

Figure 10.14 shows the behavior of specific indicated power, $W_{ind}/(P_{e,mean} \cdot V_e \cdot n)$, for the two matrices. In the figure, each data set is fitted with a curve to clarify the behavior, and the data are plotted versus mean expansion space helium pressure at 900 rpm engine speed. The figure shows that the specific indicated power of sheet 4 is ~2% higher than that of sheet 3 over the operating range.

10.2.6.1.7 Pressure-Loss Comparisons: Mesh Sheets 3
(Stainless Steel) and 4 (Nickel)

The specific pressure losses of the two mesh sheets, $\Delta W_p/(P_{e,mean} \cdot V_e \cdot n)$, are shown in Figure 10.15. This loss is caused by internal frictional resistance in the heat

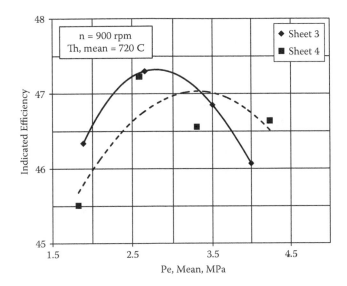

FIGURE 10.13
Indicated efficiency for mesh sheets 3 and 4 as functions of mean expansion-space helium pressure. (From Takeuchi, T. et al., 2004, Performance of New Mesh Sheet for Stirling Engine Regenerator, Paper AIAA 2004-5648, A Collection of Technical Papers, *2nd International Energy Conversion Engineering Conference*, Providence, RI, August 16–19. With permission.)

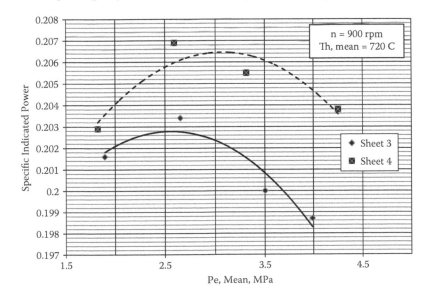

FIGURE 10.14
Specific indicated power for mesh sheets 3 and 4 as functions of mean expansion-space helium pressure. (From Takeuchi, T. et al., 2004, Performance of New Mesh Sheet for Stirling Engine Regenerator, Paper AIAA 2004-5648, A Collection of Technical Papers, *2nd International Energy Conversion Engineering Conference*, Providence, RI, August 16–19. With permission.)

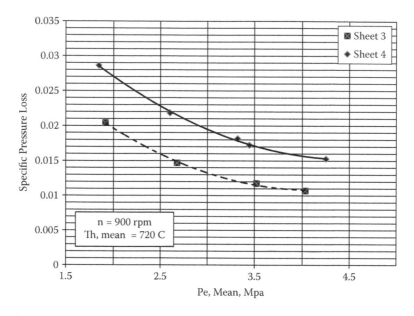

FIGURE 10.15
Specific pressure losses as functions of mean expansion space helium pressure. (From Takeuchi, T. et al., 2004, Performance of New Mesh Sheet for Stirling Engine Regenerator, Paper AIAA 2004-5648, A Collection of Technical Papers, *2nd International Energy Conversion Engineering Conference*, Providence, RI, August 16–19. With permission.)

exchangers, including the heater, the cooler, and the regenerator. The loss reduces indicated power directly. In Figure 10.15, the specific pressure loss of each matrix decreases with increasing mean expansion space pressure. Sheet 4 has larger pressure losses than Sheet 3. The absolute difference between sheets 3 and 4 is about 45 W, in spite of the same opening size for the two sheets. The reason is the larger number of layers of sheet 4, whose thickness is 20% less than sheet 3.

10.2.6.1.8 Regenerator-Loss Comparisons: Mesh Sheets 3 (Stainless Steel) and 4 (Nickel)

In Figure 10.16, the specific regenerator loss, $Q_{rloss}/(P_{e,mean} \cdot v_e \cdot n)$, of sheet 4 shows lower values than sheet 3, especially in the high pressure range. The absolute difference of specific regenerator loss is 140 W at a mean pressure of 4 MPa. The reason for the decreased regenerator loss of sheet 4 is the larger heat capacity of nickel. Taking the data scattering into consideration, it seems reasonable to state only that the specific regenerator losses of the two matrices are almost constant under the operating conditions.

10.2.6.1.9 Additional Comments on Comparisons of Mesh Sheets 3 (Stainless Steel) and 4 (Nickel)

When the internal conversion efficiencies of mesh sheet 3 and 4 regenerators were examined, it was found that mesh sheet 3 had about 1.5% higher

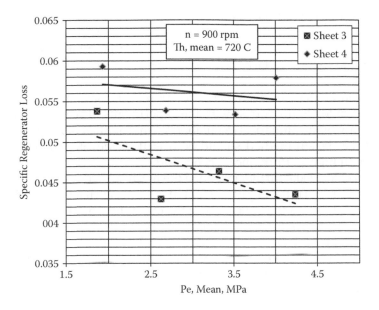

FIGURE 10.16
Specific regenerator loss for mesh sheets 3 and 4 as functions of mean expansion space helium pressure. (From Takeuchi, T. et al., 2004, Performance of New Mesh Sheet for Stirling Engine Regenerator, Paper AIAA 2004-5648, A Collection of Technical Papers, *2nd International Energy Conversion Engineering Conference*, Providence, RI, August 16–19. With permission.)

internal efficiency than mesh sheet 4 (Takeuchi et al., 2004). It was also found that the temperature efficiencies at the hot and cold ends of the regenerator were about 1% lower for sheet 4 than for sheet 3 because of the higher conductivity of nickel compared to stainless steel. Definitions and additional plots in Takeuchi et al. (2004) support these conclusions. It was suggested that the heat conduction loss in the nickel regenerator might be reduced by (1) attaching a thin layer of stainless steel to the surface of the nickel, which would be expensive, or (2) stacking stainless steel wire screens or mesh sheets between the nickel mesh sheets.

10.2.6.2 Comparisons of NS03T Engine Performance for Grooved Mesh Sheet 3, Nongrooved Mesh Sheet 5, and Mesh Sheets 6 and 7 (Furutani et al., 2006; Matsuguchi et al., 2005, 2008)

As discussed earlier, it was decided to eliminate the grooves from the mesh sheets to reduce regenerator porosity and engine dead volume. To optimize the nongrooved mesh sheets, FIDAP was used for regenerator simulation, as discussed earlier. Friction coefficient, Nusselt number, and a performance ratio (or figure of merit) defined as friction factor divided by Nusselt number were examined for ranges of mesh sheet dimensions. Via this approach, a new nongrooved mesh sheet, sheet 5, was defined. Table 10.6 compared the dimensions of grooved mesh sheet 3 and nongrooved mesh sheet 5. Mesh sheet 5 porosity is about 7% lower than that of mesh sheet 3. Mesh sheet 5's opening area ratio is also substantially larger than that of mesh sheet 3, which should tend to reduce the pressure drop loss of sheet 5 compared to sheet 3.

It was found that the increments of engine performance estimated by the FIDAP code almost agreed with the experiments. It was concluded that the reliability of the numerical calculation seemed sufficient for practical use, and that the design method should be useful for development of Stirling engines with high-performance regenerators.

Furatani et al. (2006) further extended the mesh sheet development to measure the effects of (1) the angular orientation of mesh sheets 3 and 5, and (2) addition of a different type of groove to the mesh sheet 5 design to reduce the contact area between the mesh sheets, yielding mesh sheet 6.

In the case of mesh sheet 5, it was found that random angular orientation of the mesh sheets was better than regular advances of 45° or 60° in angular orientation of adjacent mesh sheets. In the case of mesh sheet 3, whether random angular orientations or 45° advances in angular orientation were used had no significant effect (60° advance was not considered due to a 60° symmetry in the pattern of mesh sheet 3). Thus, in testing mesh sheet 6, random angular orientation of adjacent mesh sheets was used.

The design of mesh sheet 6 was optimized by using FIDAP, to examine friction factors, Nusselt numbers, and the ratio of friction factor to Nusselt number. Smaller ratios correspond to lower friction loss and higher heat transfer. Thus, small ratios of friction factor to Nusselt number should produce better performance.

A modification was later made to the mesh sheet 6 geometry to produce mesh sheet 7 (Matsuguchi et al., 2008). Table 10.6 shows that the changes in mesh sheet 6 produced larger flow openings in mesh sheet 7, resulting in increases in opening ratio, β, and porosity, ϕ.

The performance of mesh sheets 3, 5, 6, and 7 are compared in Figures 10.17, 10.18, and 10.19, on the basis of performance in the ~3 kW NS03T engine. Figure 10.17 compares experimental pressure losses and numerical-analysis friction factors for these four mesh sheets. Figure 10.18 compares experimental regenerator losses and numerical-analysis Nusselt numbers. Figure 10.19 compares experimental indicated efficiencies of the NS03T and the numerical-analysis figure-of-merit ratio, (friction factor)/(Nusselt number).

The Figure 10.17 experimental pressure loss data show that increasing the openings in mesh sheet 6 (highest pressure losses in Figure 10.17) to produce mesh sheet 7, resulted in mesh sheet 7 pressure losses being the lowest of the four mesh sheets compared in Figure 10.17, but mesh sheet 7 pressure losses were only slightly lower than those of mesh sheet 3. The numerical-analysis friction factor curves seem almost consistent with the experimental

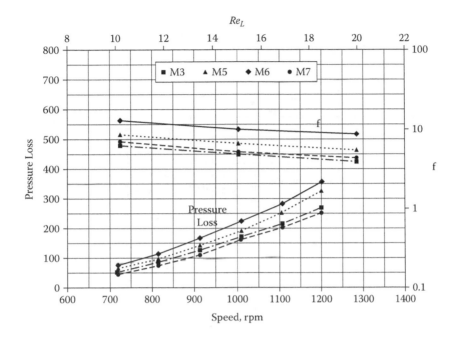

FIGURE 10.17
Experimental pressure loss as a function of engine speed and numerical-analysis friction factor, f, as a function of Reynolds number. (From Matsuguchi, A. et al., 2008, Performance Analysis of New Matrix Material for the Stirling Engine Regenerator [in Japanese, except for abstract, figure, and table captions], *Proceedings of the Japan Society of Mechanical Stirling*, Vol. 11, 2008. With permission.)

pressure losses, except that the mesh sheet 3 friction-factor curve is very slightly lower than the mesh sheet 7 friction-factor curve.

Figure 10.18 shows that the experimental regenerator losses for mesh sheet 3 are substantially lower than those of the other mesh sheets. Mesh sheet 7 regenerator losses were slightly lower than those of 5 and 6, at the highest and lowest engine speeds. However, the numerical-analysis Nusselt number for mesh sheet 3 seemed only slightly higher than for the other mesh sheets.

Figure 10.19 shows that mesh sheet 3 had the best indicated efficiencies over the entire speed range, apparently due to its superior, low, regenerator loss, and also its relatively low pressure loss. Mesh sheet 7 had the next best efficiency curve. Mesh sheets 5 and 6 had the worst, and about the same, experimental efficiencies. The numerical-analysis figure-of-merit, f/Nu, was not entirely consistent with the experimental efficiencies, because the figure-of-merit for mesh sheet 6 was substantially above that for 5.

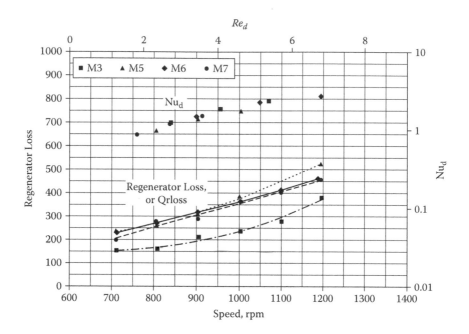

FIGURE 10.18
Experimental regenerator loss as a function of engine speed and numerical-analysis Nusselt number as a function of Reynolds number. (From Matsuguchi, A. et al., 2008, Performance Analysis of New Matrix Material for the Stirling Engine Regenerator [in Japanese, except for abstract, figure, and table captions], *Proceedings of the Japan Society of Mechanical Stirling*, Vol. 11, 2008. With permission.)

10.2.6.3 Results of Testing with Combined Mesh Sheet Types in the SERENUM05 and NS03T Engines

10.2.6.3.1 SERENUM05 Engine

The NS03T engine was described earlier in this chapter. So, this section will concentrate on a brief description of the SERENUM05 engine.

Kagawa et al. (2007) describe in considerable detail the design process used for the SERENUM05 engine, assisted via a Stirling Engine Thermodyamic and Mechanical Analysis (SETMA) mathematical model.

The SERENUM05 was also designed to be a ~3 kW engine (as was the NS03T), but was developed for use as a portable generator to be applied for military and conventional uses. However, the SERENUM05 is a double-acting engine with four pistons, U-shaped cylinders, a Z-crank mechanism, and a compact AC generator. Instead of the canister regenerators of the NS03T, the regenerators and coolers of the SERENUM05 are annular and are located around the cylinders; the annular regenerators have inner and outer diameters of 45 and 65 mm, respectively, length of 55 mm, and contain 500 layers of annular mesh sheets. The SERENUM05 was originally designed by the National Defense Academy, and was built by Toshiba in 2005.

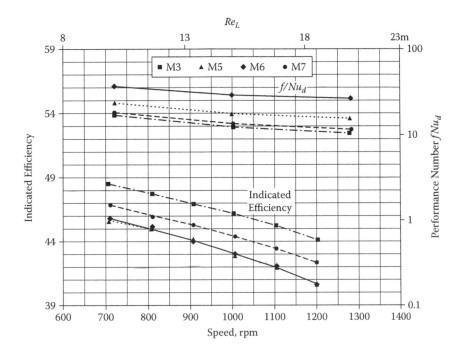

FIGURE 10.19
Experimental indicated efficiency as a function of engine speed and numerical-analysis performance number (figure of merit) as a function of Reynolds number. (From Matsuguchi, A. et al., 2008, Performance Analysis of New Matrix Material for the Stirling Engine Regenerator [in Japanese, except for abstract, figure, and table captions], *Proceedings of the Japan Society of Mechanical Stirling*, Vol. 11, 2008. With permission.)

The design parameters of the SERENUM05 engine are given in Table 10.9, and a schematic of the engine is given in Figure 10.20. Detailed engine specifications for various components can be found in Kagawa et al. (2007) and Matsuguchi et al. (2009).

10.2.6.3.2 Combination Mesh-Sheet Regenerators Tested in NS03T and SERENUM05 Engines

Figure 10.21 shows schematics of the mesh sheets that were combined for these tests. Table 10.10 compares some of the parameters for these mesh sheets. Table 10.11 shows the operating conditions used in the two engines for these tests.

Note that mesh sheets 3 and 4 are the original grooved mesh sheets, and mesh sheet 5 has no grooves. Mesh sheets 3 and 5 are made of stainless steel 304, and mesh sheet 4 is made of nickel. The two combinations used were (1) M3 + M4, with M3 and M4 types alternating throughout the regenerator stack, and (2) M3 + M5, with M3 and M5 types alternating through the stack.

TABLE 10.9

Design Parameters of SERENUM05 Engine

Main Fuel	Natural Gas
Working fluid	Helium
Mean pressure	3–5 MPa
Maximum expansion space temperature	975 ± 50 K
Compression space temperature	<323 K (water cooling)
Engine speed	500–1500 rpm
Maximum output power	>3 kW
Maximum generating efficiency	30%
NOx	<150 ppm
Noise level	<60 dB(A)
Mass	<80 kg
Height	<700 mm

Source: From Matsuguchi, A., Kagawa, N., and Koyama, S., 2009, Improvement of a Compact 3-kW Stirling Engine with Mesh Sheet, *Proceedings of the International Stirling Engine Conference (ISEC).* Reprinted by permission of the International Stirling Engine Conference.

10.2.6.3.3 Combination-Mesh-Sheet Test Results

Results of the combination-mesh-sheet tests in the NS03T and SERENUM05 engines are shown in Figure 10.22 through Figure 10.25.

Figure 10.22 shows the relationship between engine speed and indicated power. In this figure, indicated power increases with engine speed. W_{ind} of the NS03T and SERENUM05 are equivalent around 600 rpm. However, the indicated power for the SERENUM05 becomes larger than for the NS03T in the higher speed range, and the difference is from 600 to 800 W around 1200 rpm.

Figure 10.23 shows indicated efficiency as a function of engine speed. The indicated efficiency of the SERENUM05 increases with engine speed, but the indicated efficiency of the NS03T decreases with increasing engine speed. Also, in spite of the heater temperature of the SERENUM05 being lower, the indicated efficiency of the SERENUM05 is equivalent to that of the NS03T around 1200 rpm. Also, the M3 + M4 combination has the highest indicated efficiency for the SERENUM05, and the M3 + M5 combination has the highest for the NS03T.

Figure 10.24 shows the results for mechanical loss. In this figure, the mechanical loss of the NS03T is almost constant in the measurement range. However, that of the SERENUM05 increases with increasing engine speed. It appears that the friction loss of the Z-crank mechanism may be larger than for more conventional mechanisms.

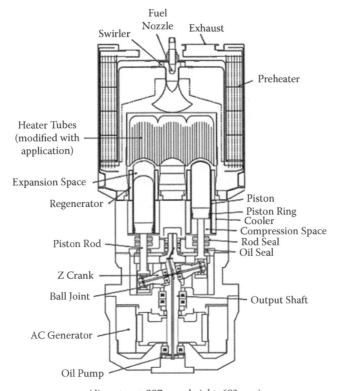

(diameter: φ 337 mm, height: 692 mm)

FIGURE 10.20
SERENUM05 double-acting, U-shaped cylinder, Z-crank mechanism, annular regenerator, and cooler, engine. (From Matsuguchi, A., Kagawa, N., and Koyama, S., 2009, Improvement of a Compact 3-kW Stirling Engine with Mesh Sheet, *Proceedings of the International Stirling Engine Conference* [ISEC]. With permission.)

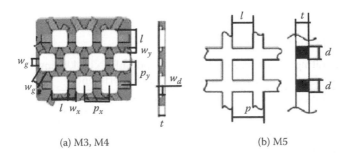

(a) M3, M4   (b) M5

FIGURE 10.21
Mesh sheets used in the combination-mesh-sheet tests of the NS03T and SERENUM05 engines. (From Matsuguchi, A., Kagawa, N., and Koyama, S., 2009, Improvement of a Compact 3-kW Stirling Engine with Mesh Sheet, *Proceedings of the International Stirling Engine Conference* [ISEC]. With permission.)

TABLE 10.10

Comparisons of Some of the Mesh Sheet 3, 4, and 5
Parameters

Dimension	M3	M4	M5
Opening l (mm)	0.300	0.300	0.650
Thickness t (mm)	0.120	0.100	0.100
Open area ratio β	0.487	0.487	0.660
Porosity ϕ	0.711	0.711	0.660

Source: From Matsuguchi, A., Kagawa, N. and Koyama, S.,
2009, Improvement of a Compact 3-kW Stirling
Engine with Mesh Sheet, *Proceedings of the
International Stirling Engine Conference (ISEC).*
Reprinted by permission of the International Stirling
Engine Conference.

Figure 10.25 shows the experimental results for regenerator loss. The
regenerator loss of the SERENUM05 is 1000 to 1500 W larger than that of the
NS03T.

10.2.6.3.4 Conclusions for the Combination Mesh-Sheet Engine Tests

1. The M3 + M4 regenerator has the highest indicated efficiency for
   the SERENUM05, and the M3 + M5 regenerator has the highest for
   the NS03T

2. The SERENUM05 has larger indicated power over most of the speed
   range; its indicated efficiency is less than that of the NS03T, until
   engine speed reaches about 1200 rpm, where the indicated efficiencies of both engines are about the same.

3. The SERENUM05 has larger mechanical loss and regenerator loss
   than the NS03T.

TABLE 10.11

Operating Conditions of the NS03T and SERENUM05 Engines
Used for the Combination-Mesh-Sheet Experiments

Conditions	SERENUM05	NS03T
Engine speed (rpm)	600–1200	600–1200
Heater tube mean temperature (°C)	550	725
Charged helium pressure (MPa)	2.0–2.5	2.0–2.5

Source: From Matsuguchi, A., Kagawa, N., and Koyama, S., 2009,
Improvement of a Compact 3-kW Stirling Engine with Mesh
Sheet, *Proceedings of the International Stirling Engine Conference
(ISEC).* Reprinted by permission of the International Stirling
Engine Conference.

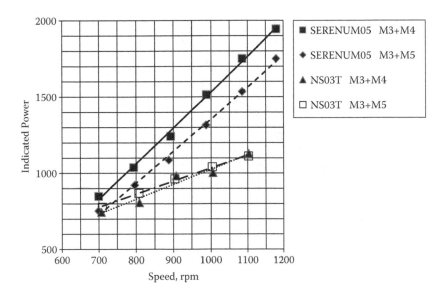

FIGURE 10.22
Indicated power as functions of engine speed, for mesh sheet combinations M3 + M4 and M3 + M5, as tested in the NS03T and SERENUM05 engines. (From Matsuguchi, A., Kagawa, N., and Koyama, S., 2009, Improvement of a Compact 3-kW Stirling Engine with Mesh Sheet, *Proceedings of the International Stirling Engine Conference* [ISEC]. With permission.)

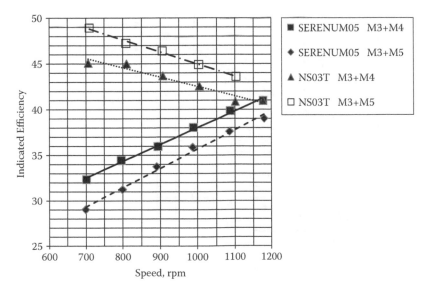

FIGURE 10.23
Indicated efficiency as functions of engine speed, for mesh sheet combinations M3 + M4 and M3 + M5, as tested in the NS03T and SERENUM05 engines. (From Matsuguchi, A., Kagawa, N., and Koyama, S., 2009, Improvement of a Compact 3-kW Stirling Engine with Mesh Sheet, *Proceedings of the International Stirling Engine Conference* [ISEC]. With permission.)

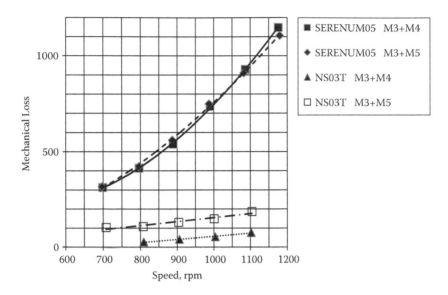

FIGURE 10.24
Mechanical loss as functions of engine speed, for mesh sheet combinations M3 + M4 and M3 + M5, as tested in the NS03T and SERENUM05 engines. (From Matsuguchi, A., Kagawa, N., and Koyama, S., 2009, Improvement of a Compact 3-kW Stirling Engine with Mesh Sheet, *Proceedings of the International Stirling Engine Conference* [ISEC]. With permission.)

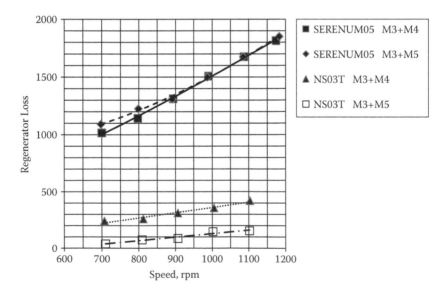

FIGURE 10.25
Regenerator loss as functions of engine speed, for mesh sheet combinations M3 + M4 and M3 + M5, as tested in the NS03T and SERENUM05 engines. (From Matsuguchi, A., Kagawa, N., and Koyama, S., 2009, Improvement of a Compact 3-kW Stirling Engine with Mesh Sheet, *Proceedings of the International Stirling Engine Conference* [ISEC]. With permission.)

10.3 Matt Mitchell's Etched-Foil Regenerators

Matt Mitchell fabricated three patterns of etched regenerator foil to be tested in the NASA/Sunpower oscillating-flow test rig (Mitchell et al., 2005, 2007). Two of those patterns allow essentially straight-through flow in the regenerator. One straight-through pattern had a porosity of about 76%, the other of about 60%. The third etched foil, a zigzag pattern, had a porosity of about 59%.

All three etched foils were approximately 49.53 mm long by 16.59 mm wide and 0.076 mm thick. The nominal etched depth was about 0.045 mm. All are etched from 316 L stainless steel. Where etched at the same locations from both sides, holes appear, leaving a lacework of parallel plates that form the primary heat-transfer surfaces. All three resulting regenerators have the same number of plates (52), but the widths of the plates and intervening spaces are different.

The etched foil labeled "1A" in Figure 10.26 is a magnified photograph of a portion of a piece of low-density straight-through foil. It is the most porous of the three. In this pattern, the plates are nominally 0.483 mm wide interspersed with gaps of the same dimension. Flow channels between parallel plates are thus about 0.045 mm thick by 0.56 mm wide by 0.48 mm long in the flow direction. The ratio of thickness to width is 12.4 to 1, which qualifies the arrangement as parallel plates as opposed to a rectangular tube. The length of the flow channel is slightly more than 10 times the thickness of the flow channel, which is of the order of the entry length for the anticipated Reynolds number. Average weight for these low-density, high-porosity foils is about 0.141 g.

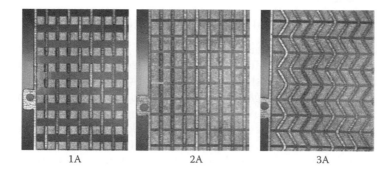

1A   2A   3A

FIGURE 10.26
Three Mitchell etched foils. A portion of the front side of the low-density, high-porosity etched foil is labeled "1A". Flow is left to right. Positioning tab is at top. The foil labeled "2A" is the high-density low-porosity foil, which was tested. A portion of the front side of the zigzag foil is labeled "3A." (From Mitchell, M. P. et al., 2007, Results of Tests of Etched Foil Regenerator Material. *Cryocoolers 14, Springer U.S. Proceedings of International Cryocooler Conference, Inc.* Boulder, CO. With permission.)

The second straight-through flow pattern, the high-density pattern, is "2A" in Figure 10.26. It differs from the first only in the relative area occupied by the plates and the spaces between them. The nominal width of the plates (or length in the direction of flow) is 0.787 mm, and the width of the spaces between the plates is 0.178 mm. The wider plates and smaller intervening spaces produce a heavier foil, averaging about 0.222 g per foil.

The third pattern, with zigzag spacer-bridges, is labeled "3A" in Figure 10.26. It has the same plate spacing as the heaver straight-through pattern. It is the heaviest of the three patterns due to the somewhat greater length of its zigzag spacer-bridges. Average weight is about 0.230 g. The rationale for the zigzag spacers is for the grooves on the back sides of the foil to redistribute part of the flow laterally at each change of direction of the zigzag flow pattern—to improve flow distribution. The zigzag foil was reported to perform substantially better than a straight-through foil in a coaxial pulse tube cryocooler equipped with an annular regenerator (Mitchell and Fabris, 2003).

Time and funding finally permitted testing only one of the three Mitchell etched foil patterns in the NASA/Sunpower oscillating-flow test rig. On close examination, one of the three stacks of regenerator foils appeared to be of substantially better quality than the other two. This foil, the high-density low-porosity foil labeled "2A" in Figure 10.26, was chosen for testing in the oscillating-flow rig. Figure of merit test results of this Mitchell etched foil are compared with some other regenerator test results in Figure 8.34. It is identified in the figure as "etched foil 2A sample."

10.4 Sandia National Laboratory Flat-Plate Regenerator

The completed parallel-plate regenerator was a cylinder 88.9 mm in diameter and 73 mm long. It was constructed from a stack of alternating stainless-steel sheets. Both sheets (A and B types) were fabricated by photochemical milling (PCM) of 316L stainless steel. After completion of the PCM process, the next steps were cleaning, sorting, stacking, and diffusion bonding of the stacked sheets. The bonded stack of sheets was machined into a cylinder by electric discharge machining (EDM). Much more of the geometric and fabrication detail is given in Backhaus and Swift (2001). It was reported that the power output of the thermoacoustic engine for which the parallel plate regenerator was fabricated almost doubled compared to performance with the earlier wire-screen regenerator; efficiency also significantly increased at the highest acoustic power input (Backhaus and Swift, 2001).

David Gedeon later reported, following a conversation with Backhaus, that the flat-plate regenerator, or a portion of it, was offered for testing in the NASA/Sunpower oscillating-flow test rig. However, by that time, it was reported that the parallel plates had apparently experienced some distortion, and the dimensional tolerances would no longer have matched the original fabricated values. The Sandia flat-plate regenerator was not tested in the oscillating-flow rig.

11

Applications Other Than Stirling Engines

11.1 Introduction

This chapter explains how the information related to Stirling regenerators is applicable also to other areas. These areas includes applications where porous media are used, such as combustion processes, catalytic reactors, packed-bed heat exchangers, electronics cooling, heat pipes, thermal insulation engineering, nuclear waste repository, miniature refrigerators, characterization of heat transport through biological media, and porous scaffolds for tissue engineering, to name a few. Only the first few applications will be discussed in this chapter.

11.2 Use of Porous Material in Combustion Processes

Porous media represent a large range of pore sizes, pore connectivity, porosities, and so forth. The solid phase could be metallic, ceramic, or from organic materials. The fluid phase could have properties that range from liquid macromolecules to low-pressure gases. On the other hand, physical and chemical processes have different length and time scales phenomenologically associated with each one. Thermal penetration depth, residence time, flame length, and flame thickness are examples of those scales to name few. The combination of the different length scales and materials leads to a wide range of applications for which those porous media can be used.

Examples of systems that involve a wide range of length scales are combustion of synthesis, combustion of solid fuels, catalytic convertors, and catalytic reactors (Figure 11.1). Oliveira and Kaviany (2001) show length scales of reactant transport in porous media.

More efficient heat exchangers are in great demand in processes involving combustion.

Via this technique, low-pollutant emissions as well as increased heat-transfer rates are accomplished. Figure 11.2 shows a double-tube air/porous

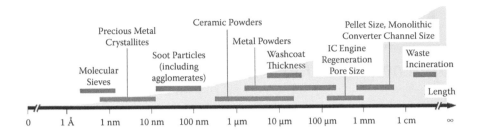

FIGURE 11.1

Length scales and characteristic dimensions of porous media for various systems. (Reprinted from *Progress in Energy and Combustion Science*, 27(5), A. A. M. Oliveira and M. Kaviany, Nonequilibrium in the Transport of Heat and Reactants in Combustion in Porous Media, 523–545. Copyright 2001, with permission from Elsevier.)

combustor heat exchanger (Moraga et al., 2009) where porous media is utilized in the inner tube. The outer annulus tube is used as a preheater for the air before entering the combustion zone.

Some of the main reasons for incorporating porous media in the combustion process (Zhdanok et al., 2000) are (1) the higher surface area (between the solid and fluid) provided by the porous media which improves both the combustion and heat-transfer processes, (2) the gas preheating (by both convection and radiation) results in a complete combustion and much higher temperature (compared to the adiabatic flame temperature) after

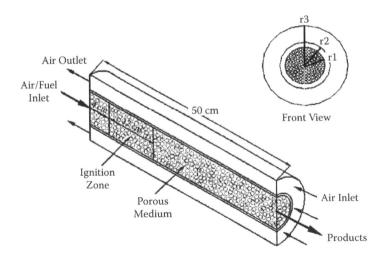

FIGURE 11.2

Double-tube air/porous combustor heat exchanger. (Reprinted from *International Journal of Heat and Mass Transfer*, 52(13–14), N. O. Moraga et al., Unsteady Fluid Mechanics and Heat Transfer Study in a Double-Tube Air–Combustor Heat Exchanger with Porous Medium, 3353–3363. Copyright 2009, with permission from Elsevier.)

combustion, and (3) the solid material (in the porous media) thermal conductivity is about 100 times that of the gases which helps improve temperature uniformity and avoid hot spots (could be a source of NOx production). The combustion process is greatly affected by the design characteristics of the porous media, such as geometric dimensions, permeability, and the pressure difference between the air-fuel supply and the combustion chamber. Several studies have focused on the implementation of a physical mathematical model that allows the development and optimization of the design of burners with porous media (Bouma et al., 1995; Hsu and Matthews, 1993; Hsu et al., 1993; Malico et al., 2000; Sathe et al., 1990).

In the area of the internal combustion (IC) engine, a similar principle to the one described earlier (see Chapter 4) for the Stirling engine is used for diesel engines (Ferrenberg, 1994; Hanamura et al., 1997). In this design (shown in Figure 11.3), an in-cylinder porous foam regenerator is used as a thermal energy storage (similar to Stirling) as well as a catalytic convertor in the

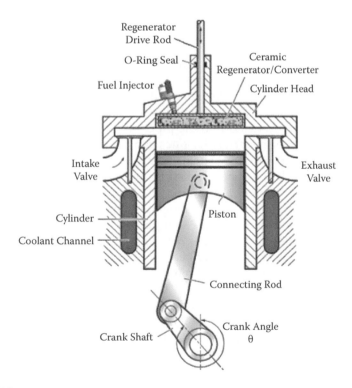

FIGURE 11.3
In-cylinder porous foam used in diesel engines (see Ferrenberg, 1994). (Reprinted from *Progress in Energy and Combustion Science*, 27(5), A. A. M. Oliveira and M. Kaviany, Nonequilibrium in the Transport of Heat and Reactants in Combustion in Porous Media, 523–545. Copyright 2001, with permission from Elsevier.)

diesel engine. The regenerator is close to either the piston top or the cylinder head most of the time. However, during the suction stroke (regenerative cooling), the regenerator moves downward toward the piston top, and the cold gases coming into the chamber get heated up by the regenerator. On the other hand, during the exhaust stroke (regenerative heating) the regenerator moves upward toward the cylinder head, and the hot gases leaving the chamber cool down through the regenerator. Utilizing this design resulted in a 50% increase in fuel efficiency for the same air/fuel ratio and using SiC 12 ppi foam (Oliveira and Kaviany, 2001).

11.3 Use of Porous Materials to Enhance Electronic Cooling

There are several ways in which porous materials might be used to enhance electronic cooling. But, generally, insertion of porous materials into a flow path increases heat transfer, and tends to increase pressure drop. Thus, in studying/designing a particular geometrical arrangement, it appears important to do some optimization of the heat transfer to pressure drop ratio to determine suitable values that might increase heat removal to a desired value, without increasing the energy required to drive the cooling flow too much.

Pavel and Mohamad (2004) experimentally and numerically studied heat-transfer enhancement in a pipe subjected to uniform heat flux, when metallic porous materials are inserted into the pipe. The results obtained led to the conclusion that higher heat-transfer rates can be achieved by using porous inserts, but at the expense of reasonable pressure drop. It was also shown that for an accurate simulation of heat transfer when a porous insert is employed, its effective thermal conductivity should be carefully evaluated. Such a pipe might be thought of as generally analogous to a heat exchanger flow path through an electronic component—though such an electronics flow path would not be expected to have a uniform cross section like that of a pipe. It does suggest that a heat exchanger heat sink using porous materials might be used to enhance electronics heat removal, but at the expense of increased drive power for the cooling flow.

Ould-Amer et al. (1998) numerically investigated a geometry that is more like a real electronic cooling geometry, as shown in Figure 11.4. The heat-generating blocks shown in the figure are analogous to electronic components like microchips. The numerical results were generated for laminar forced convection cooling of heat-generating blocks mounted on a parallel-plate channel. The flow in the porous medium was modeled using the Brinkman-Forchheimer extended Darcy model. Mass, momentum, and energy equations were solved. The local Nusselt number at the walls of the blocks, the mean Nusselt numbers, and the maximum temperature in the blocks were examined for a wide range of Darcy numbers and thermal

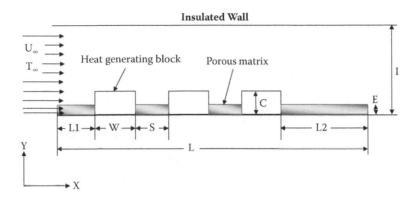

FIGURE 11.4
The physical domain used in Amer-Ould et al. (1998). (Reprinted from *International Journal of Heat and Fluid Flow*, 19(3), Y. Ould-Amer et al., Forced Convection Cooling Enhancement by Use of Porous Materials, 251–258. Copyright 1998, with permission from Elsevier.)

conductivity ratio. The computations were first conducted for a single block, then for evenly mounted blocks. The results showed that insertion of porous material between the blocks may enhance the heat-transfer rate on the vertical sides of the blocks. (The effective conductivity of the porous material must be much higher than the fluid thermal conductivity.) Although the porous matrix reduces the heat transfer on the horizontal face of the blocks, significant increases in the mean Nusselt number (up to 50%) were predicted, and the maximum temperatures within the heated blocks were reduced in comparison to the pure fluid (no porous material) case.

Ko and Anand (2003) investigated the use of porous aluminum baffles to enhance heat transfer in a rectangular channel, for the experimental geometry shown in Figure 11.5. (This is a short section taken from the experimental section which was longer and had more baffles). The experiments were conducted in the Reynolds number range of 20,000 to 50,000. The experimental procedure was validated by comparing the data for the straight channel with no baffles with those in the literature. Experiments showed that the flow and heat transfer reached periodically fully developed states downstream of the seventh module. For the range of independent parameters examined in this study, it was felt that the following conclusive statements could be made: The heat-transfer enhancement ratio decreases with increase in Reynolds number and increases with increases in pore density. The heat-transfer enhancement ratio reaches a maximum value of 300% for the range of parameters studied in the investigation. The heat-transfer enhancement ratio was found to be higher for taller and thicker baffles. The ratio of heat-transfer enhancement of per unit increase in pumping power was less than one. The friction factor slightly decreased with increase in Reynolds number, and increased with baffle thickness and pore density. Note that using porous instead of

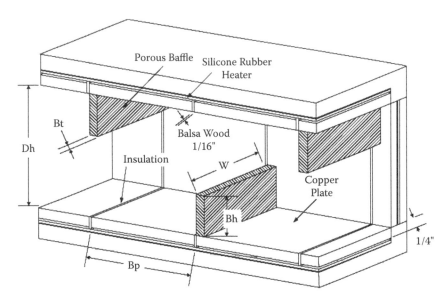

FIGURE 11.5
The test section used in Ko and Anand (2003). (Reprinted from *International Journal of Heat and Mass Transfer*, 46(22), K. Ko and N. K. Anand, Use of Porous Baffles to Enhance Heat Transfer in a Rectangular Channel, 4191–4199. Copyright 2003, with permission from Elsevier.)

solid baffles of the same size, in general, decreased pressure drop losses, but also decreased heat transfer.

11.4 Use of Porous Materials in Heat Pipes

Heat pipe is an efficient two-phase heat-transfer device (Dunn and Reay, 1994). It utilizes the latent heat of vaporization/condensation to transport heat from a hot source to a cold sink. The working fluid is pumped under capillary forces.

Heat pipes are widely used in cooling electronic equipment (e.g., laptop). However, using a conventional heat pipe design in the area of electronic cooling application has its limitations. The reason behind this is that both the evaporator and condenser are mostly located in a horizontal position and very close to each other. These, in turn, result in having most of the pressure losses associated with what is called "entrainment losses" (Faghri, 1995). Those losses are associated with the liquid flow through the porous wick (present along the whole length of the heat pipe) as well as the viscous interaction between the two phases (liquid and vapor).

Further developments in heat pipes have been achieved (Ku, 1999; Maydanik and Fershtater, 1997) by introducing what is known as loop heat pipe (LHP). This new design separates the liquid from the vapor phases and localizes the capillary structure in the evaporative section only. Thus, the wick is integrated with the evaporator and should be designed well enough to provide the capillary forces needed to get the LHP operational.

In the area of compact cooling devices, miniature LHP (mLHP) designs have been studied (Bienert et al., 1999; Hoang et al., 2003; Kiseev et al., 2003; Singh, 2006; Singh et al., 2007). Several parameters should be considered in the design of the mLHP that will affect the operational characteristics of the loop system, such as material used being plastic or metal, pore size, permeability, porosity, and so forth. Singh et al. (2004) examined polyethylene wicks and showed that they exhibit a pore size in the range of 8 to 20 micron a porosity value of less than 38%. This low porosity presents a limit on the heat capacity for the mLHP. Metal wicks, nickel (Maydanik et al., 1992), titanium (Baumann and Rawal, 2001; Pastukhov et al., 1999), stainless steel (Khrustalev and Semenov, 2003), and copper (Maydanik et al., 2005) are the other ones frequently used in LHPs. These metals are attractive due to their low thermal conductivity and the ability to obtain small pore diameters (as small as 2 microns) with high porosity (55% to 75%). In this case, monoporous wicks were used (i.e., their pore size distribution is similar to the Poisson distribution, and they are characterized by single average pore size).

In the design of a porous wick, for a high-performance heat pipe, we need high porosity to minimize the parasitic heat leaks from the evaporator zone, high permeability to reduce the pressure losses through the porous media, and the wick should be able to generate sufficient capillary head to keep the fluid in continuous circulation. Most recently, biporous wicks have been introduced (Wang, and Catton, 2004) (i.e., the wick is made from large porous particles that have in them small pores). It was confirmed via experimental data (Yeh et al., 2008) that using biporous wick will improve the heat-transfer characteristics for the LHP evaporator irrespective of how the device has been oriented w.r.t. the gravity field.

The thermophysical properties of the porous wicks, used in heat pipes, have received a lot of research attention recently (Semenic et al., 2008; Singh et al., 2009). These porous wicks are made of small particles (in the range from 30 to 100 micron) and from either copper or nickel. In this area, slugs sintered only from particles are referred to as "monoporous slugs," and those slugs sintered from clusters as "biporous slugs." Figure 11.6 shows a scanning electron microscope (SEM) photograph for a biporous wick with a cluster size of 586 micron and a powder size of 74 micron.

Different researchers were able to improve the heat rate removal of a monoporous wick by using two characteristic pore size biporous wick (Cao et al., 2002; Konev et al., 1987; North et al., 1995; Semenic and Catton, 2006).

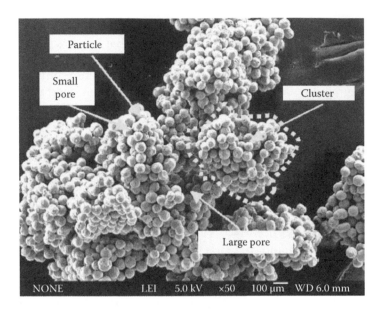

FIGURE 11.6
Scanning electron microscope (SEM) photograph of a 586/74 biporous evaporator, magnification of 50 times. (From Semenic, T., Lin, Y.-Yu, and Catton, I., 2008, Thermophysical Properties of Biporous Heat Pipe Evaporators, *ASME J. Heat Transfer*, Vol. 130, 022602-1-to-10. With permission from ASME.)

Different parameters (Semenic et al., 2008) can affect the thermal performance of a heat pipe, namely, mass flow rate of the liquid through the clusters, evaporator radius, particle diameter, pores between particles, and cluster diameter. Changing these parameters will result in changing liquid and vapor permeability and porosity, the capillary pressure, and particle-to-particle bonding area. Semenic et al. (2008) examined 15 different combinations of three cluster diameters and five particle diameters. Their work is the first, of its type, to measure thermal conductivities of biporous media for a wide range of particles and clusters and compare measured thermal conductivities to thermal conductivities of materials with similar heat conduction pathways. They obtained correlations for liquid and vapor permeabilities, capillary pressure, and the average thermal conductivity for monoporous and biporous media. These correlations can be used in optimization models that relate thermophysical properties to the critical heat flux of the biporous evaporators.

Singh et al. (2009) tested different types of wick structures in a prototype mLHP for an electronic cooling application. They found that high performance was achieved using copper wicks as compared to nickel ones, and biporous wicks showed much better heat-transfer performance compared to the monoporous ones. The small pores provided a flow path for the liquid, while the large ones provided a flow path for the vapor. Thus, the evaporative characteristics for such a wick improved greatly.

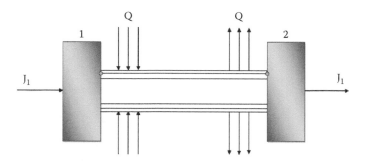

FIGURE 11.7
Longitudinal section of the open-type micro heat pipe. (With kind permission from Springer Science+Business Media: *Journal of Engineering Physics and Thermophysics*, Open-Type Miniature Heat Pipes, 65(1), 1993, L.L. Vasiliev.)

Micro heat pipe is a term that was used starting from 1984 by Cotter (1984). It represents a device that its diameter ranges from 30 micron to 1 mm and whose length varies within 1 to 5 mm. The concept behind those devices exists in nature where it plays an important role in the thermoregulation processes in plant leaves (Reutskii and Vasiliev, 1981) and the sweat glands of animals (Dunn and Reay, 1976), among others. These devices can be applied successfully in thermoregulation systems, microelectronics devices, heat pumps, and refrigerators.

Open-type micro heat pipes, in contrast to classical ones, can be dozens and hundreds of centimeters long (Vasiliev, 1993). Gravity plays an important role in this case where capillary forces take place. An open-type micro heat pipe transports a portion of the liquid through porous membranes, and the rest of the liquid circulates inside the body in the form of vapor and film. Figure 11.7 shows a longitudinal cross section of the open-type micro heat pipe. It shows the membranes-hydroseals located at the two ends (1) and (2). The liquid gets through in and out (J_1), while heat is added and rejected at (Q).

The concept of this open-type micro heat pipe has been verified experimentally (Luikov and Vasiliev, 1970). Vasiliev et al. (1991) discussed more detailed analyses of the fluid flow and heat transfer with phase change (vaporization/condensation) in microchannels with porous membranes. In their work, they showed how the porous structure intensified the condensation/vaporization heat-transfer processes.

12

Summary and Conclusions

Radioisotope-powered Stirling engine/alternators are being developed by the National Aeronautics and Space Administration (NASA) and the U.S. Department of Energy (DOE) for long-term (up to 14 years) planetary fly-by and space science missions. DOE has been supporting development of Stirling engines for solar dish-Stirling systems for terrestrial applications for many years. The regenerator is one of three heat exchangers (in addition to the heat acceptor and heat rejector) in Stirling engines and coolers. Excellent regenerator performance is crucial to good engine performance (and even more important for cooler performance). First DOE, and then NASA, provided funding for regenerator research to first learn more about the basic principles of good regenerator performance, and then to try to develop a new regenerator with superior performance to older designs. The work began with investigations based on testing and computational fluid dynamics (CFD) simulations of current-technology regenerators (i.e., wire-screen and random-fiber regenerators).

Regenerator thermal performance is critical for Stirling engine performance, because the energy stored in the regenerator matrix and removed each engine cycle may be, for example, four times that entering the cycle via the acceptor/heater per cycle. Therefore, near perfect regenerator thermal performance (~99% or better thermal effectiveness) is required. One-hundred percent regenerator effectiveness would require very large or infinite matrix heat capacity, and very large or infinite heat transfer coefficient between the oscillating gas flow and the regenerator matrix. One-hundred percent regenerator effectiveness would also imply zero integrated regenerator enthalpy flux over the cycle (integrated with time over the cycle, and with space over the regenerator cross-sectional area). Nonzero regenerator enthalpy flux, which occurs in practical regenerators, is caused by other imperfections in addition to having only finite matrix heat capacity and finite gas-to-solid heat transfer coefficients. Among nonideal regenerator features contributing to nonzero enthalpy flux losses are: nonuniform flow across the cross-section perpendicular to the main flow axis, for various reasons; blow-by at the regenerator wall due to lower matrix porosity at the wall than in the core; and different flow resistance at the wall than in the core due to differences in turbulent eddies near the wall and in the core. And there are other thermal losses in addition to those that contribute to nonzero enthalpy flux losses: Axial conduction losses due to conduction through the gas and the metal are direct heat-leaks through the regenerator that add to the engine heat accepted and rejected; there is conduction through the regenerator canister or container, in addition to that through the matrix and gas; vortex shedding from matrix wires or structures perpendicular to the flow, and any turbulent eddies, cause thermal dispersion, which enhances the axial conduction through the gas. In addition to regenerator thermal losses, there are fluid viscous or pressure drop losses due to flow through the regenerator. As for all heat exchangers involving fluid-flow, design involves a trade-off between viscous and thermal losses to minimize overall regenerator losses.

Wire screens tend to provide a bit better regenerator performance than random fibers but are more expensive to fabricate; in many modern Stirling designs developed for commercialization purposes, random-fiber regenerators have been used to minimize cost. In both wire-screen and random-fiber regenerators, there are many layers of wires that lie primarily normal to the main flow path; thus, flow separations on the downstream sides of the wires tend to enhance pressure drop losses. This was realized during the DOE-funded phase of the research.

It has been known for a long time that, theoretically, flat-plate regenerators oriented parallel to the flow path have superior ratios of heat transfer to pressure drop compared to wire screens and random fibers. However, attempts to implement flat-plate regenerators, or approximations thereof, encountered some practical problems. When flat-plate regenerators are inserted into circular or annular canisters, the geometrical details where the plates encounter, or come close to, the wall vary around the circumference of the wall(s); this tends to produce nonuniformity of flow, which reduces regenerator performance. Also, thin metal foils were wrapped in a helical manner, to provide wrapped-foil regenerators. Unfortunately, these wrapped metal foils were found to be subject to deformation due to both steady-performance and transient-performance temperature gradients. Engine performance with wrapped metal foils tended to end up worse than with random-fiber regenerators. Flat-plate regenerators have also been found to be subject to deformation due to these temperature gradients. (The reasons for such deformations are discussed in this book.)

Based on what was learned about random-fiber/wire-screen regenerator performance during the DOE phase of the research, and a follow on NASA grant, the Cleveland State University led regenerator-research team proposed to NASA to design and microfabricate a new type of regenerator matrix. Upon receiving an award, the team considered a number of concepts and discussed fabrication ideas with various vendors. Eventually, proposals were solicited from the vendors, were evaluated, were narrowed down to two, and eventually a vendor was chosen to microfabricate their proposed design.

One of the initial microfabricated concepts seemed a bit like an evolution of the wrapped-foil concept, except it consisted of concentric foils with walls fabricated to stiffen them so that they would not be subject to deformation. However, further investigation/consideration led to conclusions that this structure was not practical to fabricate, and if fabricated, would be subject to relative large conductions losses along the flow axis.

Eventually, the concept of a segmented involute-foil was developed. It is actually segmented in two directions. Several involute-foil segments in the radial direction are separated by concentric rings. And the foils are segmented in the axial direction by being formed into disks of about 250 microns in thickness. (Later, some ~500 microns disks were fabricated and tested to save funding, but the 250 micron disks performed better.)

Testing of a stack of segmented-involute-foils in a NASA/Sunpower (Athens, Ohio) oscillating flow rig determined that they performed substantially better, via having a higher figure of merit (or ratio of heat transfer to pressure drop loss) than current random-fiber designs. Later, a segmented-involute-foil regenerator was fabricated and tested in a Sunpower free-piston engine. Although an attempt was made to modify the engine to make it more suitable for use with the involute-foil regenerator, the engine/involute-foil regenerator did not constitute an optimized design. As a result, the performance of the engine with the segmented-involute foil was about the same as a similar version of the engine tested with a random fiber regenerator. Earlier computer simulations of an engine/involute-foil regenerator suggested that a completely optimized engine/involute-foil regenerator might have produced an improvement of ~6% to 9% relative to an optimized engine/random-fiber regenerator.

In addition to likely engine performance improvements achievable with involute-foil regenerators, relative to random-fiber regenerators, it is believed that the segmented-involute-foils should be less subject to the possibility of regenerator fragments breaking off and escaping from the regenerator into the engine working space. Fiber breakage and escape from the regenerator could plug the gas bearing pads, cause friction between pistons and cylinders, and seriously compromise engine performance.

Based on the superior oscillating-flow-rig results for the segmented involute foils, and the promising, nonoptimal-engine/regenerator-design, engine test results, it appears that the segmented-involute-foil should be capable of improving Stirling engine performance. However, the LiGA technique that was planned to be used was relatively expensive, and it was decided to fabricate the involute-foils from nickel, only, when it was found how extremely slow the process of electric discharge machining was for machining of stainless steel. (By using nickel, the electric discharge part of the LiGA microfabrication process could be eliminated.) Thus, for application of segmented-involute-foils to practical, particularly for commercial, designs, a cheaper process for fabricating them from the desired materials needs to be developed.

Nickel was OK, though not optimal, for the 650°C (hot end) Stirling engine in which the involute foil regenerator was tested. It could not be used for the 850°C enhanced version of the engine. And nickel, due to its high thermal conductivity, has higher axial conduction losses than is desirable.

Also reviewed in the book, in Chapter 10, is the regenerator research and development effort of Norboru Kagawa and his staff and students at the National Defense Academy of Japan. They used chemical etching to fabricate "mesh sheet" regenerators. Mesh sheet regenerator geometry is somewhat similar to that of wire screens, with their square holes etched into metal disks, and with various types of grooves (and no grooves) used to try to enhance engine/regenerator performance. The benefit of chemical etching of mesh sheets, compared to wire screens, is that the various microdimensions of the mesh sheets are capable of optimization, which is being attempted in the National Defense Academy effort. In contrast, wire screens are available in only a limited number of wire sizes and configurations. The results of this research, development, and engine testing program are reviewed in this text. We attempted to accurately portray Kagawa's effort, based on published papers, and Kagawa's review comments on a couple of drafts; however, the authors are responsible for any errors incorporated in the description of the mesh sheet development at Japan's National Defense Academy.

Engine tests were conducted with wire screens and mesh sheets. Some mesh sheets performed better than others. Tests were also conducted in two different engines using combinations of two different types of mesh sheets. Please see Chapter 10 for an extensive discussion of the different types of mesh sheets and the engine test results.

In the initial investigations of the Cleveland State–led team, chemical etching was one of the processes considered. At the time, that process was rejected because it was believed that the etched microfabricated walls of the vendors under consideration would be at an angle not suitable for the concepts under consideration. This might not be a problem for the relatively thin (250 micron) involute-foil disks that were eventually fabricated and tested. Perhaps chemical etching of the type used by Kagawa and his staff and students might also be applicable to segmented-involute-foil disks.

13

Future Work

The following are recommendations for future work in the area of Stirling regenerators.

13.1 Developing New Stirling Engine/Coolers

There is a need to design a space-power engine/alternator and cooler from the ground up to employ an involute-foil regenerator. This approach is needed to realize the full benefit of involute-foil regenerators.

13.2 Developing a New Regenerator Design

There are several design options that could potentially produce improvements in regenerator and engine/cooler performance, such as:

1. Using variable diameter regenerators could possibly provide some benefits, if they could be manufactured.
2. Using variable porosity matrices in both the axial and radial directions might produce performance improvement. Some research has been attempted already both computationally and experimentally. It is believed that this territory has not yet been fully explored.
3. Utilizing nanotechnology in getting specific thermal properties material (e.g., higher thermal conductivity in the radial direction compared to the axial, flow, direction) might result in performance improvements.
4. Development of Stirling regenerators for high-temperature applications (such as Venus environment) is needed.

13.3 Further Investigations in the Regenerator

In this category, we are proposing additional study of the following:

1. Further study of "Blowby." Some research effort has already been made experimentally at the University of Minnesota.

2. What is the effect of plenum thickness (at the end of the regenerator) for different matrix porosities? Our finding (for high porosity) was that the effect is negligible. This is opposite other investigator's findings for lower-porosity matrices; for example, the designers of the automotive Stirling engines, which used hydrogen as the working fluid, thought that the proper design of the regenerator plenums (volumes at the end of the regenerator) was critical for good engine performance. The porosity of those regenerators was lower than the modern, small Stirlings that tend to have porosities of about 90%.

3. Examine the effect of the regenerator porosity on the jet spreading into the matrices.

4. Directly measure eddy transport in a porous media.

5. Develop two-point radial spatial correlations for $\langle \tilde{u}\tilde{T} \rangle$ (at the exit plane of the regenerator matrix) from measurements. This will enable us to find a mathematical connection between a spatially averaged dispersion term (porous media theory) and a time-averaged dispersion term (turbulence theory).

13.4 Computational Fluid Dynamics (CFD) Modeling of the Regenerator

Several CFD models have been developed to simulate the Stirling regenerator. However, there are still some issues that have not yet been explored, such as:

1. Modeling porous media under oscillatory flow conditions with more realistic boundary conditions and/or two-dimensional (2-D)/three-dimensional (3-D) modeling,

2. Applying nonequilibrium models for the porous media used in simulating the regenerator

13.5 Microfabrication of New Regenerators

The microfabrication process needs to be further developed to permit microfabrication of higher-temperature materials than nickel; and a less-expensive process needs to be developed. NASA and Sunpower are currently developing an 850°C engine for space-power applications. A potential power/cooling system for Venus applications could need regenerator materials capable of temperatures as high as 1200°C. Early attempts by International Mezzo Technologies (Baton Rouge, Louisiana) to electric discharge machining (EDM) stainless steel, using a LiGA-developed EDM tool, involved a burn time (dependent on EDM machine setting) that was much too large to be practical. Some possible options for further development of a microfabrication process for high-temperature involute-foils are:

1. Optimization of an EDM process for high-temperature materials that cannot be processed by LiGA only. Burn times can be greatly reduced by higher-power, EDM machine settings than originally used, by Mezzo, but "overburn" (i.e., the gaps between the EDM tool and the resulting involute-foil channels) increases with higher powers.

2. Development of a LiGA-only process for some high-temperature alloy or pure metal that would be appropriate for the regenerator application. Pure platinum would work but has very high conductivity, which would tend to cause larger axial regenerator losses and is very expensive.

3. Microfabrication of an appropriate ceramic material for high-temperature regenerators. Structural properties of ceramics, which tend to be brittle, would be a concern. Matching of ceramic-regenerator and metal-regenerator-container coefficients-of-thermal-expansion would also likely be a problem area.

4. It may be worthwhile to revisit the possibility of chemically etching involute-foil regenerators. This approach was initially not chosen because of concerns about variation in wall thickness via the etching process. However, that decision was made before it was decided to use thin involute-foils on the order of 250 microns thick. For such thin involute-foils, variation in wall thickness due to etching might not be a problem. Also, Noboru Kagawa of the National Defense Academy of Japan and his staff and students have had good success in the chemical etching of mesh sheet regenerators. Those are also rather thin. There is no indication, in their many papers, that variation in wall thickness of the mesh sheets was a problem.

5. A breakthrough in materials and manufacturing processes is needed that will make producing such devices (involute-foil or similar) cost effective.

6. The spatter-like debris and edge roughness need to be eliminated in further development of the LiGA manufacturing process for invloute foil. Also, notch defects extending completely through or mostly through channel walls need to be eliminated.

Appendix A: NASA/Sunpower Oscillating-Flow Pressure-Drop and Heat-Transfer Test Rig

A.1 General Description (*Chapman et al., 1998*)

Figure A.1 illustrates the apparatus for measuring heat-transfer and pressure-drop characteristics of porous plug specimens in oscillating flows. The apparatus is built around an oscillating-flow test rig that was originally designed for pressure-drop (but not heat-transfer) measurements and has since been modified and refined. The flows and specimens are chosen to be representative of those encountered in the regenerators of Stirling engines, or Stirling coolers.

The apparatus includes an assembly of a piston cylinder, cooling section, specimen holder, and heating section aligned sequentially from bottom to top along a vertical flow path. The foregoing assembly is contained in a pressure vessel to enable testing at specified elevated pressures. The oscillating flow is generated by the piston, which fits closely in the cylinder and is driven by a variable-stroke, variable-frequency linear motor.

The bottom end of the heating section opens to the top of the specimen holder, and the top end of the heater opens to a relatively large, fixed, thermally insulated buffer volume. A capillary tube vents the buffer volume to the surrounding space within the pressure vessel.

The heating section is composed of a copper cylinder with drilled flow passages and with electrical band heaters clamped to its outer surface. The electrical power to the heaters is regulated by a commercial temperature controller. The cooling section is a shell-and-tube heat exchanger, with circulating water at a controlled temperature serving as the coolant.

Five fine-wire thermocouples are installed on each face of a specimen. The specimen with thermocouples attached is sandwiched between flow-diffuser disks. The sandwich is installed in the specimen holder.

Because of the large volume and other aspects of the design, the pressure swing in the piston cylinder is attributable mostly to frictional pressure drop in the specimen and in the heating and cooling sections, rather than to compression effects. Therefore, the mass flow rate is very nearly sinusoidal and spatially uniform in the specimen and in the heating and cooling sections (Figure A.1).

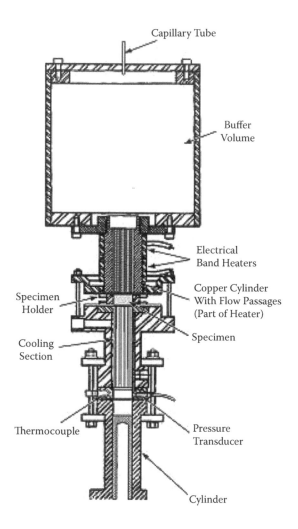

FIGURE A.1

National Aeronautics and Space Administration (NASA)/Sunpower Inc. (Athens, Ohio) oscillating-flow heat-transfer and pressure-drop test rig. Provides measurement of thermal and flow properties of specimens representative of regenerator matrices of Stirling engines and coolers, or of other general porous materials.

In its heat-transfer mode, the apparatus is used to measure the net thermal-energy flux, which is the quantity of "bottom-line" importance in a Stirling engine. During operation, the piston is actuated, while the heating section is maintained at a temperature about 200°C above that of the cooling section, giving rise to a temperature gradient in the specimen. The resulting axial conduction and imperfect heat transfer give rise to a net thermal flux along the specimen; this flux is ultimately rejected to the cooler, where it is measured. Static-conduction losses through cylinder walls and flanges and piston work due to pressure drop also contribute to cooler heat rejection, but these quantities can be calibrated out, so that it is possible to infer the thermal performance of the specimen in isolation.

Specimens tested thus far have been made of woven metal screens and metal felts/random fibers with a wide range of porosities all representative of Stirling-engine regenerator matrices. Experiments on these specimens have yielded generic correlations for friction factors, Nusselt numbers, axial-conductivity-enhancement ratios, and overall-heat-flux ratios. Recently, two new involute-foil matrices were tested; development, fabrication, and test results for these involute-foil matrices are discussed in Chapters 8 and 9.

A.2 Modifications Made to Oscillating-Flow Test Rig (*Ibrahim et al., 2004a*)

A number of modifications were made to the National Aeronautics and Space Administration (NASA)–owned oscillating-flow rig located at Sunpower Inc. (Athens, Ohio). We designed and fabricated a new shell-and-tube water cooler in order to reduce the time required to achieve thermal equilibrium between data points and eliminate a problem with a trapped air pocket in the previous cooler. We also designed and fabricated new thermocouple/diffuser-disk assemblies. The thermocouples (0.005 inch wire chrome-alumel) measure temperature at either face of the test matrix. The diffuser disks (0.7 mm thick stack of 200 × 200 mesh stainless screens) between the cooler and heater passages and the matrix serve to diffuse incoming jets before they impinge on the regenerator matrix.

Rig modifications were performed with the permission of NASA, although without any NASA funding. Gedeon Associates (Athens, Ohio) did the design work and preliminary assembly work under U.S. Department of Energy (DOE) funding. Sunpower manufactured the necessary parts, paid for outside brazing operations, and performed the final assembly into the overall test rig under internal funding.

Figure A.2 shows the new diffuser-disk and thermocouple assembly. There is one such assembly on each face of the regenerator matrix. The assembly allows the option of removing the diffuser screens in order to duplicate the jet penetration aspect of the University of Minnesota (UMN) large-scale rig. In the previous assembly, the diffuser disks were glued into place.

The regenerator test canister fits into the outer recess, held by clamping pressure and sealed with an O-ring in the canister. The actual matrix is directly above the inner hole. Below the hole is the diffuser disk assembly

FIGURE A.2
Close-up of thermocouple/diffuser-disk assembly and its torlon housing.

and a sintered stack of 200 mesh screens, 0.7 mm thick. Five 0.005 inch diameter wire thermocouples monitor the time mean gas temperature between the regenerator matrix and the diffuser disks. One such assembly bolts to the cooler entrance, the other to the heater entrance, both sealed by O-rings.The thermocouples are permanently glued to the torlon, but the diffuser disks are removable.

The computer-aided design (CAD) drawings in Figures A.3 and A.4 show an overall view and details of all the parts of the oscillating-flow test rig that were modified.

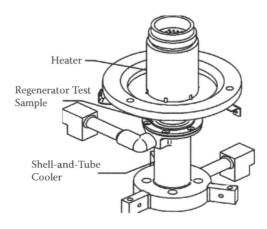

FIGURE A.3
An overall view of a portion of the oscillating-flow test rig.

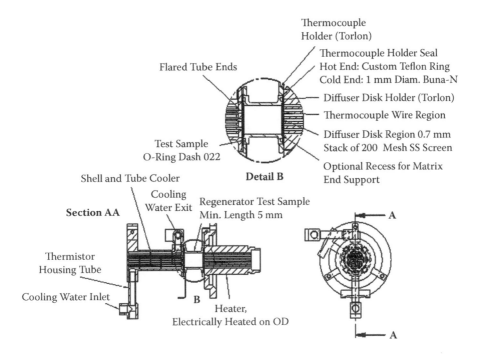

Thermocouple
Holder (Torlon)

Thermocouple Holder Seal
Hot End: Custom Teflon Ring
Cold End: 1 mm Diam. Buna-N

Flared Tube Ends

Diffuser Disk Holder (Torlon)

Thermocouple Wire Region

Diffuser Disk Region 0.7 mm
Stack of 200 Mesh SS Screen

Test Sample
O-Ring Dash 022

Optional Recess for Matrix
End Support

Shell and Tube Cooler   **Detail B**

Cooling
Water Exit

Section AA

Regenerator Test Sample
Min. Length 5 mm

A

Thermistor
Housing Tube

Cooling Water Inlet

B

Heater,
Electrically Heated on OD

A

FIGURE A.4
Details of all the parts of the oscillating-flow test rig that were modified.

A.3 Some Additional Modification Made to Test Rig (*Ibrahim et al., 2009a*)

A.3.1 New Water Flow Meter for Oscillating-Flow Test Rig

In our ongoing series of rig improvements, we upgraded the water-flow meter electronics and installed a length of straight tubing downstream of the turbine flow meter (itself unchanged). The new electronics makes it possible to total the water flow over a period of time rather than just display an instantaneous flow rate. This makes it is easier to calibrate the meter. As a result, we learned that the previous water-flow calibration was somewhat off at low flow rates. See Ibrahim et al. (2009a) for additional information.

A.3.2 Revised Heat-Transfer Test Procedure

The original 2006 data were logged under a heat-transfer test procedure that did not allow as much time between data points for the test rig to reach thermal equilibrium compared to the 2008 practice.

Appendix B: Some Comments on Dynamic Similitude as Applied to Physical Simulations of Stirling Engines

Prepared by T. W. Simon, Yi Niu, and L. Sun

B.1 Fluid Mechanics

The National Aeronautics and Space Administration (NASA) sponsored a "Loss Understanding" program in the late 1980s with the objective of enhancing the understanding of fluid mechanics and heat transfer of oscillatory flows as applied to Stirling engine heat exchangers. An initial contribution by Simon and Seume (1988) was a literature review. Similarity parameters, Re_{max}, Va, and AR were discussed. They were taken from the literature on unsteady fluid mechanics in Stirling engine heat exchangers. The following shows how the Re_{max}, Va are derived from the momentum equation.

$$\frac{\partial \vec{u}}{\partial t} + \vec{u} \cdot \ \vec{u} = -\frac{p}{\rho} + v \ ^2\vec{u} \tag{B.1}$$

If the hydraulic diameter of the porous matrix, d_h, the maximum mean velocity within the porous matrix, u_{max}, and $1/\omega$ are chosen as the length scale, the velocity scale, and the time scale, respectively, the normalized equation can be written as:

$$\frac{\omega d^2 h}{4v} \frac{\partial \vec{u}}{\partial t} + \frac{1}{2} \frac{u_{max} d_h}{v} \vec{u}^* \cdot \ ^*\vec{u}^* = -\frac{1}{2} \frac{u_{max} d_h}{v} \frac{^*p^*}{\rho} + v^* \ ^{*2}\vec{u}^* \tag{B.2}$$

Here, * denotes a dimensionless quantity. So, we define dimensionless numbers as:

$$Re_{max} = \frac{u_{max} d_h}{v} \tag{B.3}$$

$$Va = \frac{\omega d_h^2}{4v} \tag{B.4}$$

The dimensionless frequency, Va, is called the Valensi number or the kinetic Reynolds number. It represents the unsteady inertial effect relative to the viscous effect. The parameter Re_{max} is the Reynolds number, which represents the steady (quasi-steady) inertial effect relative to the viscous effect. In this report, Simon and Seume (1988) gave ranges of representative values for Stirling engine heat exchangers in terms of those parameters. They were based upon a survey of 12 regenerator operating points and the simple Schmidt analyses.

$$Va = 0.06\text{\textasciitilde}5$$

$$Re = 20\text{--}\text{\textasciitilde}2000$$

For an incompressible fluid, the continuity equation is:

$$\cdot \vec{u} = 0 \tag{B.5}$$

or

$$\frac{\partial u}{\partial x} + \frac{\partial v}{\partial y} + \frac{\partial w}{\partial z} = 0 \tag{B.6}$$

The normalized equation can be written as:

$$\frac{\partial u^* u_{max}}{\partial x^* d_h} + \frac{\partial v^* u_{max}}{\partial y^* d_h} + \frac{\partial w^* u_{max}}{\partial z^* d_h} = 0 \tag{B.7}$$

Therefore,

$$\frac{\partial u^*}{\partial x^*} + \frac{\partial v^*}{\partial y^*} + \frac{\partial w^*}{\partial z^*} = 0 \tag{B.8}$$

which means the nondimensionalized continuity equation is the same for geometrically similar systems, regardless of the length scale and the velocity scale chosen. If the differential equations (including boundary conditions and initial conditions), expressed in dimensionless terms, are identical, the solutions will be identical. The continuity equation automatically satisfies this requirement, but the momentum equation will be identical only if the Valensi number and the Reynolds numbers for the engine and the model are the same. Thus, dynamic similarity will hold when Va and Re_{max} of the model match the real engine values.

B.2 Heat Transfer

With regard to heat transfer, in the Simon and Seume (1988) report, the Eckert number was derived by normalizing the energy equation. The Eckert number:

$$Ec = \frac{U_\infty^2}{Cp(T_s - T_\infty)} \qquad (B.9)$$

is a ratio of thermal energy generated by viscous dissipation to thermal energy convected in a flow stream, based upon the overall temperature difference in that stream. Stirling engine Eckert numbers ranged from 0.0066 to 0.025 (Simon and Seume, 1988). Such low values indicate that viscous dissipation does not contribute significantly to regenerator heat transfer or the regenerator temperature field. Thus, we need not consider the Eckert number (the effect of viscous dissipation) in the present research. Recently, we extended the list of dimensionless numbers to include parameters particularly relevant to regenerators. One is the Fourier number, Fo, chosen for the purpose of simulating heat transfer behavior within a Stirling regenerator. The Fo number is defined as:

$$Fo = \frac{\alpha \tau}{\delta^2} \qquad (B.10)$$

It represents the ratio of the cycle time to the time required for thermal penetration to a distance δ.

According to information provided by Cairelli (2002), values of Fo number for the solid material in the regenerators of some free piston Stirling engines are those in Table B.1.

The large Fourier numbers indicate that the wires are always uniform in temperature, though they are changing in temperature temporally. This is discussed in more detail in the Section B.5.

TABLE B.1

Values of Fo Number for Some Free-Piston Stirling Engine Regenerators

Engine	Fo (Solid Phase)
RE1000	80.4
Space power research engine (SPRE)	236.0
Component test power convertor (CTPC)	112.8
Stirling Technology Co. (STC) (now Infinia)	544.0

The Fourier number could also be applied to temperature distribution within a fluid space. It can be defined as:

$$Fo = \frac{\alpha_f \tau}{\delta^2}$$ (B.11)

where δ represents a characteristic length of the fluid space. This also will be discussed in Section B.5.

B.3 Free Molecular Convection

When the length scales of the flow become very small, one must be concerned that the flow and thermodynamics are not in equilibrium on the molecular level. When the length scales of the flow space approach the mean free path of molecules, the concept of continuum cannot be accepted, and free molecular convection, where the kinetic energy of each molecule must be considered, must be applied. Below, we take a closer look at diffusion from the molecular level. From Maxwell (1860) (Bird et al., 1960):

$$\frac{\tau}{\rho} = \frac{y - \text{diffusion}}{\rho} = -\frac{\bar{c}\,\lambda}{3}\frac{\partial u}{\partial y}$$ (B.12)

where $\bar{c}$ is the mean molecular speed (related to probability of collision), and λ is the mean free path (represents distance between collisions), so the diffusivity is:

$$v = \frac{\mu}{\rho} = \frac{\bar{c}\,\lambda}{3}$$ (B.13)

for a low-pressure gas. Using the Boltzmann relationships for the speed and mean free path,

$$\mu = \rho v = \frac{2}{3\pi^{3/2}}\sqrt{\frac{mkT}{d^2}}$$ (B.14)

where m is the molecule mass, k is Boltzmann's constant, T is temperature, and d is σ which is effective molecular diameter (collision diameter).

For a monatomic gas, this expression can be written in terms of the characteristic collision diameter, $\sigma(\text{Å})$; the molecular mass, m (g/mol); the temperature, T (K); and a parameter Ω_μ which is a weak function of kT/ε. as:

$$\mu = 2.6693 \times 10^{-5}\frac{\sqrt{mT}}{\sigma^2 \Omega_\mu}$$ (B.15)

The viscosity, μ, is in gm/cm·sec. The quantity ε is the characteristic energy of interaction in the Lennard-Jones model:

$$(r) = 4\varepsilon \left[\left(\frac{\sigma}{r} \right)^{12} - \left(\frac{\sigma}{r} \right)^{6} \right] \tag{B.16}$$

(r) is the repulsive force.

This relationship is also successfully applied to modeling of polyatomic gases, as discussed by Bird et al. (1960). Note that μ rises with T and is independent of p, so long as p is not large enough to transition the gas out of the "low pressure" or "ideal gas" state of behavior. Under such simple conditions, $\mu = \rho v$, and $k = \rho c_p$ α can all be computed. Again, note that these diffusion coefficients depend upon the fluid and its state and are not dependent upon the fluid flow situation, as would be true if the "diffusion" were by turbulent eddies.

When pressure is large and the above model is violated, the analysis becomes far more difficult, if not impossible. One must then resort to empirical relations formulated on the macroscopic level, such as $\frac{\mu}{\mu_0} = f\left(\frac{p}{p_c}, \frac{T}{T_c} \right)$ (Figure 1.3.1, P.16 of Bird et al., 1960). The subscript 0 indicates low-pressure behavior, and c indicates "critical."

Of course, the problem at higher pressures is that the molecules are not "behaving." They are interacting with one another in ways other than with perfectly elastic collisions.

Now, let's use that Boltzmann relationship to discuss the continuum condition. For helium, the molecular collision diameter is given to be:

$$\sigma = 2.576 \text{ Å}$$

(Table B.1, Bird et al., 1960). The molecular density at engine conditions, $T = 800$ K and $p = 150$ atm is $n = 1 \times 10^{27}$ molecules/m³. The mean free path length, $\lambda = \frac{1}{\sqrt{2}\pi\sigma^2 n} = 2.5$ nm. According to Bird et al. (1960), continuum is expected where $Kn = \frac{\lambda}{L} \ll 1$. Thus, continuum holds when the feature size, L, is $\gg 2$ nm. This should be the case for all flow features of Stirling engines where the smallest feature size is the wire size of the regenerator, which is around 20,000 nm.

We also expect a statistically large number of molecules within our volume of interest if the flow is to behave as a continuum. If we need 10^6 molecules to meet this, then our above analysis indicates a cubic volume holding the molecules must be at least 90 nm on a side.

Finally, we should use the above empirical relationship for viscosity to see if our high pressure is violating ideal gas behavior from a diffusion (viscosity) perspective. For conditions of a representative engine, $\frac{p}{p_c} = 66$, $\frac{T}{T_c} = 152$ and $\frac{\mu}{\mu_c} = 1.0$ (with extrapolation). Also, computing the compressibility factor, $Z = \frac{pv}{RT}$, for this condition, $Z = 1.00$ (again, with extrapolation). So, the fluid is behaving as an ideal gas and a continuum fluid.

Another means of expressing the Knudsen criterion is $Kn = \frac{\lambda}{L} = \sqrt{\frac{\lambda \pi}{2}} \frac{Ma}{Re}$. Our very low Ma numbers and high Reynolds numbers lead again to the conclusion that the fluid is behaving as a continuum.

B.4 Regenerator Figures of Merit

Ruhlich and Quack (1999) gave a figure of merit as the ratio of heat transfer to dimensionless pressure drop for internal flow, which can be reduced to:

$$F_R = \frac{4St}{f} \quad \text{or} \quad \frac{St}{C_f} \tag{B.17}$$

where St is the Stanton number, f is the Darcy-Weisbach friction factor, and C_f is the Fanning skin friction coefficient. We note that $f = 4C_f$ for either a round pipe or flow between two infinite walls. They compare different heat transfer surfaces and find an optimum. This optimum consists of slim elements oriented in the flow direction in a staggered overlapping arrangement. Such regenerators can provide equal thermal performance to that of the stacked screens, with up to one-fifth the pressure drop.

Gedeon (2003a) proposed an extended figure of merit that includes the effects of thermal dispersion:

$$F_R^* = \frac{1}{f\left(\frac{1}{4St} + \frac{N_k}{RePr}\right)} \tag{B.18}$$

where N_k is the ratio of the effective conductivity due to thermal dispersion to the gas molecular conductivity. (See Figure 8.34 and Appendix D.)

B.5 The University of Minnesota (UMN) Test Section Design

A representative Stirling engine was chosen for selecting the test section design parameters of the UMN facility. The intention of the design is to make the dimensions as large as possible (without changing the thermal-hydraulic

TABLE B.2

Dimensionless Parameters Based on the Hydraulic Diameter
of the Respective Components

	Regenerator		Cooler	
	University of Minnesota (UMN)	A Real Engine	UMN	A Real Engine
Re_{max}	800	248	20,700	20,700
Va	2.1	0.23	14.3	14.3
Amplitude ratio, AR	3.8	1.0	15.5	15.5
F_o	59.6	136		

behavior) in order to allow taking detailed measurements within the regenerator matrix, while simulating the temporal and spatial characteristics of the real engine. Thus, we wished to maintain dynamic similitude of all the parameters that were relevant to the study.

In designing the regenerator, the matrix wire diameter was scaled up to 0.81 mm (0.032 inch) for the 6.3 mm × 6.3 mm square mesh screen layers, which resulted in a hydraulic diameter, dh, of 7.3 mm on our 90% porosity stacked screen geometry. This was a larger scale than strict dynamic similarity would have given ($d_h \approx 2.5$ mm if scaled from Table B.2), but a scale that was suitable for the measurements. The following momentum and heat transfer analyses were done in support of that particular regenerator design choice.

One can rewrite the Valensi number as $1/2(d_h/\delta_v)^2$, where δ_v is viscous penetration depth, $\delta v = \sqrt{2v/\omega}$. The viscous penetration depth is the thickness of the viscous layer over a surface. Therefore, the Valensi number (actually $\sqrt{2Va}$) measures the ratio of hydraulic diameter to viscous penetration depth. It represents the time required for establishing uniformity of flow by viscous (and eddy transport if v is replaced with $v + \varepsilon_M$ where ε_M is the turbulent eddy diffusivity) within a pore of characteristic dimension, d_h, relative to the cycle period. As noted, the flow in the regenerator pore of the real engine is expected to be influenced by both eddy and molecular transport. To accurately apply to turbulent flow, the Valensi number should be $1/2(d_h/\delta v)^2$, where $\delta_v = \sqrt{2(v + \varepsilon_M)/\omega}$. Our previous experiments (Niu et al., 2003a) show that the eddy viscosity in the regenerator matrix would be 550 times the molecular viscosity in our experiment. Using this value, the viscous penetration depth within one cycle would be 11.4 hydraulic diameters. Thus, we conclude that the flow within an interstitial space is well mixed when the flow is turbulent, both in the test and in the operating engine. The selected wire size and mesh size give a hydraulic diameter within the matrix of three times that which one would compute by simply scaling the regenerator with the same scale factor as used on the other features of the engine. We did this to create feature sizes that would allow temperature and flow (such as turbulence) measurements within the regenerator. This analysis shows that we can do so without jeopardizing the hydraulic similarity of the test when the eddy transport is active.

Moreover, Gedeon and Wood (1992) conducted experiments on woven screens at low and high *Va* values. It was observed that friction factors show little dependence on Valensi number' and no essential difference compared to steady flow. This assumption has been used throughout the years for getting regenerator design information. Thus, our analysis essentially showed only that this could be extended to our larger-than-scaled-size regenerator when the eddy transport is active. Gedeon and Wood (1992) concluded that as long as Valensi number is smaller than 3.72, the maximum value in their experiments, the Valensi number dependence is negligible when the eddy transport is active, we meet that test criterion. When we have only molecular transport, we would expect to see a *Va* effect.

Another important consideration in scaling the regenerator is the transient heat transfer within the regenerator. As in the analysis of the Valensi number, the Fourier number could be rewritten in the form of $\pi(\delta_\alpha/d_w)^2$, where δ_α is the thermal penetration depth into the wire in a given time interval, $\delta_\alpha = \sqrt{2\alpha/\omega}$. Therefore, the Fourier number represents the thermal penetration during a cycle relative to the wire diameter. So long as F_o is large enough, the interior of the wire will remain nearly isothermal during a cycle. In the present design, F_o is 14.9. This is considered to be large enough to maintain thermal quasi-equilibrium. Thus, whether 59.6 of the model or 136 of the engine (Table B.2), the solid is always isothermal.

Another important engine design parameter is the length of the regenerator relative to the displacement that a fluid particle would have traveling through the regenerator (if it were continuous in geometry over that length), the amplitude ratio. The length of the regenerator in the experiment is about 1/10 the length it should have if scaled up from the engine. However, because our intent is to study penetration of the jets into the regenerator and flow within the regenerator after end effects have disappeared, all that is needed is a regenerator of sufficient length to have the jets from the cooler merge completely before reaching the hot end (~25 mm) and the flow from the heater end to become fully established (~4 mm).

For the cooler, all characteristic dimensions were scaled by a factor of around 12, consistent with scale-up from the dimensions of the engine. The operating frequency was reduced to 1/150 of the selected engine operating frequency, for matching the Valensi number of the cooler tubes. Such dimensions and operating frequency keep the Reynolds number and the Valensi number of the cooler section almost the same as those of the actual engine cooler. Dynamic similarity of the cooler section is important for this study because it determines the behavior of the jets which enter from the cooler to the regenerator.

To recap, the cooler is dynamically similar to an actual engine. The regenerator matrix features (interstice size and wire diameter) are larger than a scaled version of the engine regenerator would have, but not so much that thermal-hydraulic quasi-equilibrium within the regenerator is violated so long as the turbulent eddies are active.

In testing, we discovered signs of nonequilibrium during a portion of the cycle in the current experiments. It was first found in the heat transfer coefficient measurements within the regenerator matrix. At the beginning of each half cycle (from 0° to about 40°), the measured *Nu* number is negative and infinity (see Figure B.1, same as Figure 7.4). This leads us to a closer look. We began with velocity measurements of the flow departing the regenerator, as measured within the plenum (Figure B.2, same as Figure 7.7).

It is observed in Figure B.2 that the root-mean-square (rms) fluctuation suddenly jumps to a high value (>0.3 m/s) at the crank angle of about 45°. This is the time within the cycle that the temperature difference changes sign. Prior to 45°, the rms fluctuation $\sqrt{\overline{u'^2}}$ is low and remains about 0.1 m/s. We surmise that at the beginning of the half cycle, when the acceleration is large, the eddy transport is relatively weak compared to values during the part of the cycle from 45° to 150°. Therefore, the momentum and thermal transport

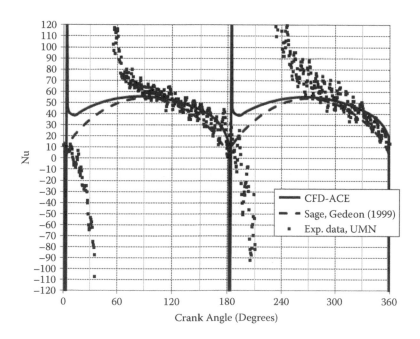

FIGURE B.1
Instantaneous Nusselt number, Nu, during one cycle.

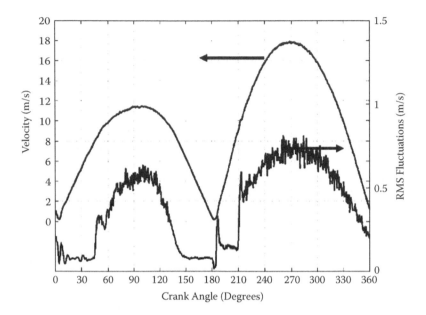

FIGURE B.2
Instantaneous velocities and rms within the plenum during one cycle.

are hypothesized to be merely due to molecular transport, which is a less effective transport mechanism than if the transport were by turbulent eddies. As a result, the momentum of the flow passing over the screen wires and through the pores in the matrix is not mixed well over the pore. Thus, it is not in equilibrium in terms of momentum and thermal effects on the pore scale. One may conclude that the difference between the wire temperature and the measured temperature at a certain distance away from the screen wire does not represent the temperature gradient of the film on the wire surface, the gradient that is directly participating in heat transfer with the screen wire. This might be the reason for the negative heat transfer coefficients during the beginning of acceleration and the lag in the temperature difference relative to the heat flux. If we were capable of locating a thermocouple close enough to the wire to take the temperature of the film, we should not get this lag. Thus, for this portion of the cycle, the experiment does not represent the engine under study. The study with this difference did afford us the opportunity to study this effect and therefore was quite valuable.

In Table B.3, dimensionless numbers used in heat transfer, fluid mechanics, and Stirling engine design are employed. Many are not relevant to our work because the nondimensionalized parameters are not important to our studies. Examples are Mach number (our Mach numbers are low and propagation to pressure information can be considered to be infinite) or Rayleigh number (gravitational forces are considered to be negligible). Nevertheless, they are included for completeness. Some of these numbers are relevant to

TABLE B.3

A List of Dimensionless Numbers that May Apply

Amplitude ratio, AR	$\dfrac{2 \times \text{max fuid particle displacement}}{\text{section length}}$	Two times the amplitude of fluid particle displacement during a cycle relative to the component length
Axial conduction enhancement ratio Nu_k,	$\dfrac{\text{Nu}_{\varepsilon_H + \alpha}}{\text{Nu}_\alpha}$	Ratio of Nusselt number based on eddy and molecular diffusion of heat to Nusselt number based solely on molecular diffusion of heat
Beale number, B	$\dfrac{P_{\text{brake}}}{p_{\text{ref}} V_{\text{sw}} f}$	Engine power relative to a form of PV power
Bejan number, Be	$\dfrac{P L^2}{\mu \alpha}$	Parameter of the optimal spacing for heat-generating boards under forced convection
Biot number, Bi	$\dfrac{hL}{k_s}$	Ratio of internal thermal resistance of a solid body to its surface convective thermal resistance
Brinkman number, Br	$\dfrac{\mu U_\infty^2}{k(T_s - T_\infty)}$	Thermal energy converted by dissipation relative to energy transported by conduction
Characteristic Mach number, Ma_ε	$\dfrac{\omega V_{\text{sw}}^{\frac{1}{3}}}{\sqrt{RT_c}}$	Piston speed based upon length scale computed from swept volume, V_{sw}, relative to sound speed based on cold end temperature, T_c
Colburn j factor, j_H	$St\, Pr^{\frac{2}{3}}$	Dimensionless heat transfer coefficient
Compression ratio, r_v	$\dfrac{\text{max volume}}{\text{min volume}}$	Ratio of volumes of the working fluid over the cycle, maximum to minimum
Dissipation Reynolds number	$\dfrac{v\eta}{v}$	(= 1) Based on Kolmogorov length, η, and velocity, v, scales
Darcy-Weisbach friction factor, f	$\dfrac{P}{\frac{L}{D}\frac{1}{2}\rho U_\infty^2}$	Pressure drop per dimensionless length of duct relative to the free-stream dynamic head
Dean number, De	$Re_d \sqrt{\dfrac{d}{2R}}$	Curvature effect on stability relative to the flow viscous effect; flow in a curved duct
Drag coefficient, C_D	$\dfrac{D}{\frac{1}{2}\rho U_\infty^2}$	Ratio of drag force (pressure forces plus shear forces with respective areas) to free-stream dynamic head
Eckert number, Ec	$\dfrac{U_\infty^2}{c_p(T_s - T_\infty)}$	Thermal energy converted by dissipation relative to energy convected in the free stream
Euler number	$\dfrac{P}{\rho U_\infty^2}$	Ratio of pressure forces to inertial forces

(continued)

TABLE B.3 (CONTINUED)

A List of Dimensionless Numbers that May Apply

Fanning friction factor, C_f	$\dfrac{\tau}{\frac{1}{2}\rho U_\infty^2}$	Ratio of shear stress to free-stream dynamic head
Finkelstein number, F	$\dfrac{P_{computed}}{P_{ref} V_{sw} f}$	Computed engine power relative to a form of PV power
Flush volume, FL	$\dfrac{\delta M_{regen,cycle}}{M_{regen}}$	Mass of fluid flushed through the regenerator in one cycle relative to mass of fluid within the regenerator matrix
Fourier number, Fo	$\dfrac{\alpha t}{L^2}$	Time of event relative to thermal penetration time by conduction
Froude number	$\dfrac{U_\infty}{\sqrt{gL}}$	Inertial force relative to gravitational force
Grashof number, Gr_L	$\dfrac{g\beta(T_s - T_\infty)L^3}{v^2}$	Ratio of buoyancy to viscous forces (g might be replaced with another acceleration, as appropriate)
Gortler number, Gr	$\dfrac{U\theta}{v}\sqrt{\dfrac{\theta}{R_0}}$	Curvature effect on stability relative to the viscous effect; boundary layer on a curved surface
Knudsen number, Kn	$\dfrac{\lambda}{L}$	Ratio of mean free path length to characteristic length
Mach number	$\dfrac{U_\infty}{a}$	Ratio of flow speed to sound speed
Momentum thickness Reynolds number, Re_θ	$\dfrac{U_\infty\theta}{v}$	Ratio of momentum advection rate to momentum diffusion rate (based on gradient u_∞/θ); used in boundary layer flow
Number of transfer units, NTU	$\dfrac{St\, L_{regen}}{r_h}$	Regenerator number of transfer units (from Kays and London, 1964)
Nusselt number, Nu	$\dfrac{hL}{k_f}$	Dimensionless heat transfer coefficient; ratio of conduction resistance to convective layer resistance
Peclet number, Pe_L	$Re_L Pr$	Ratio of streamwise thermal advection to cross-stream thermal diffusion
Porosity, β		Ratio of void volume to total volume
Prandtl number, Pr	$\dfrac{v}{\alpha}$	Ratio of molecular diffusivity of momentum to thermal molecular diffusivity
Pressure ratio, P_r	$\dfrac{\text{max pressure}}{\text{min pressure}}$	Ratio of pressures of the working fluid over the cycle, maximum to minimum

TABLE B.3 (CONTINUED)

A List of Dimensionless Numbers that May Apply

Rayleigh number, Ra	$\dfrac{g\beta(T_s - T_\infty)L^3}{v\alpha}$	Ratio of thermal advection due to buoyancy to thermal molecular diffusivity
Regenerator heat transfer scaling parameter, Tr	$\left(\dfrac{L_r}{r_{hr}}\right)^{3/2}\left(\dfrac{p_{ref}}{\omega\mu}\right)^{-1/2}$	Regenerator heat transfer scaling parameter
Reynolds number, Re	$\dfrac{U_\infty L}{v}$	Ratio of advection of momentum to molecular diffusion of momentum
Schmidt number, Sc	$\dfrac{v}{D_{12}}$	Momentum diffusion rate relative to mass diffusion rate of species 1 in a 1–2 binary mixture
Sherwood number, Sh	$\dfrac{h_m D}{c D_{12}}$	Dimensionless mass transfer coefficient; ratio of molecular diffusion resistance to mass transfer convective layer resistance
Stanton number, St	$\dfrac{h}{\rho U_\infty c_p}$	Ratio of heat transfer by convection to thermal transport by streamwise advection
Stirling number, SG	$\dfrac{p_{ref}}{\omega\mu}$	Ratio of charge pressure to shear stress
Strouhal number	$\dfrac{fL}{U_\infty}$	Ratio of transit time to event time
Temperature ratio, TR	$\dfrac{T_E}{T_C}$	Expansion to compression space temperature ratio
Thermal capacitance ratio, C	$\dfrac{\rho_w c_w}{\rho c}$	Thermal capacitance ratio between solid and fluid
Tortuosity	$\dfrac{A_e}{A_{cs}}$	Ratio of effective area for solid conduction relative to the cross-sectional area
Turbulence Reynolds number, Re_λ	$\dfrac{\sqrt{\overline{u'^2}}\,\lambda_g}{v}$	Ratio of advection on the dissipation length scale to molecular diffusion
Valensi number, Va	$\dfrac{\omega d_h^2}{4v}$	Time for momentum to diffuse d_h relative to cycle time
Wormsley parameter, Wo	$\sqrt{Va}$	Root of the Valensi number

TABLE B.4

Some Parameters from Stirling Engines, and Our Experiment

Application	Porosity	Wire Diameter (mils)	Hydraulic Diameter (mils) Dh	Cooler Tube Diameter (mils)	Heater Tube Diameter (mils)	dh d_{cooler}	dh d_{heater}
Experiment	0.90	32	288	750		0.38	
GPU-3	0.71	1.6	3.9	40	119	0.098	0.033
MOD-I	0.68	1.96	4.2	39.4	677+	0.11	0.006
MOD-II	0.68	2.2	4.7	39.4	79.5	0.12	0.06
RE1000	0.76	3.5	11.1	35	125	0.310	0.09
SPRE (space power research engine)	0.84	1.0	5.25	60	50	0.09	0.10
CTPC (component test power convertor)	0.728	2	5.35	30.8	40	0.174	0.09
Stirling Technology Co. (STC) (now Infinia)	0.90 (estimated)	0.5 (estimated)	4.5	63 (estimated)	212	0.07	0.02

Stirling engine systems but are not relevant to the "isolated effects" studies such as in our test section. Some are not addressed above because they are combinations of other effects. An example is the Strouhal number that can be written in terms of the Reynolds number and Valensi number. Finally, some of these numbers can be assumed to scale on the Reynolds number. An example is the dissipation number.

Table B.4 shows values of some dimensional parameters (for our test and for some engines) relevant to our study.

Appendix C: Radiation-Loss Theoretical Analysis

C.1 Summary of Results

During the phase I final review (of the NRA effort), an evaluation of the effects of radiation through the regenerator was requested. This was done by a simplified theoretical analysis of radiation down a long thin tube (the results were later confirmed by a Cleveland State University computational fluid dynamics [CSU-CFD] analysis). A long thin tube overestimates radiation through a stack of involute-foil disks because there is a clear sight path down the whole length of the tube. In the actual involute-foil stack, it is impossible to see any light passing through it when held up to a bright light source.

Looking down a long thin tube—focused to the far end—one sees mostly wall. So, too, radiation emitted at the hot end of such a tube sees mostly wall, where it is absorbed before it gets too far, provided those walls are diffuse gray absorbing surfaces (not highly reflecting). Emitted radiation gets about 3.5 tube diameters before 99% of it hits the tube wall (easy to work out by comparing the surface area of a hemisphere with radius 3.5 tube diameters with that of the tube cross-sectional area). Some fraction of that radiation is reflected, but it, too, cannot get very far before multiple reflections eventually absorb practically all of it. The walls of a long, thin diffuse-gray tube act like a sort of distributed radiation shield.

To the extent that the microfabricated regenerator (stack of involute-foil disks) looks like a bundle of long thin tubes, it will also block radiation transmission. The analogy is not too far fetched. A view at the hot end of the regenerator looking toward the cold shows mostly foil surfaces, except for a tiny view angle where the cold end is visible. So, a quantitative estimate of radiation loss down a long thin tube can serve as a basis for estimating the radiation loss in a microfab regenerator. Perhaps it is not too unreasonable to substitute passage hydraulic diameter for tube diameter.

The remainder of this section considers the limiting case of diffuse black regenerator walls, which is arguably a reasonable approximation for long, narrow passages. The results are summarized in Table C.1, which shows that radiation loss under these assumptions is negligible compared to other regenerator losses.

TABLE C.1

Relative Loss Estimates for a 100 W Class
Space-Power Stirling Regenerator

Hot temperature	850 °C (1120 °K)
Cold temperature	100 °C (370 °K)
Passage aspect ratio L/d	300
Passage wall emissivity ε	0.5
Radiation flow at cold end	10 mW
Radiation flow at hot end	200 mW
Time-average enthalpy flow	13,000 mW
Solid conduction	7,000 mW

C.1.1 Radiation in Long, Thin Tube

A tube of radius a (diameter d) and length L, as shown in Figure C.1, was used in this study. Coordinate $\xi = x/d$ is the dimensionless axial coordinate. The tube has open ends at $\xi = 0$ and $\xi_L = L/d$. The tube wall is presumed to be a diffuse black surface (emissivity $\varepsilon = 1$) with a fixed wall temperature $T(\xi)$ that varies linearly with ξ between T_0 and T_L. The tube terminates in black cavities at the two ends at temperatures T_0 and T_L. Of interest is the radiation heat flux $q(\xi)$ through the tube section $A(\xi)$.

If radiation flux $q(\xi)$ is small compared to the helium time-average enthalpy flux and solid thermal conduction flux down the regenerator (when installed in a running engine), then it will have a small effect on the usual regenerator temperature distribution, and the assumption of a linear temperature distribution is valid.

The radiation flux depends on position and can be represented as a fraction R of the worst-case radiation flux:

$$q(\xi) = R q_{\max} \tag{C.1}$$

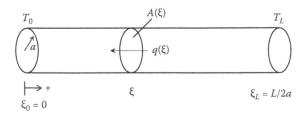

FIGURE C.1
The tube used to study radiation heat transfer in the regenerator.

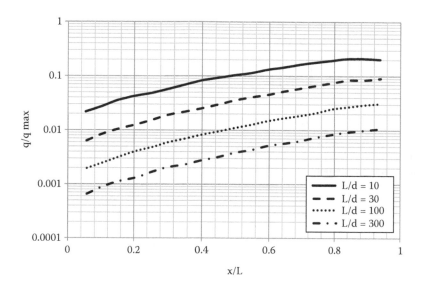

FIGURE C.2
Radiation loss estimates.

where q_{max} is the black-body radiation exchange between two parallel planes at temperatures T_0 and T_L (flux limit as tube length approaches zero):

$$q_{max} = -\sigma\left(T_L^4 - T_0^4\right)$$ (C.2)

constant σ is the Stefan-Boltzmann constant 5.670E-8 W/(m^2 °K^4).

In general, multiplication factor R depends on the position ξ, temperature ratio T_0/T_L and tube aspect ratio L/d. Numerical calculation in a custom-written Delphi Pascal program for the representative case $T_0/T_L = 1/3$ (typical for Stirling engine) and various values of L/d gave the results shown in Figure C.2. L/d_f for the involute-foil regenerator is ~350 for a total regenerator length of 60 mm. In Figure C.2, the cold end is at $x/L = 0.0$.

C.1.2 Justifying the Black-Wall Assumption

Even though the actual wall emissivity for a metallic regenerator material is probably closer to $\varepsilon = 0.5$ than $\varepsilon = 1$, multiple reflections in a long, thin tube render the apparent tube-wall emissivity near one at any location. The apparent emissivity is the value ε_a for which the total outgoing radiation (emitted + reflected) is $\varepsilon_a \sigma T^4(\xi)$. Figure 8.9 on p. 257 of Siegel and Howell (2002) shows that the apparent wall emissivity for an isothermal tube approaches 1 within a few diameters of the tube entrance, regardless of actual emissivity. In particular, for actual emissivity $\varepsilon = 0.5$, the apparent emissivity is nearly 1 at a distance of only two tube diameters from the entrance. For the present

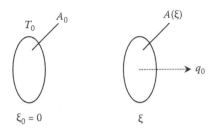

FIGURE C.3
Negative end contribution.

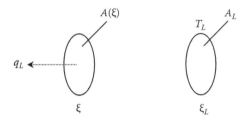

FIGURE C.4
Positive end contribution.

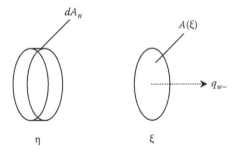

FIGURE C.5
Negative wall contribution.

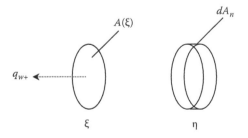

FIGURE C.6
Positive wall contribution.

analysis, this means that it is not unreasonable to consider the tube walls to be black surfaces, which greatly simplifies the analysis because reflected radiation need not be considered. This black-wall assumption is arguably valid so long as the wall temperature does not change much over a distance of several diameters, which implies that the local radiation environment is similar to that of an isothermal tube.

C.1.3 Applied to a Regenerator

For the temperatures of a space-power engine, T_L might be on the order of 1120 °K, and T_0 might be on the order of 370 °K ($T_L/3$), so the worst-case parallel-plate radiation heat flux q_{max} works out to 8.9 W/cm². A regenerator void frontal area (corresponding to the tube interior of the above analysis) for a 100 W class Stirling engine is on the order of 2 cm². So, the total worst-case radiation flow would be about 18 W. The actual radiation flow down the regenerator would be reduced by a fraction corresponding to the curve $R(\xi)$ in the above plot for $L/d = 300$. The aspect ratio for our current microfab regenerator design (based on length-to-hydraulic-diameter ratio) is about 350 for a total regenerator length of 60 mm.

The conclusion is that the radiation heat flow down the regenerator would be very small. Near the cold end about 6×10^{-4} of 18 W, or about 10 mW. Near the hot end about 1×10^{-2} of 18 W, or about 200 mW. In terms of the time average enthalpy flux (13 W) and the solid thermal conduction (7 W), the radiation loss is smaller by two orders of magnitude.

C.2 Detailed Derivation of Radiation Heat Flux through a Tube with a Small Cross Section (*Gedeon Associates, Athens, Ohio*)

The problem is as follows: Evaluate the radiation heat flux $q(\xi)$ through the tube cross section $A(\xi)$ as the sum of the heat flows from the two ends and the two wall surfaces before and after point ξ. For each part, base the calculations on the configuration factors for radiation heat transfer tabulated in Appendix C of Siegel and Howell (2002). With the assumption that ends and walls are black-body emitters, there is no reflected radiation to consider and the analysis is relatively straightforward.

C.2.1 Negative End Contribution (*See Figure C.3*)

q_0 is the radiation flux leaving surface A_0 that passes through surface $A(\xi)$. It may be written as:

$$q_0 = \sigma T_2^4 F_{0-\xi} \tag{C.3}$$

where $F_{0-\xi}$ is the configuration factor for the fraction of the total radiation leaving A_0 that arrives at $A(\xi)$. From Case 21, p. 826, Siegel and Howel (2002):

$$F_{0-\xi} = \tfrac{1}{2}(G - \sqrt{G^2 - 4}) \tag{C.4}$$

where

$$G = 2 + 4\xi^2 \tag{C.5}$$

C.2.2 Positive End Contribution (*See Figure C.4*)

q_L is the radiation flux leaving surface A_L that passes through surface $A(\xi)$. It may be written as:

$$q_L = -\sigma T_L^4 F_{L-\xi} \tag{C.6}$$

where $F_{L-\xi}$ is the configuration factor for the fraction of the total radiation leaving A_L that arrives at $A(\xi)$. Similar to the previous case:

$$F_{L-\xi} = \tfrac{1}{2}\left(H = \sqrt{H^2 - 4}\right) \tag{C.7}$$

where

$$H = 2 + 4(\xi_L - \xi)^2 \tag{C.8}$$

C.2.3 Negative Wall Contribution (*See Figure C.5*)

q_{w-} is the radiation flux leaving the wall surface $\eta < \xi$ that passes through surface $A(\xi)$. Integrating the contributions of differential elements dA_η, it may be written as:

$$q_{w-} = \frac{1}{\pi a^2}\sigma \int_0^\xi T(\eta)^4 \, F_{d\eta-\xi} \, dA_\eta \tag{C.9}$$

Substituting for the wall area element:

$$dA_\eta = 4\pi a^2 \, d\eta \tag{C.10}$$

this becomes:

$$q_{w-} = 4\sigma \int_0^\xi T(\eta)^4 \, F_{\overline{d\eta-\xi}} \, d\eta \tag{C.11}$$

$F_{\overline{d\eta-\xi}}$ is the configuration factor for the fraction of the total radiation leaving dA_η that arrives at $A(\xi)$. From Case 30, p. 829, Siegel and Howell,

$$F_{\overline{d\eta-\xi}} = \frac{(\xi-\eta)^2 + \frac{1}{2}}{\sqrt{(\xi-\eta)^2 + 1}} - (\xi-\eta) \tag{C.12}$$

C.2.4 Positive Wall Contribution (*See Figure C.6*)

q_{w+} is the radiation flux leaving the wall surface $\eta > \xi$ that passes through surface $A(\xi)$. Similar to the previous case:

$$q_{w+} = -4\sigma \int_\xi^{\xi_L} T(\eta)^4 \, F_{d\eta-\xi}^+ \, d\eta \tag{C.13}$$

$F_{d\eta-\xi}^+$ is the same configuration factor as the previous case except with ξ and η switched:

$$F_{d\eta-\xi}^+ = \frac{(\eta-\xi)^2 + \frac{1}{2}}{\sqrt{(\eta-\xi)^2 + 1}} - (\eta-\xi) \tag{C.14}$$

C.2.5 Normalization

Dividing the four heat flux contributions by $q_{max} = -4\sigma(T_L^4 - T_0^4)$ converts them to normalized radiation heat fluxes:

$$\frac{q_0}{q_{max}} = \frac{-(T_0/T_L)^4 F_{0-\xi}}{1 - (T_0/T_L)^4} \tag{C.15}$$

$$\frac{q_L}{q_{max}} = \frac{F_{L-\xi}}{1 - (T_0/T_L)^4} \tag{C.16}$$

$$\frac{q_{w-}}{q_{max}} = \frac{-4 \int_0^\xi (T(\eta)/T_L)^4 F_{\overline{d\eta-\xi}} \, d\eta}{1 - (T_0/T_L)^4} \tag{C.17}$$

$$\frac{q_{w+}}{q_{max}} = \frac{4 \int_0^{\xi_L} (T(\eta)/T_L)^4 F_{d\eta-\xi}^+ \, d\eta}{1 - (T_0/T_L)^4} \tag{C.18}$$

For a linear temperature variation, the temperature ratio $T(\eta)/T_0$ may be expressed as:

$$\frac{T(\eta)}{T_L} = \frac{T_0}{T_L} + \left(1 - \frac{T_0}{T_L}\right)\frac{\eta}{\xi_L} \qquad (C.19)$$

C.2.6 Programming

Custom Delphi program RadiationDownTube.pas performs the above calculations and sums all the radiation contributions to produce the results plotted above. (Excel does the actual plotting of data.) An adaptive quadrature routine does the required integrations. The program was tested for two test cases with known solutions: (1) the radiation flow in the limit of zero tube length and (2) the radiation flow for a step temperature distribution with $T = T_0$ up to point ξ and $T = T_L$ beyond. For both cases $q/q_{max} = 1$.

Appendix D: Regenerator Figure-of-Merit Degradation with Intra-Regenerator Flow-Gap Variations

Gedeon Associates, Athens, Ohio

To: Regenerator Research Team

From: D. Gedeon

January 7, 2004

The intra-regenerator flow streaming results for parallel foil regenerators are used to estimate the degradation in our regenerator figure of merit as a result of gap variations between sections.

D.1 Background

We have settled on the following formula for calculating the figure of merit for preliminary ranking of regenerator matrices (Gedeon, 2003a):

$$F_M = \frac{1}{f\left(\frac{R_e P_r}{4 N_u} + \frac{N_k}{R_r P_r}\right)} \tag{D.1}$$

for a foil-type regenerator $f = 96/R_e$, $N_u = 8.23$, and $N_k = 1$. When plotted as a function of Reynolds number (assuming $P_r = 0.7$), the figures of merit for a foil regenerator, as well as a few other regenerators of interest, look like those shown in Figure D.1.

D.2 Effects of Gap Variations

A foil regenerator is split into two parts. The flow gap in one is increased while decreased in the other. This gives rise to DC flow circulations between the two parts and a degradation in overall engine efficiency, as previously documented (Gedeon, 2004a).

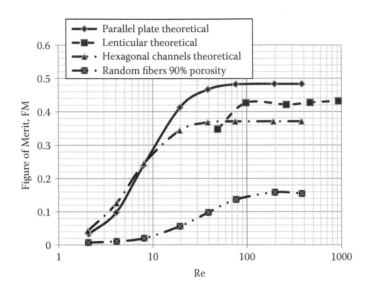

FIGURE D.1
Figures of merit for several matrices.

Not previously reported was the effect of the gap variations on combined-regenerator pumping loss W_f (Sage output AEfric) and cycle-average enthalpy flow $\dot{H}$ (Sage output HNeg or HPos). Because W_f relates to friction factor and $\dot{H}$ relates to Nusselt number, this information can be used to estimate the effective figure of merit as a function of gap variations. Specifically, friction factor is directly proportional to W_f. So, the effective friction factor for the gap-perturbed regenerator is greater (or less) than the friction factor for the baseline regenerator by the factor:

$$\frac{f}{f_0} = \frac{W_{fA} + W_{fB}}{(W_{fA} + W_{fB})_0} \tag{D.2}$$

where the 0 subscript refers to the baseline (equal flow gap) regenerator, and subscripts A and B refer to the two regenerator parts. Because the Nusselt number is inversely proportional to $\dot{H}$, the effective Nusselt number for the gap-perturbed regenerator is related to Nusselt number for the baseline regenerator by the factor:

$$\frac{N_u}{N_{u0}} = \frac{(\dot{H}_A + \dot{H}_B)_0}{\dot{H}_A + \dot{H}_B} \tag{D.3}$$

The effective f and N_u thereby computed can be substituted into the formula for figure of merit with Reynolds number taken as the baseline regenerator

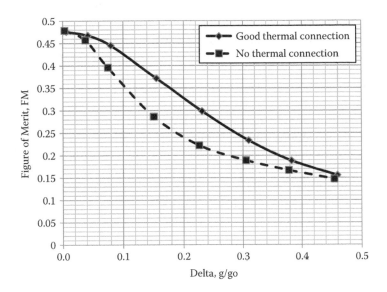

FIGURE D.2
Figures of merit as functions of flow-gap variation.

mean value $R_e = 62$. The result is two curves of figure of merit degradation as a function of relative gap variation, one for the case where the two regenerator parts are in good thermal contact and one for the case of no thermal contact. The latter case is worse because of the temperature skewing effect of DC flow which amplifies the DC flow if not suppressed by thermal contact between the two parts, as shown in Figure D.2.

The "Delta g/g_0" represents the amount by which the gap is higher in part A and simultaneously lower in part B, compared to the baseline gap. A gap-variation *amplitude*, in other words. The curve labeled "good thermal connection" corresponds to gap variations between regenerator parts in good transverse thermal contact—for example, between adjacent foil layers or nearby in the same layer. The curve labeled "no thermal connection" corresponds to gap variations between distant parts of the regenerator, as might occur when all the layers on one side of an annulus are crammed together and spread apart on the other side due to some systematic assembly error or misalignment between inner and outer canisters.

It is seen that in either case, it does not take too much gap variation to significantly reduce a foil regenerator figure of merit. Increased cyclic enthalpy flow (sum for both parts) is mainly responsible for the degradation. Flow frictional loss hardly changes, even decreases a bit. At a gap variation of ±45%, the figure of merit is down in the vicinity of random fibers at 90% porosity ($F_M = 0.14$). This is for a Stirling engine regenerator. The figure of merit degradation for a cryocooler regenerator would be much faster, based on the conclusions of the January 5, 2004, memorandum (Gedeon, 2004a).

It is reasonable to apply the above plot to the general case of any channel-type regenerator, such as the honeycomb type. In that case, hydraulic-diameter variation would replace flow-gap variation, but the resulting curves should be quite similar.

So, based on the above, what should we specify for an allowable gap variation (hydraulic diameter variation) for a microfabricated regenerator? Maybe a ±10% variation would be reasonable. That should keep us above $F_M = 0.4$ for both localized variations and systematic gap variations. If systematic variations (over large distances) are not a problem, then we might go as high as ±15%, or maybe a bit more, but not too much more lest we wind up in the unenviable position of producing an "improved" regenerator that performs as well as random fibers at many times the cost.

Appendix E: Potential 6% to 9% Power Increase for a Foil-Type "Microfab" Regenerator in the Sunpower ASC Engine

Gedeon Associates, Athens, Ohio

To: Regenerator Research Team
From: D. Gedeon
March 26, 2004

E.1 Introduction

Gary Wood of Sunpower Inc. has been wondering if the previous estimates for the benefits of a microfabricated regenerator (in an August 25, 2003, memo) might have been optimistic considering the regenerator length was excessive (133 mm) and foil solid conduction was discounted. Solid conduction multiplier *Kmult* was set to 0.1 as an approximation to the interrupted and convoluted solid conduction path we were considering for the lenticular matrix design.

This memo addresses these concerns. A foil type is inserted into the canister of the most recent Sunpower ASC engine Sage model (D25B3.stl), and the foil spacing is optimized, with length constrained first to 70 mm, then 60 mm. The foil is assumed to be 15 micron thick stainless steel with the full solid conductivity accounted for in the model.

After reoptimization, the result is a 6.6% increase in pressure-volume (PV) power for the same heat input for 70 mm long foil, compared to the baseline random fiber regenerator—a 5.5% increase for 60 mm long foil. Details follow.

E.2 Details

One of the two Sage files created for this study is named *D25B3FoilRegen*. It has an optimization structure much like that described in a December 23, 2003, memo "Sage Model for ASC Optimization" (SunpASCOptimizationFile.doc),

TABLE E.1

Key Results and Dimensions for Two Foil Regenerator Optimizations

	Baseline Random Fiber Regenerator D25B3	Foil Regenerator 70 mm long D25B3FoilRegen	Foil Regenerator 60 mm Long D25B3FoilRegenL60
Heat input (W)	230	230	230
PV power (W)	111.8	119.2	118
Percent increase	—	6.6	5.5
Foil gap (microns)	—	97.2	92.6
Canister area (cm²)	2.83	2.25	2.06

except the acceptor length is constrained to 25 mm and the regenerator matrix type is foil. Regenerator length is constrained to 70 mm. A second Sage file named *D25B3FoilRegenL60* is identical except regenerator length is constrained to 60 mm.

When these Sage files are optimized, all of the basic engine dimensions get adjusted to best serve the foil regenerator, subject to all the dimensional constraints that have so-far evolved for the advanced Stirling convertor (ASC) engine design. The objectives were to see how well a foil regenerator would do compared to the baseline random fiber regenerator and get some idea of the trades for reduced regenerator length. Foil gap was optimized, but foil thickness held constant at 15 microns.

Table E.1 provides some key results and dimensions for the two foil regenerator optimizations, compared to the baseline random-fiber regenerator.

Once again, there are significant benefits to foil regenerators, even at reduced length. Note that the regenerator canister area went down for the two foil cases resulting in a slightly more compact engine with reduced pressure-wall conduction loss.

For the 60 mm case, the foil solid conduction averaged over the regenerator represents about 6.7 W out of 230 W heat input. This would be worth another 3% of efficiency (power increase for given heat input) if eliminated. For the regenerator plates we are considering, there would be an interrupted solid conduction path allowing us to eliminate some of the 6.7 W conduction loss. So, the original estimate of a potential benefit on the order of 9% for a microfabricated regenerator continues to appear reasonable.

Appendix F: Electric Discharge Machining (EDM) Regenerator Disks—Concentric Involute Rings (In Which Involute Foils Are Packaged in a Cylindrical Form within Concentric Rings)

Gedeon Associates, Athens, Ohio

To: Microfab Team
From: D. Gedeon
April 13, 2004

F.1 Packing a Cylinder

Thinking ahead to the test regenerator we are planning to build in year 3, we need to come up with a matrix that fits within a 19 mm diameter cylindrical form. What should it look like?

The involute-foil idea is tempting, except that the only way to *completely* fill a cylinder with an involute is with the extreme case of a single foil, wound in spiral fashion. This will not do because it is not geometrically similar to the multifoil involute structure we would use for a thin annular regenerator (space power engine), and it results in a long unsupported length of foil, more like a watch spring than a regenerator matrix.

Both problems can be solved by packing the involute elements into concentric rings, each ring separated by a thin wall. Each ring may contain a different involute family (different generating circle), so they may all be geometrically similar to "thin annular" involutes. The ring walls also serve as support points for individual involute elements, thereby increasing their stiffness.

Figures F.1 and F.2 show two ways one might package involutes into concentric rings. In the picture labeled "geometric spacing," the rings are defined by circles with successive diameters in the same ratio. The advantage of geometric spacing is that the involutes in all rings have about the same angle relative to the cylinder radius and are therefore fluid-dynamically and

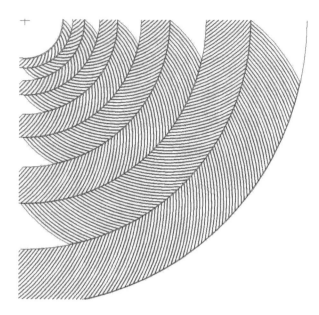

FIGURE F.1
Geometrically spaced involute rings.

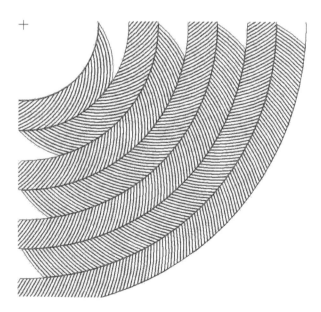

FIGURE F.2
Equally spaced involute rings.

structurally similar insofar as those things matter. The disadvantage is that the ring spacing decreases toward the center, resulting in shorter elements (shorter aspect ratios for the flow channels). In the picture labeled "uniform spacing," the rings are defined by circles with successive diameters in constant increments. The advantage of uniform spacing is that the lengths of the involute elements do not vary as much. The disadvantage is that the radius-angle of the involute elements increases toward the center, possibly resulting in fluid-dynamic or structural differences between inner and outer rings.

F.2 Dimensions

The illustrations show cross-sectional views of regenerator disks created with the Solid-Edge computer-aided design (CAD) software. Only about one-quarter of the disks are cut into the involute patterns so as not to bog down the tiny brain of Solid-Edge in computing all the little features. Common dimensions for the two regenerator disks are as listed in Table F.1.

The involute channel widths are exact, as these things go, with the wall thickness varying slightly to accommodate the approximate circular profiles used instead of true involute profiles. The error so created is estimated later on.

F.3 Central Holes

In each case, there comes a point toward the center of the matrix where continuing the pattern becomes absurd and one can either change to another type of foil pattern or just stop and fill the central hole with an insulating stuffer. For purposes of the test regenerator, it should be acceptable to go

TABLE F.1

Common Dimensions for Two Types of Regenerator Disks (See also Figures F.1 and F.2)

Outer diameter	19.05 mm
Involute channel width (gap)	86 microns
Wall thickness between involutes	≈14 microns
Wall thickness between rings	20 microns
Disk thickness	500 microns

with the insulating stuffer. For the geometrically spaced matrix, the central hole diameter is 2.6 mm. For the equally spaced matrix it is 5 mm.

F.4 Alternating Sense Involutes

As illustrated, the angular sense of the involutes alternates in successive rings. This results in a herringbone pattern. It would also be possible to maintain the same angular sense in each ring. In either case, the idea would be to flip alternating regenerator disks so the angular sense changes with each layer. This way there should be no need for rotational alignment between layers, and the solid conduction path between involute wall elements is interrupted. The opportunity for flow distribution between layers is also maximized.

The structural and flow advantages to the herringbone pattern, if any, are not completely clear. One possible item of consequence is that any radial flow in a plenum at the discharge end of such a herringbone matrix would tend to get mixed because the flow coming from successive involute ring would tend to swirl in opposite directions. This is a good thing, I suppose.

F.5 EDM

Based on my current understanding of the International Mezzo Technologies (Baton Rouge, Louisiana) EDM process, the illustrated regenerator disks should be possible to make. One concession to EDM might be to round the corners where the involute elements attach to the circular walls between rings which would result in somewhat greater axial thermal conduction loss.

F.6 Structural Analysis

In principle, one could perform a finite-element stress analysis of a complete involute-ring regenerator disk, but I have not done that. The important things to understand would be the axial load required to buckle a regenerator comprising a stack of such disks and the resistance to deformation of the individual involute wall elements.

The axial buckling load is important if we decide to hold the stack in place by compression. I have some hope that the axial bucking strength will be adequate because each involute flow channel forms a structural cell consisting of side walls integrally connected to end walls. Each structural cell should be relatively stiff, compared to foil layers unconnected at the ends. And there are a large number of such cells to distribute the load.

The resistance to deformation of the individual wall elements is important to maintaining uniform spacing. If nothing else, the walls should be stiff enough so that internal stresses relieved by the EDM process do not result in significant spacing variations. I am not sure, but it seems that the curved shapes of the involute walls will help in this regard. The radius of curvature cannot change too much without affecting the separation between endpoints, which is constrained by the end walls (inter-ring walls). Were the elements straight (e.g., radial spokes), they could deflect much more for the same end constraints. Obviously, we need to think more about this.

Appendix G: Involute Math

Written by David Gedeon of Gedeon Associates, Athens, Ohio

Gathered together here are some useful formulas and suggestions for generating involute patterns in computer-aided design (CAD) software.

G.1 Involute Mathematics

Each ring of flow channels is a circular pattern of elemental flow channels. At the center of each elemental flow channel is a segment of an involute curve defined by a generating circle and two boundary circles, as illustrated in Figure G.1.

The generating circle has radius R_0 and the two boundary circles have radii R_1 and R_2. The involute segment is the curved arc between R_1 and R_2. By definition, it is generated by the mathematical equivalent of a string unwinding from the generating circle. Dotted lines indicate the position of the string at the beginning and ending of the arc. The exposed lengths of the string at the two endpoints of the involute segment are just the arc lengths along the generating circle subtended by angles θ_1 and θ_2 or

$$r_1 = R_0\theta_1 \tag{G.1}$$

and

$$r_2 = R_0\theta_2 \tag{G.2}$$

r_1 and r_2 are also the local radii of curvature of the involute segment at the two endpoints. By drawing some right triangles, it is easy to conclude that r_1 and r_2 are geometrically related to the various circle radii by

$$r_1\sqrt{R_1^2 - R_0^2} \tag{G.3}$$

and

$$r_2 = \sqrt{R_2^2 - R_0^2} \tag{G.4}$$

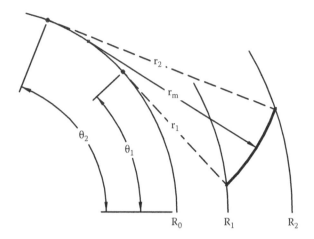

FIGURE G.1
Generating and boundary circle defining an involute curve.

A circular pattern of N involute segments can be made by rotating the original involute by angular increments $2\pi/N$ (in radians) about the center of the generating circle. This pattern can also be thought of as being made by shortening the string by increments s_0 equal to the circumference of the generating circle divided by N, or

$$s_0 = \frac{2\pi R_0}{N} \tag{G.5}$$

The advantage of this point of view is that s_0 is evidently the normal spacing between involute elements. Solving the previous equation for N, the number of involute elements spaced by distance s_0 is:

$$N = \frac{2\pi R_0}{s_0} \tag{G.6}$$

The arc drawn in the above illustration is not actually a true involute curve but rather just an ordinary circular arc centered on the generating circle with radius

$$r_m = (r_1 + r_2)/2 \tag{G.7}$$

The normal spacing between the circular arcs is therefore not exactly s_0. It varies as:

$$s \approx s_0 \cos \alpha \tag{G.8}$$

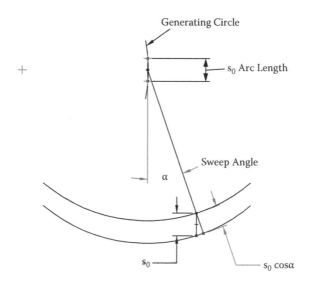

FIGURE G.2
Two successive circular "involute" arcs drawn with centers vertically aligned on the generating circle.

where α is the rotation angle of radius r_m relative to its angle at mid segment. This can be seen from Figure G.2, which shows two successive circular "involute" arcs drawn with their centers vertically aligned on the generating circle. The reason for the approximately equals symbol in the previous equation is that s_0 is the arc length between two successive center positions along the generating circle, not exactly the cord length.

G.2 Involute Cutouts

As mentioned, the "involute" arc segments so generated lie at the centers of the flow channels. The way the flow channels are generated is to offset the arc segments toward each side by small increments (concentric arcs) to create cutout boundaries, as in Figure G.3 that illustrates a single-flow channel between two boundary circles.

In this case, the involute arc segment is offset 43 microns toward each side forming a channel of width 86 microns. The cutout ends are offset 10 microns from the inter-ring boundary circles, consistent with 20 micron thick walls between ring sections. The normal distance separating true involute arc segments in the circular pattern is 100 microns. With 86 micron channel widths, this leaves 14 microns for the side-wall thickness. Actually, the side walls vary

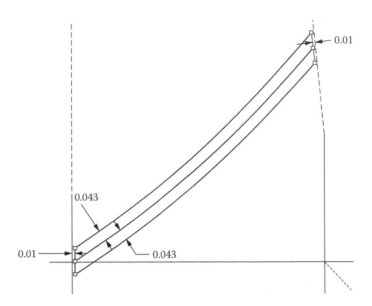

FIGURE G.3
Offsetting of arc segments to create cutout boundaries.

in thickness slightly because of the circular-arc approximation to true involute segments and the resulting error in normal spacing that results. That error affects only the side wall thickness. The channel width is exact, to within the limits of the CAD software.

G.3 Spreadsheet Calculations

Generating the involute channels requires a sequence of boundary circles and a sequence of generating circles. In principle, the two sequences could be different, but it is easiest if the same sequence of circles is used for both purposes. In other words, given an arbitrary sequence of circles C_i, for $i = 0...M$, with diameters D_i in increasing order, circle C_0 is the generating circle for the involute arc segments between C_1 and C_2, circle C_1 is the generating circle for the involute arc segments between C_2 and C_3, and so forth. Then the only remaining choice is the spacing between the diameters D_i. The options of geometrical spacing (D_i / D_{i-1} = constant) or arithmetic spacing ($D_i - D_{i-1}$ = constant) have already been illustrated and their properties discussed. We might eventually want to look at other options, too.

To automate the process, I wrote a spreadsheet that calculates circle sequences and other useful information for the 19.05 mm diameter regenerator disks illustrated earlier. There are actually two spreadsheets: *Involute19DiamCyl.*

TABLE G.1

Spreadsheet Calculated Values for the Equal-Space Rings

Circle Diameters D_i (mm)	Number Elements in Pattern N	Radius of Involute Arcs r_m (mm)	Sweep Angle α (radians)	Relative Spacing Error at Endpoints	Relative Spacing Error Due to N Roundoff
19.05					
17.05	472.00	4.92	0.24	0.007	−0.002
15.05	409.00	4.62	0.27	0.009	−0.002
13.05	347.00	4.29	0.30	0.011	0.000
11.05	284.00	3.94	0.34	0.014	−0.001
9.05	221.00	3.55	0.40	0.020	−0.002
7.05	158.00	3.11	0.51	0.033	−0.004
5.05	95.00	2.60	0.76	0.072	−0.009
3.05					

xls for geometrically spaced rings and *Involute19DiamCylEqualSpaced.xls* for equally spaced rings. The spreadsheets generate a sequence of circles spanning the desired diameter range, then for each concentric ring of the structure calculate the following items, based on the desired normal spacing s_0 between involute channels:

- Number N of elements in the circular pattern (Equation G.1 to the nearest integer)
- Radius of curvature of the approximate-involute arc segment from Equation (G.2)
- Sweep angle α in radians $(\theta_2 - \theta_1)$
- Relative spacing error at endpoints $1 - \cos\alpha/2$ based on Equation (G.3)
- Spacing error due to roundoff of N

The last two items are handy for estimating the variation in wall thickness between involute channels. For example, the spreadsheet-calculated values for the equal-spaced rings are given in Table G.1.

Note that the relative spacing error for the inner ring is relatively large (0.072). This times s_0 is the approximate amount by which the wall thickness is thinned down near the channel endpoints. For the present case of $s_0 = 100$ microns, the absolute wall thickness error is about 7 microns. When everything is accounted for in solid edge, the actual wall thickness for the inner ring varies from about 10 microns at the ends to 19 microns at mid-chord. The wall thickness is much more uniform for subsequent rings.

Appendix H: Implications of Regenerator Figure of Merit in Actual Stirling Engines

Gedeon Associates, Athens, Ohio

Re: NASA/CSU Regenerator Microfabrication Contract NAS3-03124
From: David Gedeon
January 27, 2006

H.1 Introductory Discussion

Our regenerator figure of merit measures the heat transfer per unit flow resistance of a regenerator matrix. But what does that mean in the context of an actual Stirling engine (or cooler)? The question can be answered by imagining a fixed Stirling engine into which regenerators of variable figure of merit F_M are substituted. It turns out (derived below) that the figure of merit is inversely proportional to the product of regenerator pumping loss W_p, thermal loss Q_t, and the square of regenerator mean flow area A_f:

$$F_M \propto \frac{1}{W_p Q_t A_f^2}$$

For regenerators with the same flow areas, the figure of merit is inversely proportional to the product of pumping loss and thermal loss. So, a high figure of merit will correspond to a low pumping loss, a low thermal loss, or both. But depending on the relative sizes and importance of the two losses in an actual engine, the overall benefit to engine efficiency will vary. It is even logically possible that a regenerator with a higher figure of merit will result in lower actual engine efficiency if it reduces one of the two losses that is not very important while allowing the other, that *is* important, to increase a bit, or if it reduces the engine power density because of a larger void volume.

H.2 Figure of Merit Reformulation

The figure of merit we adopted comes from a memorandum (Gedeon, 2003a):

$$F_M = \frac{1}{\left(\frac{R_e P_r}{4N_u} + \frac{N_k}{R_e P_r}\right)} \tag{H.1}$$

where

f = Darcy friction factor
N_u = Nusselt number $h d_h / k$
N_k = effective gas conductivity due to thermal dispersion as a fraction of molecular conductivity
R_e = Reynolds number $\rho u d_h / \mu$
P_r = Prandtl number $c_p \mu / k$
d_h = hydraulic diameter

In Equation (H.1), the two terms in the denominator measure the effects of heat transfer and thermal dispersion, respectively. Another memorandum (Gedeon, 2003b) discusses the equivalence between mean-parameter enthalpy flows produced by heat transfer and microscopic enthalpy flows produced by thermal dispersion.

So what does the figure of merit have to do with the losses in a regenerator? To answer that question, it is convenient to start with the expressions for time average thermal energy transport (enthalpy + dispersion) per unit void flow area q_t and pumping power per unit regenerator void volume w_r given in the 1996 regenerator test-rig contractor report (Gedeon and Wood, 1996).

$$q_t = -k \frac{\partial T}{\partial X} \left\{ \frac{p_e^2}{4N_u} + N_k \right\} \tag{H.2}$$

and

$$w_r = \frac{1}{2d_h} \{ f \rho u^2 |u| \} \tag{H.3}$$

where {} stands for time average, and Peclet number P_e is shorthand for $R_e P_r$. Multiplying q_t by regenerator void flow area A_f converts it to total thermal energy transport Q_t:

$$Q_t = -kA_f \frac{\partial T}{\partial x} \left\{ \frac{P_e^2}{4N_u} + N_k \right\} \tag{H.4}$$

Multiplying w_r by the regenerator void volume $A_f L$ converts it to total pumping power W_p:

$$W_p = \frac{A_f L}{2d_h} \{f \rho u^2 |u|\} \tag{H.5}$$

For reasons that are about to become clear, it is convenient to express pumping power in terms of F/R_e by introducing the factor 1 in the form $\rho |u| d_h / (\mu R_g)$, which results in

$$W_p = \frac{A_f L \rho^2}{2\mu} \left\{ \frac{f}{R_e} u^4 \right\} \tag{H.6}$$

One can already see signs of the figure of merit (Equation H.1) in the preceding equations. It is even clearer by writing the figure of merit in the form

$$F_M = \frac{P_r}{\left(\frac{f}{R_e}\right)\left(\frac{R_e^2 P_r^2}{4 N_u} + N_k\right)} \tag{H.7}$$

Ignoring time averages and constants (because we are only interested in proportionalities), representing the temperature gradient as T/L, and after a little simplification, the figure of merit can be reduced to the form:

$$F_M \propto \frac{\rho^2 c_p}{A_f^2} \frac{T(u A_f)}{} \frac{1}{W_p Q_t} \tag{H.8}$$

What does it mean? In the first factor on the right, all of the quantities in the numerator are constant for any given Stirling machine.

Fixed gas and charge pressure $\Rightarrow c_p$ and ρ are fixed

Fixed hot and cold temperatures $\Rightarrow \Delta T$ is fixed

Fixed piston, displacer volumetric flow rates $\Rightarrow u A_f$ is fixed

So assuming all of the above, the figure of merit reduces to:

$$F_M \propto \frac{1}{W_p Q_t A_f^2} \tag{H.9}$$

Or solving for the pumping-work thermal-energy-transport product,

$$W_p Q_t \propto \frac{1}{F_M A_f^2} \tag{H.10}$$

This interesting result suggests that large regenerator flow areas are good. For two regenerators with the same F_M, the one with the larger flow area will have a lower $W_p Q_t$ product. I suppose that makes sense given the dependence of pumping power on velocity to the fourth power in Equation (H.6). This may be one reason F_M does not track overall engine efficiency so closely when comparing regenerators of different matrix structures (e.g., parallel plates versus random fibers).

A cautionary remark in applying Equation (H.10) too literally is that it does not take into account total void volume of the regenerator. The pressure amplitude and therefore power density of the engine will vary inversely with the regenerator void volume, though not in direct proportion because there are other volumes in the total engine. The implication is that if one measures the $W_p Q_t$ product relative to the square of the engine power, then the A_f^2 factor may be somewhat canceled out. But if one is comparing two regenerators with the same void volume, then this concern vanishes.

Appendix I: Random-Fiber Correlations with Porosity-Dependent Parameters— Updated with New 96%-Porosity Data

Gedeon Associates, Athens, Ohio

Re: NASA/CSU Regenerator Improvement Grant NNC04GA04G

From: D. Gedeon

November 1, 2006

Since the previous memo on this topic (Gedeon, 2006a), additional testing of a 96%-porosity random-fiber half-length sample "A" produced a lower figure of merit compared to the tests of the full-length 96% porosity sample. Recent full-length retesting suggested that the original full-length data was bad for the 50-bar helium test case (Gedeon, 2006b).

This memo updates the porosity-dependent master correlations for f, N_u, and N_k based on heat transfer data for the 96%-porosity half-length sample, which is believed to be more reliable.

I am planning to go ahead and insert these latest master correlations into the development version of Sage, because the peak figure of merit for 96%-porosity random-fibers has dropped significantly from about 0.43 down to about 0.28. This will have an effect for *far-out* regenerator investigations but should have minimal impact for analysis of conventional regenerators with porosities on the order of 90%. There is a table of relative errors produced by the master correlations near the end of this memo.

I.1 Master Correlations

The updated correlations are in the same form as before:

$$f = \frac{a_1}{R_e} + a_2 R_e^{a3} \tag{I.1}$$

$$N_u = 1 + b_1 P_e^{b2} \tag{I.2}$$

$$N_k = 1 + b_3 P_e^{b2} \tag{I.3}$$

R_e is the Reynolds number based on hydraulic diameter. $P_e = R_e P_r$ is the Peclet number. What has been updated is the porosity dependence embedded in the individual coefficients, as follows $(x = \beta /(1 - \beta)$, where β is porosity):

$$a_1 = 22.7x + 92.3$$

$$a_2 = 0.168x + 4.05 \tag{I.4}$$

$$a_3 = -0.00406x - 0.0759$$

$$b_1 = (0.00288x + 0.310)x$$

$$b_2 = -0.00875x + 0.631 \tag{I.5}$$

$$b_3 = 1.9$$

I.2 Figure of Merit

The overall figure of merit still increases with increasing porosity, but not as much as before. It now peaks at a value of about 0.28 at 96% porosity, down from about 0.43 previously.

I.3 Correlating Porosity

The F_M plot in Figure I.1 and the material in this section are derived from the revised spreadsheet *RandomFiberPorosityDependence.xls*.

The data-modeling parameters for the individual regenerator tests are listed in Table I.1. The only changes compared to the table in the August 19, 2006, memo (Gedeon, 2006a) are for the 0.96 porosity N_u and N_k parameters, which are now based on tests for half-length sample "A." The individual parameters are plotted in Figure I.2 along with their trend-lines.

The previous quadratic curve fit (trend line) for the b_1 heat-transfer parameter (Nusselt number coefficient) is no longer obviously necessary, but it does reduce relative errors for the individual correlations, compared to a linear trend line, and also results in a more uniform progression of increasing figure of merit with porosity. The y-intercept remains zero so that the implied heat-transfer coefficient does not diverge at zero porosity ($h \propto Nu$ $(1 - \beta)/\beta$) as argued in the August 19, 2006, memo (Gedeon, 2006a).

TABLE I.1

Friction-Factor and Heat-Transfer Correlation Parameters for Different Porosities

β	β /(1 – β)	f Parameters			N_u, N_k Parameters		
		a1	a2	a3	b1	b2	b3
0.688	2.205	128.8	3.858	–0.063	0.499	0.635	3.787
0.820	4.556	248.5	4.889	–0.071	0.945	0.632	2.157
0.850	5.667	233.8	4.15	–0.082	1.552	0.539	1.113
0.897	8.709	211.2	5.139	–0.151	1.287	0.600	1.026
0.900	9.000	321.4	5.138	–0.108	2.323	0.534	0.583
0.930	13.29	380.3	9.906	–0.195	7.447	0.424	1.983
0.960	24.00	651.5	6.627	–0.135	8.600	0.461	2.498

Table I.2 gives the updated table of correlated versus individual ratios for various quantities of interest, averaged over the Reynolds number range from 10 to 1000. Values for individual correlations are indicated by zero subscripts. F_M is the overall figure of merit.

Compared to the table in the August 19, 2006, memo (Gedeon, 2006a) (including 0.93 porosity data), the figure-of-merit ratios in the range of porosities 0.688 to 0.85 are substantially higher. There is not much change at porosity 0.90 and a bit more pessimism at 0.93. The correlation at porosity 0.96 is based on new data.

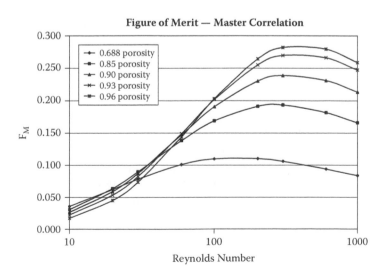

FIGURE I.1
Master correlation figures of merit (2006) as functions of porosity.

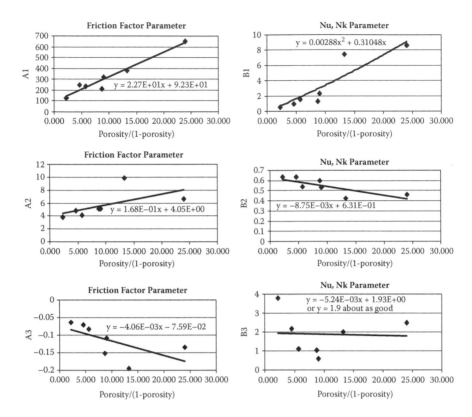

FIGURE I.2

Trend lines for friction-factor and heat-transfer correlation parameters as functions of porosity/(1-porosity). (Data points are also given in Table I.1.)

TABLE I.2

Master Correlation Relative Errors

B	f/f_0	N_u/N_{u0}	N_k/N_{k0}	F_M/F_{M0}
	Master Correlation Relative Errors Averaged Over R_e = 10 to 1000			
0.688	1.05	1.22	0.47	1.35
0.820	0.83	1.28	0.75	1.52
0.850	1.03	1.40	1.95	1.09
0.897	1.35	1.80	1.47	0.99
0.900	0.99	1.39	3.09	0.99
0.930	0.97	0.93	1.40	0.94
0.960	0.98	0.89	0.66	1.12

TABLE I.3

Gedeon Associates' Derived Data Files for Different Regenerator Samples

Sample	Tested	Porosity	DP file	HX file
2 mil Brunswick	1992–1993	0.688	P06-29Scaled	H11-21Scaled
1 mil Brunswick	1992–1993	0.82	P11-04Scaled	H11-18Scaled
30 micron Bekaert	2006	0.85	DP_85Porosity	HX_85PorosityTrunc
12 micron Bekaert	2003	0.897	BekD12P90DPRetestScaled	BekD12P90HXScaled
30 micron Bekaert	2006	0.90	DP_90Porosity	HX_90PorosityTrunc
30 micron Bekaert	2006	0.93	DP_93Porosity	HX_93PorosityTrunc
30 micron Bekaert	2006	0.96	DP_96Porosity	HX_96PorosityHalfA

I.4 Data Files

The derived data files used for parameter modeling are define in Table I.3. (Gedeon Associates created and maintains these files.)

Appendix J: Regenerator Final Design

Summarized by Gedeon Associates, Athens, Ohio

J.1 Detailed Specifications

According to the Sage computer simulation, Table J.1 shows the regenerator design that produces the best efficiency under the constraints of the frequency-test bed (FTB) installation.

The involute geometry is composed of primary and alternate disks, designed to be alternated in the stack-up assembly. Computer-aided design (CAD) drawings of the two disks are shown in Figures J.1a and J.1b

Hydraulic diameter and porosity for the above disks are $D_h = 0.159$ mm and $\beta = 0.837$, respectively, which are close to the optimized values as determined with the Sage model.

These drawings do not show any rounded corners where foil elements meet partition circles. We decided to round the corners in the production disks to facilitate the manufacturing process, and because rounded corners had produced good results in the phase II prototype regenerator. There was some concern over nonuniform flow patterns in sharp corners and also structural weakening due to stress concentrations there.

J.2 Jet Diffuser Design

Random-fiber flow diffusers are located at either end of the regenerator for purposes of spreading the incoming flow jets from the narrow channels of the acceptor or rejector heat exchangers. The diffuser concept is sketched in Figure J.2 with the arrows attempting to convey the idea of the gas flow field upstream and downstream of the diffuser. The diffuser design was backed by two-dimensional (2-D) computational modeling at Cleveland State University, as will be shown in Appendix L.

The 600 g/m^3 material density for the random-fiber material refers to the density of a material previously supplied by Bekaert Corporation (Belgium) which was readily available for use. That material compressed to 0.6 mm thickness results in a porosity of $\beta = 0.88$.

TABLE J.1

Frequency-Test Bed (FTB) Regenerator Dimensions

Channel gap (mm)	0.086 (0.001, –0.001)
Web wall thickness (mm)	0.014 (0.001, –0.001)
Inner and outer wall thickness (mm)	0.030 (0.005, –0.005)

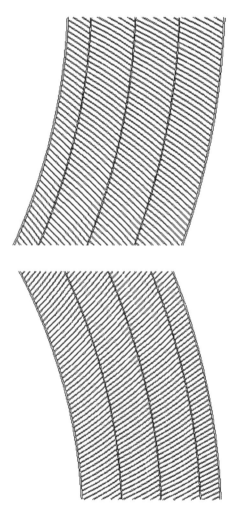

FIGURE J.1
A computer-aided design (CAD) drawing of the involute disk: (a) primary (top) and (b) alternate (bottom).

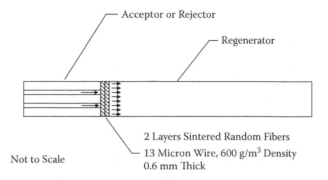

FIGURE J.2
Sketch of the jet diffuser model.

In addition to diffusing jets, the diffusers accommodated irregularities in the FTB regenerator cavity (region between displacer cylinder outer diameter [OD] and pressure-wall inner diameter [ID]) at the two ends). At the rejector end, part of the piston cylinder extends into the regenerator space by 0.8 mm, resulting in a tapered regenerator cavity there. At the acceptor end, there may have been a small braze fillet where the acceptor heat exchanger joins to the pressure wall.

The diffuser disks have a nominal thickness of 0.6 mm, but we intended to adjust thickness as necessary by compressing the random fiber material more or less. The nominal regenerator stack-up height would be achieved by stacking an integral number of precisely made 0.500 mm thick regenerator disks. Our original intention was to use two diffuser disks at the rejector end of the regenerator (to clear the piston cylinder intrusion) and one at the acceptor end. As a result of quality control issues during the polishing process, it turned out the regenerator disks were not precisely 0.5 mm thick (see below, section V), and the assembly process was not as smooth as we had hoped. The clearance between the regenerator disk ID and displacer hot cylinder was too tight, and the disks did not slide onto the cylinder smoothly. We stacked the disks onto the displacer cylinder starting from the rejector end. When we got to the acceptor end, we found it necessary to leave out the last regenerator disk and use two diffuser disks in order to best fill the remaining gap between the regenerator face and cavity end. As a result, we have one extra regenerator disk that was not installed in the engine (total number of disks installed was 126).

J.3 Thermal Expansion and Assembly Issues

The length of the space occupied by the regenerator is defined by the outer pressure wall of the heater head. A calculation shows that the relative thermal expansion between the nickel regenerator and stainless-steel pressure

TABLE J.2

Thermal Expansion Coefficient for Nickel and SS

Material	Coefficient of Thermal Expansion (ppm per °C)
High purity nickel (70–1000 F)	15.5
304 stainless steel (32-212 F)	17.3

wall is only about 0.030 mm in heating from room temperature to operating temperature, with the pressure wall expanding more. The thermal expansion calculations were based on the values given in Table J.2.

After installation, there was easily this amount of resilience in the regenerator assembly to accommodate the anticipated expansion. The regenerator disk stack was not rigid as one might expect had all the disks been precisely 0.500 mm thick with flat faces. Instead, the local variations in disk thickness resulted in a large number of small random gaps between disks, so that the regenerator could be compressed elastically on the order of 0.1 mm or more,

TABLE J.3

Jet Boundary Conditions

Rejector	
Pressure (Pa), mean, amplitude, and phase angle	3.10E6 + 4.1E5 at –22°
Mass flow rate (kg/s), amplitude, and phase angle	4.6E-3 at 44°
Mean temperature (°C)	43
Mean density ρ_m (kg/m^3)	4.8
Velocity amplitude u_1 (m/s)	9.6
Pressure head amplitude $\rho_m u_1^2/2$ (Pa)	2.2E2
Mean jet spacing $(A_{regen}/N_{jets})^{0.5}$ (mm)	1.8
Acceptor	
Pressure (Pa), mean, amplitude, and phase angle	3.10E6 + 4.0E5 at –24°
Mass flow rate (kg/s), amplitude, and phase angle	1.8E-3 at –12°
Mean temperature (°C)	624
Mean density ρ_m (kg/m^3)	1.7
Velocity amplitude u_1 (m/s)	17
Pressure head amplitude $\rho_m u_1^2/2$ (Pa)	2.5E2
Mean jet spacing $(A_{regen}/N_{jets})^{0.5}$ (mm)	1.3
Rejector jet diffuser layer	
Pressure drop amplitude (Pa)	1.2E3
Acceptor jet diffuser layer	
Pressure drop amplitude (Pa)	1.8E3
Microfab regenerator	
Pressure drop amplitude (Pa)	16.3E3

depending on applied pressure. There is also some resiliency in the random-fiber material. Room-temperature experiments at Sunpower showed sintered random-fiber samples rebound elastically by about 2% after removal of the applied force. For 2 mm total thickness of random fiber material, this suggests an elastic rebound of 0.040 mm, which would accommodate the required 0.030 mm due to thermal expansion.

J.4 Jet Boundary Conditions

Table J.3 shows the jet boundary conditions as obtained from Sage.

Appendix K: Estimated Alternator Efficiency

Written by David Gedeon of Gedeon Associates, Athens, Ohio

In Table 8.28a, the alternator efficiency is not measured directly but rather estimated from a simple alternator loss model calibrated to the data. The estimated pressure-volume (PV) power output is then the measured electrical output divided by the estimated alternator efficiency.

The reason for this approach is that the engine PV power measurements are not very accurate and not available for the microfabricated regenerator tests. For the random-fiber regenerator tests, the piston PV power output was measured, but the electrical power measurements are more accurate because they are based on true integrated electrical power calculations performed by a dedicated electrical power meter (Yokogawa) designed for that task. PV power, on the other hand, is calculated in terms of pressure and piston amplitudes and their relative phase angle. It is affected by transducer errors and is a "phasor-math" calculation rather than an actual time integration. Neither piston position nor pressure is recorded by a fast-sampling data acquisition system.

For the purpose of this memo, the PV power calculations are only used to calibrate a formula for alternator efficiency as a function of current phase angle, which is available for all the tests. Actually, only one PV power calculation is used for that purpose, the one for the July 12, 2004, data point (see Tables 8.28 and 8.30). That point was chosen because alternator efficiency was lowest for that point, thereby producing the biggest difference between PV and electrical power and arguably the most accurate measurement of alternator efficiency.

The simple alternator-loss model amounts to the observation that the alternator electrical loss, W_{loss}, scales as the square of the length of the alternator force phasor imposed on the piston—the length of arrow F in Figure K.1. This follows because the alternator force is proportional to electrical current, and the electrical losses grow as current squared ($W_{loss} \propto I^2 R$). The useful electrical output, W_e, on the other hand, is proportional to the square of the length of force component, F_d, in phase with piston velocity, because that is the component absorbing power from the piston. For a given power output, the electrical loss is smallest when the current phase angle, θ (relative to the piston motion) is 90°. When the current phase differs from 90°, the alternator force is also helping to resonate the piston by providing a force component, F_s, in phase with the piston spring. Figure K.1 illustrates the resultant alternator force phasor, F, and its drive and spring components, F_d and F_s, for the case when current phase, θ, is greater than 90°.

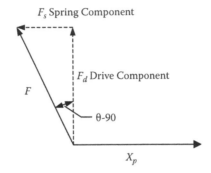

Alternator Force Phasor Diagram

FIGURE K.1
Alternator force phasor diagram.

Applying some trigonometry to this model, the ratio W_{loss}/W_e is proportional to $1/\cos^2(\theta\text{-}90)$. Introducing a calibration parameter c, it follows that alternator electrical efficiency, η_e, may be written

$$1-\eta_e = \frac{W_{loss}}{W_e} = \frac{c}{\cos^2(\theta-90)} \tag{K.1}$$

According to the measured efficiency of the July 12, 2004, data point (electrical output/calculated PV power), calibration parameter c has the value 0.086 (see Table 8.28a). This formula, applied to the other measured current phase angles, gives the alternator electrical efficiency values in the table.

Appendix L: Computational Fluid Dynamics (CFD) Results, Simulating the Jet Diffuser (Cleveland State University)

L.1 Computational Fluid Dynamics (CFD) Geometry

A two-dimensional (2-D) geometry with parallel plates was chosen to simulate the jet diffuser model discussed earlier (see Figures 7.34 and 7.35). Simulating the actual geometry would require a three-dimensional (3-D) geometry and thus more central processing unit (CPU) and memory allocation. Figure L.1 shows the 2-D geometry used to model the jet flow (from the frequency-test-bed [FTB] acceptor) into a random fiber matrix (porous media) separating the acceptor and the involute-foil regenerator. The dimensions were selected based on data provided by Gedeon Associates (Athens, Ohio) to match the FTB design (see Figure J.2 and Table J.1, Appendix J). With 1.3 mm mean jet spacing (from Table J.3), 650 microns (see Figure L.1) equals one-half of the mean jet spacing. Also, the distance from the jet exit to the involute-foil inlet (600 microns shown in Figure L.1) corresponds to the 0.6 mm porous-material thickness (Figure J.2). Porosity is 0.9. It should be noted that the jet enters from the west side with a half-width of 133 microns, and the upper and lower boundaries of the CFD domain were chosen to be symmetric (as shown in the figure). With the dimensions given, six parallel plates were placed at 0.6 mm from the jet exit with metal thicknesses of 14 micron and 86 micron gaps.

L.2 CFD Results

The FLUENT commercial code (Fluent, 2005) was utilized to simulate the above case in order to help validate the choice of diffuser dimensions. Version 6.3.26 was used with 213,560 cells. The code ran on a Dell Precision PWS670, Intel (R) Xeon (TM) with a 2.8 GHz CPU. A steady flow with a V2f turbulence model was utilized in the simulation, and the input data shown in Table L.1 were assumed.

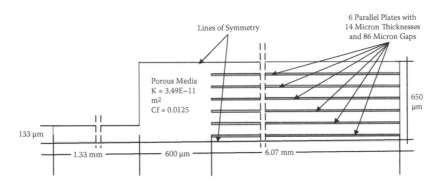

FIGURE L.1
Geometry used to model the porous media (placed in the 600 micron thickness) between the acceptor outlet (shown as slot with 133 micron half thickness) and the involute-foil entrance (shown with 6 parallel plates, 14 micron metal thickness, and 86 micron gap). The flow is from west to east, and top and bottom planes are lines of symmetry.

Figures L.2, L.3, and L.4 show CFD velocity vectors obtained for three different cases (of various gap/porous-media configurations between the jet exit and the involute-foil inlet): (1) without porous media (see Figure L.2), (2) with porous media and a 133 micron axial gap between the jet exit and the porous media (see Figure L.3), and (3) with porous media and no gap (see Figure L.4). A big recirculation area is noticed in the case without any porous media, as expected. The case with the porous media and gap shows how the flow spreads out vertically before entering the porous media. The case with no gap (and with porous media) shows no recirculation as the flow passes through the porous media.

TABLE L.1

Input Data for Computational Fluid Dynamics Simulation of Jetting through Porous-Media Diffuser into Simulated Involute-Foil Regenerator

Fluid	Air
Pressure, Pa	101,325
Temperature, °K	300
Jet velocity, m/s	35.41
Permeability, m²	3.49E-11
Inertial coefficient	0.0125
Porosity	0.9

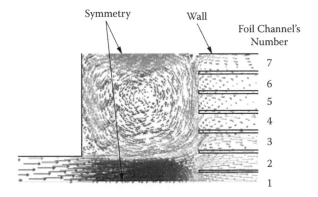

FIGURE L.2

Velocity vectors, without porous media between the jet exit and involute-foil inlet.

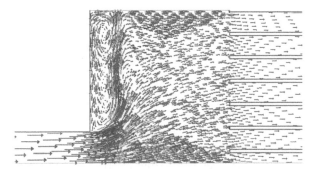

FIGURE L.3

Velocity vectors, with porous media and with a gap (133 micron) between the jet exit and the porous media.

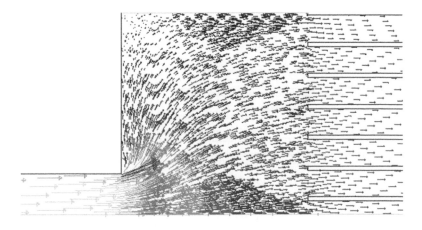

FIGURE L.4

Velocity vectors, with porous media between the jet exit and involute foil inlet and without a gap.

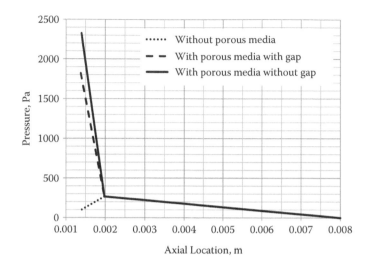

FIGURE L.5
Pressure distribution along the flow direction starting from the jet exit.

Figure L.5 shows the pressure distribution along the flow direction starting from the jet exit. There is a pressure recovery (in the space between the jet exit and the involute-foil inlet) in the case without porous material (about 156 Pa). The pressure drop is 1595 Pa for the case with a gap upstream of the porous material and 2099 Pa for the case of no gap upstream of the porous material.

Figure L.6 shows the mass flow rate in each channel of the involute foil (normalized by the maximum flow rate, which occurs in channel (1)—see

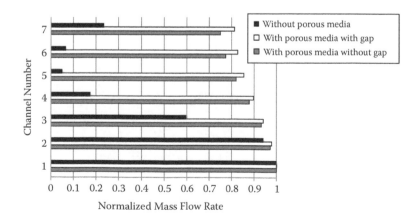

FIGURE L.6
The mass flow rate in each channel normalized by the maximum flow rate, which occurs in channel (1), for the three cases studied.

Figure L.2 for channel identification), for the three cases studied above. The best flow uniformity (which relates directly to reducing the regenerator losses) was obtained for the case with porous media and with a gap. These results combined with the results for the pressure drop shown in Figure L.5 indicate the optimum case (of the three cases examined) is the one with a 133 micron gap and porous media. This case provides the lowest pressure drop and the best velocity uniformity entering the involute foil.

There is a pressure recovery of about 156 Pa in the case without any porous media between the jet exit and involute foil inlet. The pressure drops in the cases with porous media are 1595 Pa with a gap between the jet exit and the porous media and 2099 Pa without a gap.

Nomenclature

English Symbols

A: Amplitude (m)

A_c: Open (fluid flow) area of jet generator (m²), as in Equation (9.14)

A_f: Regenerator mean-flow area (m²)

A_h: Open (fluid flow) area of slot-jet generator (m²), as in Equation (9.21)

A_{int}:   Total interface area between the fluid and solid (m²), as in Equation (7.30)

A_R: Amplitude ratio—that is, fluid displacement during half-cycle divided by tube, or other component, length (dimensionless)

A_r: Open (fluid flow) area of large-scale mock-up (LSMU) plates (m²), as in Equation (9.18)

A_{sf}:   Area of interface between solid and fluid in porous medium (m²), as in Equation (2.33)

A_T: Total, or wetted, flow area (m²), as used in calculating hydraulic diameter

A_t: Tangential area (m²)

A_o, d_h: Fluid displacement (dimensionless), $\left(\dfrac{x_{max}}{\beta d_h}\right)$

A_j: Area (m²)

a: Acceleration (m/s²)

a_{sf}:   Total interface area between the fluid and solid (m²), as in Equation (7.30)

a_{total}:   Total (solid + fluid) cross-sectional area of regenerator/porous medium (m²)

b: Jet width, m

C: Dimensionless coefficient, as in Equation (7.17); Specific heat (J/kg-°K), as in Equations (9.24) and (9.25)

C_1, C_2:   Constant coefficients in $\bar{v}^2 - f$ turbulence model, as in Equation (2.48)

C_D: Drag coefficient (dimensionless), as in Equation 3.27); pressure coefficient (dimensionless), as in Equation (3.34)

C_p: Constant pressure specific heat (J/kg-K)

C_f: Inertial coefficient (dimensionless)

427

C_L: Lift coefficient (dimensionless), as in Equation (3.28); Constant coefficient in $\bar{v}^2 - f$ turbulence model, as in Equation (2.48)

$C_{\varepsilon 1}, C_{\varepsilon 2}$:   Constant coefficients in standard $k - \varepsilon$ turbulence model, as in Equation (2.35)

C_η: Constant coefficient in $\bar{v}^2 - f$ turbulence model, as in Equation (2.48)

C_μ: Constant coefficient in turbulence models, see Equations (2.34), (2.42), and (2.45)

$C_{\omega 1}, C_{\omega 2}$:   Constant coefficients in standard $k - \omega$ turbulence model, as in Equation (2.43)

CA: Crank angle, degrees

c: Specific heat (J/kg-K)

D: Diameter (m)

D_h: Hydraulic diameter (m), as defined by Equation (8.3)

D_p: Piston bore diameter (m)

d_c: Diameter of jet channel (m)

d_h: Hydraulic diameter (m)

d_i: Regenerator fiber diameter (m)

d_e: Regenerator mean effective fiber diameter (m)

d_w: Wire diameter (m); diffusion depth (m)

E: Internal energy per unit mass (kJ/kg)

e: Internal energy per unit mass (kJ/kg)

E_t: Internal energy per unit volume (kJ/m³)

$erfc$:   Complementary error function

F: Fraction of matrix not participating fully in thermal exchange with fluid (dimensionless)

F_D: Drag force (N), as in Equation (3.27)

F_L: Lift force (N), as in Equation (3.28)

Fo: Fourier number (dimensionless)

F_M: Regenerator figure of merit (dimensionless)

f: Friction factor (dimensionless); frequency (radians/s)

f_D: Friction factor (dimensionless), as in Equations (2.59) and (8.28)

f_F: Friction factor for hydrodynamically developing flow (dimensionless), as in Equations (3.12) and (3.22)

f_K: Friction factor (dimensionless), as in Figure 7.8

f_r: Friction factor (dimensionless), as in Equation (8.1)

f_x, f_y, f_z: Friction factors for flows in x, y, and z directions (dimensionless), as in Equations (2.14)

g: Involute-foil gap (m), as in Figures 8.3 and 8.27

H: Size of structural unit (m)

h: Convective heat-transfer coefficient (W/m²-K); feature size (m); enthalpy (kJ/kg)

h_{sf}:   Convective heat-transfer coefficient across solid-fluid interface of porous medium (W/m²-°K), as in Equation (7.42)

J_0:   Zeroth Bessel J function

J_1:   First Bessel J function

K: Thermal conductivity (W/m-K); permeability (m²)

k: Thermal conductivity (W/m-°K)

k_0:   Stagnant thermal conductivity (W/m-°K)

k_{dis}:   Thermal dispersion thermal conductivity (W/m-°K), as in Equation (7.39)

k_e: Overall effective thermal conductivity (W/m-°K)

k_f: Fluid thermal conductivity (W/m-°K)

k_r: Gas thermal conductivity at rejection end of regenerator (W/m-°K), as in Equation (6.17)

k_s: Sand grain roughness (dimensionless); solid thermal conductivity (W/m-°K)

k_y: Cross-stream thermal dispersion (W/m-°K)

k_x: Streamwise thermal dispersion (W/m-°K)

k_{fe}:   Fluid effective thermal conductivity (W/m-°K)

k_{se}:   Solid effective thermal conductivity (W/m-°K)

k_{tor}:   Tortuosity thermal conductivity (W/m-°K)

$k_{eff,\,s+f}$:   Lumped effective thermal conductivity for fluid and solid of porous medium (W/m-°K), as in Equation (2.69)

$k_{f,eff}$:   Fluid effective thermal conductivity (W/m-°K)

$k_{f,stag}$:   Fluid stagnant thermal conductivity (W/m-°K)

$k_{s,eff}$:   Solid effective thermal conductivity (W/m-°K)

L: Length or characteristic length (m)

L_c: Channel flow length through segmented involute foil (m), as in Figure 8.27

L_{hy}:   Hydrodynamic entrance length for developing flow (m), as in Equation (3.11)

$L_{th,H}$:   Entrance length for thermally (or simultaneous hydrodynamically and thermally) developing flow with uniform heat flux (m), as in Equations (3.15) and (3.19)

$L_{th,T}$:   Entrance length for thermally (or simultaneous hydrodynamically and thermally) developing flow with uniform wall temperature (m), as in Equations (3.13) and (3.17)

m: Mass (kg)

N: Number of points; number of mesh layers

N_k: Effective thermal conductivity ratio due to thermal dispersion enhancement, or (molecular conduction + thermal dispersion-eddy conduction)/(molecular conduction) (dimensionless)

N_u: Nusselt number (dimensionless)

N_{ue}:   Effective Nusselt number (dimensionless)

Nu: Nusselt number (dimensionless)

Nu_H: Nusselt number for uniform wall heat flux (dimensionless)

$Nu_{x,H}$:   Nusselt number for thermally (or simultaneous hydrodynamically and thermally) developing flow with uniform heat flux (dimensionless), as in Equations (3.16) and (3.20)

$Nu_{x,T}$:   Nusselt number for thermally (or simultaneous hydrodynamically and thermally) developing flow with uniform wall temperature (dimensionless), as in Equations (3.14) and (3.18)

Nue: Effective Nusselt number (dimensionless)

Nu_m: Mean Nusselt number (dimensionless)

Nux: Local Nusslet number (dimensionless)

N_u: Nusselt number (dimensionless)

n: Vector normal to surface

p: Pressure (Pa)

P_e: Peclet number (dimensionless)

Pe: Peclet number (dimensionless)

Pr: Prandtl number (dimensionless)

P_{wet}:   Wetted perimeter (m)

Q: Heating power (W/m)

Q_{in}:   Heat into engine (W)

$\dot{q}$:   Heat transfer rate (W/m^2)

$\dot{q}_w$:   Heat flux (W/m^2)

$\dot{q}''$:   Heat flux (W/m^2)

$\dot{q}'''$:   Heat flux (W/m^3)

R: Thermal resistance (1/m^2); radial position (m); radius of pipe (m)

R_i: Inside radius (m)

R_O: Outside radius (m)

Re: Reynolds number (dimensionless)

Re_0:   Reynolds number value at which the friction factor departs from the Hagen-Poiseuille theory (dimensionless), as in Equation (8.2)

Re_d: Maximum Reynolds number based upon diameter (dimensionless), $\left(u_{max}d \middle/ v \right)$

Re_H: Maximum Reynolds number based on cell height, H, and inlet velocity (dimensionless)

Re_K: Maximum Reynolds number based upon permeability (dimensionless), $\left(u_{max}\sqrt{K} \middle/ v \right)$

Re_{max}:   Maximum Reynolds number (dimensionless), $\left(u_{max}d_h \middle/ v \right)$

Re_ω: Kinetic Reynolds number (dimensionless), same as Valensi number

Re_{ω,d_h}:   Kinetic Reynolds number based on hydraulic diameter (dimensionless), $\left(\omega d_h^2 \middle/ v \right)$

r: Distance in the radial direction (m)

r_h: Hydraulic radius, or (wetted area)/(wetted perimeter) (dimensionless)

S: Mesh wire spacing (m); spacing between two cooler tubes (m)

St: Strouhal number (dimensionless), $\left(\dfrac{L}{U\tau} \right)$

s:	Basic involute-foil element spacing, gap + wall thickness (m), see Figure 8.27
T:	Temperature (°K)
T_{ACCEPT}:	Temperature of acceptor heat exchanger (°K)
T_{REJECT}:	Temperature of rejector heat exchanger (°K)
T^+:	Dimensionless temperature, as in Equation (2.39)
T_c:	Cold end temperature (°K); temperature at a central point within regenerator (°K), as in Equation (6.17)
T_h:	Hot end temperature (°K)
T_r:	Temperature at rejection end of regenerator (°K), warmer end, as in Equation (6.17)
t:	Time (s)
U:	Darcy velocity (m/s) (Regenerator fluid velocity in absence of porous medium, or regenerator entrance velocity); velocity components in the x-direction (m/s)
U_c:	Average velocity over flow area of round-jet generator (m/s), as in Equation (9.14)
U_h:	Average velocity over flow area of slot-jet generator (m/s), as in Equation (9.21)
U_p:	Piston velocity (m/s), as in Equation (9.14)
U_r:	Flow velocity within large-scale mock-up (LSMU) plates (m/s), as in Equation (9.18)
u:	Velocity (m/s); velocity components in the x-direction (m/s)
u^+:	Dimensionless velocity in standard $k-\varepsilon$ turbulence model, as in Equation (2.37)
u_{max}:	Maximum bulk mean velocity in regenerator (m/s)
$\sqrt{u'^2}$:	rms fluctuation of axial velocity (m/s)
$u'v'$:	Turbulent shear stress (m²/s²)
V:	Elementary representative volume (m³)
V:	Volume (m³)
V_α:	Volume of integration over α-phase—that is, fluid or solid phase (m³); Valensi number (dimensionless)
Va:	Valensi number (dimensionless), same as kinetic Reynolds number, $\left(\omega d^2 \middle/ 4\nu\right)$
v:	Velocity components in the y direction (m/s)
W:	Involute-foil channel width (m), see Figure 8.27; microscopic quantity (any dimension)
w:	Velocity components in the z-direction (m/s)
W_{PV}:	PV, or indicated, power predicted for engine (W)
W_T:	Total wetted perimeter (m), as used in calculating hydraulic diameter
$Wdis$:	Excess displacer drive power (W)
X_c:	Amplitude of particle displacement within jet generator tubes (m)
X_p:	Piston displacement (m)

X_r: Amplitude of particle displacement within regenerator (m)
X_{max}: Maximum particle displacement amplitude (m)
x: Distance in streamwise direction (m)
x':   Dimensionless streamwise location, based on the hydraulic diameter
x_m: Streamwise thermal dispersion multiplier (dimensionless)
x_p: The jet penetration depth (m)
y: Distance in cross-stream direction (m)
y':   Dimensionless cross-stream location, based on the hydraulic diameter
y^+:   Dimensionless location relative to wall in standard $k - \varepsilon$ turbulence model, as in Equation (2.38)
y_T^+:   Dimensionless thermal sublayer thickness, see Equations (2.40) and (2.41) and discussion
y_v^+:   Dimensionless viscous sublayer thickness in standard $k - \varepsilon$ turbulence model, see Equations (2.37) and (2.38) and discussion
y_m: Cross-stream thermal dispersion multiplier (dimensionless)
z: Distance in streamwise direction in polar coordinates (m)

Greek Symbols

α: Thermal diffusivity (m²/s)
β: Porosity—that is, (porous-media void volume)/(total porous media volume (dimensionless)
B_{cte}:   Coefficient of thermal expansion (1/°K)
γ: Diffusivity ratio (dimensionless); ratio of specific heats ($c_p/c_v = 1.67$ for helium)
Δ: Differential, or change in, some quantity, like distance, velocity, and so forth (same dimension as quantity operated on)
ΔP_1:   Amplitude of phasor pressure drop, as in Equation (6.17)
δ: Penetration depth (m); Kronecker delta function (dimensionless); oscillating-fluid amplitude (m), as in Table 6.11; nominal thickness of plenum at end of regenerator (m)
ε: Eddy transport term, absolute roughness height (m); eddy diffusivity, dissipation of turbulence kinetic energy (N/(s-m²))
ε_M: Eddy diffusivity (m²/s), as in Equation (7.2)
ζ: Characteristic constant (dimension depends on equation usage)
η: Similarity variable (dimension depends on equation usage)
Θ: Dimensionless integration variable
θ: Crank position (degree); dimensionless temperature
κ: Integration variable (s); bulk coefficient of viscosity (kg/m-s), as in Equation (2.10)
λ: Coefficient; eigenvalues; eddy diffusivity coefficient, as in Equations (7.4) and (7.6)
$\overline{\lambda}$: Dispersion constant tensor

μ: Dynamic viscosity (kg/m-s)

μ′:   Second coefficient of viscosity (kg/m-s)

μ_r: Gas viscosity at rejection end of regenerator (kg/m-s), as in Equation (6.17)

ν: Kinematic viscosity (m²/s)

$\overline{v^2}$:   Normal stress parameter of $\overline{v^2} - f$ turbulence model, see Equation (2.47) and discussion

ξ: Characteristic constant (dimension depends on equation usage)

Π: Stress tensor (N/m²)

π: Constant ratio of circle's circumference to diameter = 3.1415……

ρ: Density (kg/m³)

σ: Stefan-Boltzman constant (W/m²-°K⁴); laminar Prandtl number (dimensionless), as in Equation (2.40)

σ_k: Constant coefficient in standard $k - \varepsilon$ and $k - \omega$ turbulence models (values different in two models), see Equations (2.34) and (2.42)

σ_t: Turbulent Prandtl number (dimensionless), as in Equation (2.41)

σ_ε: Constant coefficient in standard $k - \varepsilon$ turbulence model, as in Equation (2.35)

σ_ω: Constant coefficient in standard $k - \omega$ turbulence model, as in Equation (2.43)

τ: Shear stress tensor (N/m²); time period of vortex shedding (s); time period of oscillatory flow (s); characteristic period (s)

Φ: Dimensionless temperature; dissipation function, or rate at which mechanical energy is expended in the process of deformation of fluid due to viscosity, W

φ: Porosity, or volume of fluid phase of porous medium divided by total volume of porous medium; dimensionless temperature

ω: Angular frequency, or rotational speed (rad/sec); parameter of $k - \omega$ turbulence model, see Equation (2.43)

Dimensionless Groups

AR: Amplitude ratio

Er: Energy ratio

Fo: Fourier number

Pr: Prandtl number

Pr_t: Turbulent Prandtl number

Re: Reynolds number

Re_ω: Kinetic Reynolds number, same as Valensi number

St: Strouhal number

Va: Valensi number

Subscripts

∞:	Far away spatially, infinite distance
AC:	Alternating current, as in W_{AC} implies AC Watts
air:	Air
ambient:	Ambient, environmental, surroundings (as in ambient temperature)
cr:	Critical
d:	Dispersion
dc:	"Direct current" fluid flow, component of oscillating-flow, as in Equations (6.14) through (6.16)
dis:	Dispersion
eff:	Effective
f:	Fluid phase
g:	Fluid phase
H:	Thermal, heat
h:	Hydraulic
hx:	Heat exchanger
i:	Inner
M:	Momentum
matrix:	Solid phase of porous medium
max:	Maximum
min:	Minimum
n:	Indexing variable
o:	Outer
regen:	Regenerator
s:	Solid phase
stg:	Stagnation
thermal:	Denotes thermal, not momentum
tor:	Tortuosity
w:	Wall; wire

Superscripts

s:	Denotes solid
f:	Denotes fluid
':	Deviation of quantity, as from a mean value

Overbars, or Other Symbols Used Directly over an Algebraic Letter or Symbol

~:   Fluctuation from mean denotes a tensor quantity

¯:   Time average, as in Equation (7.9); spatial average, as in Equation (6.11); or denotes a vector

→:   Denotes a vector

Operators

$\langle\ \rangle$:   Volume-average

$\langle\ \rangle^{f,s}$:   Intrinsic average (volume average over fluid and solid part of porous medium)

•:   Dot product

Δ: Gradient

$\dfrac{\partial}{\partial x}$:   Partial derivative with respect to x direction, or one-dimensional gradient operator

Imag. ( ):   Imaginary part of complex number

Real ( ):   Real part of complex number

Abbreviations/Acronyms

1-D, 2-D, 3-D: One-dimensional, two-dimensional, three-dimensional

4-215:   Ford Phillips automotive engine

4L23:   General Motors engine

AE: Available energy

AMS-02:   Alpha magnetic spectrometer 2 (Instrument for NASA space mission)

ASC: Advanced Stirling convertor (engine and linear alternator)

CAD: Computer-aided design

CAMD: Center for Advanced Microstructures and Devices

CFD: Computational fluid dynamics

CFD-ACE: A particular commercial computational fluid dynamics (CFD) code

COP:	Coefficient of performance, heating or cooling (dimensionless figure of merit for heat pumps, refrigerators, etc.)
CSU:	Cleveland State University
CTPC:	Component test power convertor (a particular free-piston Stirling engine/linear alternator)
DI:	De-ionized
DOE:	Department of Energy
EDM:	Electric discharge machining
EFAB:	Electrochemical FABrication
EM:	Mechanical Technologies, Inc. Engineering Model
FEA:	Finite element analysis
FTB:	Frequency test bed (convertor, or engine and linear alternator)
Genset:	Sunpower Inc. 3 Kw Generator
GPU-3:	Ground Power Unit 3 (a particular rhombic-drive Stirling engine generator, developed by General Motors for the U.S. Army)
GRC:	NASA's Glenn Research Center
ID:	Inner diameter
IPA:	Isopropyl alcohol
LiGA:	Lithographie, Galvanoformung and Abformung (the German words for lithography, electroplating, and molding x-ray lithography is used here)
LSMU:	Large-scale mock-up (of involute-foils)
M1, M2, etc.:	Different types of mesh sheet regenerator elements (developed by Prof. Norboru Kagawa and students and staff of the National Defense Academy of Japan)
M87:	A particular Stirling cryocooler (developed by Sunpower Inc. of Athens, Ohio)
MEMS:	Microelectromechanical systems
MOD I:	Mechanical Technologies automotive Stirling
MOD II:	A modification of an earlier four-cylinder double-acting kinematic, or crank drive, automotive Stirling engine
MTI:	Mechanical Technology, Inc. (company that developed the MOD II, SPDE, SPRE, and CTPC for the U.S. Department of Energy and the National Aeronautics and Space Administration)
NASA:	National Aeronautics and Space Administration
NRA:	NASA Research Award
NS03T:	A particular Japanese kinematic (crank drive) Stirling engine
N-S:	Navier-Stokes equations
OD:	Outer diameter
P40:	United Stirling AB Engine

PMMA:	PolyMethyl MethAcrylate (a clear plastic, also marketed as Acrylic, Plexiglas, Lucite, and so forth. Used as a photoresist in LiGA process for microfabrication of involute-foils)
PR:	Photo resist, or photoresist, as used in LiGA procedure
RE1000:	Sunpower Inc. Early Free-Piston Engine
RTV:	Room-temperature vulcanizing (type of sealant)
SEM:	Scanning electron microscope
SERENUM05:	A particular Japanese kinematic, or crank drive, Stirling engine
SES:	Stirling Engine Systems, Inc.
SPDE:	Space power demonstrator engine (a particular free-piston Stirling engine/linear alternator)
SPDE-D:	SPDE trial parameters
SPDE-O:	SPDE intended operation
SPDE-T:	SPDE trial parameters
SPRE:	Space power research engine (a particular free-piston Stirling engine/linear alternator)
STES:	Technical University of Denmark engine
SU-8:	SU-8 is a negative, epoxy-type, near-ultraviolet photoresist used in microelectromechanical (MEMS) applications
TCR:	Thermal contact resistance
UMN:	University of Minnesota
UV:	Ultraviolet
XLRM2:	Name of one of the bending magnet beamlines at CAMD

References

Adolfson, D. (2003). Oscillatory and Unidirectional Fluid Mechanics Investigations in a Simulation of a Stirling Engine Expansion Space, M.S. Thesis, Mechanical Engineering Department, University of Minnesota.

Atwood, C. L., Griffith, M. L., Schlienger, M. E., Harwell, L. D., Ensz, M. T., Keicher, D. M., Schlienger, M. E., Romera, J. A., and Smugeresky, J. E. (1998). Laser Engineered Net Shaping (LENS): A Tool for Direct Fabrication of Metal Parts, *Proceedings of ICALEO '98*, Orlando, FL, November 16–19.

Ayyaswamy, P. S. (2004). University of Pennsylvania, Private communication.

Backhaus, S. N., and Swift, G. W. (2001). Fabrication and Use of Parallel-Plate Regenerators in Thermoacoustic Engines, *Proceedings of the 36th Intersociety Energy Conversion Engineering Conference*, Savannah, GA, 29 July–2 August.

Backhaus, S., and Swift, G. W. (2000). A Thermoacoustic-Stirling Heat Engine Detailed Study, *J. Acoust. Soc. Am.*, 107(6): 3148–3166.

Baumann, J., and Rawal, S. (2001). Viability of Loop Heat Pipes for Space Solar Power Applications, American Institute of Aeronautics and Astronautics Paper no. 2001–3078.

Beavers, G. S., and Sparrow, E. M. (1969). Non-Darcy Flow through Fibrous Porous Media, *J. Appl. Mech.*, 36(4): 711–714.

Berchowitz, D. M. (1993). *Miniature Stirling Coolers*, Sunpower Inc. Web site: www. sunpower.com/lib/site files/pdf/publications/Doc0049.pdf.

Bienert, W. B., Krotiuk, W. J., and Nikitkin, M. N. (1999). Thermal Control with Low Power Miniature Loop Heat Pipes, *Proceedings of the 29th International Conference on Environmental Systems*, Denver, CO, July 12–15, SAE Paper No. 1999-01-2008.

Bird, R. B., Stewart, W. E., and Lightfoot, E. N. (1960). *Transport Phenomena*, Wiley, New York.

Boomsma, K., and Poulikakos, D. (2001). On the Effective Thermal Conductivity of a Three-Dimensionally Structured Fluid-Saturated Metal Foam, *Int. J. Heat Mass Transfer*, 44: 827–836.

Bouma, P., Eggels, R., Goey, L., Nieuwenhuizen, J., and Van Der Drift A. (1995). A Numerical and Experimental Study of the NO-Emission of Ceramic Foam Surface Burners, *Combust. Sci. Tech.*, 108: 193–203.

Bowman, L. (1993). A Technical Introduction to Free-Piston Stirling Cycle Machines: Engines, Coolers, and Heat Pumps, Sunpower Inc. Web site: www.sunpower. com/lib/sitefiles/pdf/publications/Doc0050.pdf.

Bowman, R. (2003). Oxidation of a Type 316L Stainless Steel Stirling Convertor Regenerator, NASA, Glenn Research Center, Cleveland, OH, NASA-TM-2003-212117.

Bucci, A., Celata, G. P., Cumo, M., Serra, E., and Zummo, G. (2003). Water Single-Phase Fluid Flow and Heat Transfer in Capillary Tubes, *Therm. Sci. Eng.*, 11(6): 81–89.

Burmeister, L. (1993). *Convective Heat Transfer*, John Wiley and Sons, New York.

Cairelli, J. (2002). Information Regarding Solid Material in the Regenerators of Some Free-Piston Stirling engines, Private communication.

439

Cairelli, J. E., Thieme, L. G., and Walter, R. J. (1978). Initial Test Results with a Single-Cylinder Rhombic-Drive Stirling Engine, DOE/NASA/1040-78/1, NASA TM-78919.

Cao, X. L., Cheng, P., and Zhao, T. S. (2002). Experimental Study of Evaporative Heat Transfer in Sintered Copper Bidispersed Wick Structures, *J. Thermophys. Heat Transfer*, 16(4): 547–552.

CFD-ACE User Manual. (1999). CFD Research Corporation, 215 Wynn Drive, Huntsville, AL 35805.

Chan, Jack, Wood, J. Gary, and Schreiber, Jeffrey G. (2007). Development of Advanced Stirling Radioisotope Generator (ASRG) for Space Applications, NASA/TM 2007-214806.

Chapman, D. M., Gedeon, D., and Wood, J. G. (1998). Oscillating-Flow Heat-Transfer and Pressure-Drop Test Rig, NASA Tech Brief.

Chen, R.Y. (1973). Flow in the Entrance Region at Low Reynolds Number, *J. Fluids Eng.*, 95: 153–158.

Churchill, S. W., and Ozoe, H. (1973a). Correlations for Laminar Forced Convection with Uniform Heating in Flow over a Plate and in Developing and Fully Developed Flow in a Tube, *J. Heat Transfer*, 95: 78–84.

Churchill, S. W., and Ozoe, H. (1973b). Correlations for Laminar Forced Convection in Flow over an Isothermal Flat Plate and in Developing and Fully Developed Flow in an Isothermal Tube, *J. Heat Transfer*, 95: 416–419.

Cohen, A., Zhang, G., Tseng, F., Mansfield, F., Frodis, U., and Will, P. (1999). EFAB: Rapid Low-Cost Desktop Micromachining of High Aspect Ratio True 3-D MEMS, *Proceedings of IEEE MicroElectroMechanical Systems Workshop*, January 17–21.

Cotter, T. M. (1984). Principles and Prospects of Micro Heat Pipes, *Proceedings of the 5th International Heat Pipes Conference*, Tsukuba, Japan, pp. 328–335.

Danila, D. (2006). CFD Investigation of Fluid Flow and Heat Transfer in an Involute Geometry for Stirling Engine Applications, M.S. Thesis, Cleveland State University, OH, December.

Deckard, C., and Beaman, J. J. (1988). Process and Control Issues in Selective Laser Sintering, *Sensors and Controls for Manufacturing*, 33: 191–197.

Dhar, M. (1999). Stirling Space Engine Program, Volume 1—Final Report, NASA/ CR— 1999- 209164/VOL1.

Dunn, P. D., and Reay, D. A. (1994). *Heat Pipes*, Pergamon, London.

Durbin, P. A. (1995). Separated Flow Computations with the $k - \varepsilon - v^2$ Model, *AIAA J.*, 33(4): 659–664.

Dyson, R. W., Wilson, S. D., and Demko, R. (2005a). On the Need for Multidimensional Stirling Simulations, *Proceedings of 3rd International Energy Conversion Engineering Conference*, Paper no. AIAA-2005-5557, San Francisco, CA.

Dyson, R. W., Wilson, S. D., and Demko, R. (2005b). Fast Whole-Engine Stirling Analysis, *Proceedings of 3rd International Energy Conversion Engineering Conference*, Paper no. AIAA-2005-5558, San Francisco, CA.

Faghri, A. (1995). *Heat Pipe Science and Technology*, Taylor and Francis, London.

Ferrenberg, A. J. (1994). Low Heat Rejection Regenerated Engines a Superior Alternative to Turbocompounding, SAE Trans., SAE Paper no. 940946, DOI: 10.4271/940946.

Finegold, J. G., and Sterrett, R. H. (1978). Stirling Engine Regenerators Literature Review, Jet Propulsion Laboratory Report no. 5030-230, Pasadena, California Institute of Technology.

FLUENT, Inc. (2005). Fluent 6.3—User Guide.

Franke, R., Rodi, W., and Schönung, B. (1995). Numerical Calculation of Laminar Vortex-Shedding Flow Past Cylinders, *J. Wind Engng. Ind. Aerodyn.*, 35: 237–257.

Fraser, P. R. (2008). Stirling Dish System Performance Prediction Model, Master's Thesis, University of Wisconsin.

Fried, E., and Idelchik, I. E. (1989). *Flow Resistance: A Design Guide for Engineers*, Hemisphere, Washington, DC.

Friedmann, M., Gillis, J., and Liron, N. (1968). Laminar Flow in a Pipe at Low and Moderate Reynolds Number. *Appl. Sci. Res.*, 19: 426–438.

Furutani, S., Matsuguchi, A., and Kagawa, N. (2006). Design and Development of New Matrix with Square-Arranged Hole for Stirling Engine Regenerator, Paper AIAA 2006-4017, *Proceedings of 4th International Energy Conversion Engineering Conference*, San Diego, CA, 26–29 June 2006.

Gedeon, D. (1986). Mean-Parameter Modeling of Oscillatory Flow, *ASME J. Heat Transfer*, 108: 513–518.

Gedeon, D. (1992). Advection-Driven vs Compression-Driven Heat Transfer, NASA Report under PO C-23433-R, Task 2.

Gedeon, D. (1997). DC Gas Flows in Stirling and Pulse-Tube Cryocoolers, In R. G. Ross, Editor, *Cryocoolers 9*, Pages 385–392. Plenum, New York.

Gedeon, D. (1999). Sage Stirling-Cycle Model-Class Reference Guide, 3rd edition, Gedeon Associates, Athens, OH.

Gedeon, D. (2002). Recipes for Calculating Viscous and Thermal Eddy Transports, Internal memorandum, April 4.

Gedeon, D. (2003a). Regenerator Figures of Merit, (CSUMicrofabFiguresofMerit.tex), Unpublished memorandum to Microfabrication Team, August 6.

Gedeon, D. (2003b). Digression on Regenerator Figure of Merit Calculations, (CSUMicrofabFMeritConsistency.tex), Unpublished memorandum to Microfabrication Team, December 12.

Gedeon, D. (2004a). Intra-regenerator Flow Streaming Produced by Nonuniform Flow Channels, Unpublished memorandum to Microfabrication Team (CSUmicrofabIntraRegenFlows.doc), January 5.

Gedeon, D. (2004b). Intra-Regenerator Flow Streaming Theory, Unpublished memorandum (Csuintraregenstreamingtheory.Pdf), also See Appendix D of this Report, Gedeon Associates, Athens, OH.

Gedeon, D. (2005). Flow Circulations in Foil-Type Regenerators Produced by Non-Uniform Layer Spacing, *Cryocoolers 13*, by Springer U.S.

Gedeon, D. (2006a). NASA Random Fiber Master Correlations, 8-19-06 Unpublished memorandum (NASARandomFiberMasterCorrelations.doc).

Gedeon, D. (2006b). Two Unpublished Memoranda on NASA Random Fiber Correlations, 10-20-06 memorandum NASARandomFiberMysteryLengthDep. doc and 11-1-06 memorandum NASARandomFiberReasonsLengthDep.doc.

Gedeon, D. (2007). Private communication.

Gedeon, D. (2009). Sage Stirling-Cycle Model-Class Reference Guide, 5th edition, Gedeon Associates, Athens, OH.

Gedeon, D. (2010). Sage User's Guide, Electronic Edition for Acrobat Reader, Sage v7 Edition, available on Web site: http://sageofathens.com/Documents/SageStlxHyperlinked.pdf, Gedeon Associates, Athens, OH.

Gedeon, D., and Wood, J. G. (1992). Oscillating-Flow Regenerator Test Rig: Woven Screen and Metal Felt Results, Ohio University Center for Stirling Technology Research, Status Report for NASA Lewis contract NAG3-1269.

Gedeon, D., and Wood, J. G. (1996). Oscillating-Flow Regenerator Test Rig: Hardware and Theory with Derived Correlations for Screens and Felts, NASA Contractor report 198442, February 1996.

Hanamura, K., Bohda, K., and Miyairi, Y. (1997). A Study of Super-adiabatic Combustion Engine *Energy Conversion and Management*, 38(10–13, July–September): 1259–1266, International Symposium on Advance Energy Conversion Systems and Related Technologies.

Hargreaves, C. M. (1991). *The Phillips Stirling Engine*, Elsevier Science, New York.

Hinze, J. O. (1975). *Turbulence*, 2nd ed., McGraw-Hill, New York.

Hoang, T. T., O'Connell, T. A., Ku, J., Butler, C. D., and Swanson, T. D. (2003). Miniature Loop Heat Pipes for Electronic Cooling, ASME Paper no. 35245, 2003.

Hornbeck, R.W. (1965). An All-Numerical Method for Heat Transfer in the Inlet of a Tube. *Am. Soc. Mech. Eng.*, Paper no. 65-WA/HT-36.

Hsu, C. T. (1999). A Closure Model for Transient Heat Conduction in Porous Media, *J. Heat Transfer*, 121: 733–739.

Hsu, C. T., and Cheng, P. (1990). Thermal Dispersion in a Porous Medium, *Int. J. Heat Mass Transfer*, 33(8): 1587–1597.

Hsu, P., Evans, W., and Howell, J. (1993). Experimental and Numerical Study of Premixed Combustion within Nonhomogeneous Porous Ceramics, *Combust. Sci. Tech.*, 90: 149–172.

Hsu, P., and Matthews, R. (1993). The Necessity of Using Detailed Kinetics in Models for Premixed Combustion within Porous Media, *Combust. Flame* 93: 157–166.

Hunt, M. L., and Tien, C. L. (1988). Effects of Thermal Dispersion on Forced Convection in Fibrous Media, *Int. J. Heat Mass Transfer*, 31: 301–308.

Ibrahim, M. B., Tew, R. C., and Dudenhoefer, J. E. (1989). Two-Dimensional Numerical Simulation of a Stirling Engine Heat Exchanger, *Proceedings of the 24th Intersociety Energy Conversion Engineering Conference*, The Institute of Electrical and Electronics Engineers, 6: 2795–2802, Washington, DC, August.

Ibrahim, M. B., Bauer, C., Simon, T., and Qiu, S. (1994). Modeling of Oscillatory Laminar, Transitional and Turbulent Channel Flows and Heat Transfer, *10th International Heat Transfer Conference*, Vol. 4, pp. 247–252, Brighton, England, August 14–18.

Ibrahim, M. B., Zhang, Z., Wei, R., Simon, T. W., and Gedeon, D. (2002). A 2-D CFD Model of Oscillatory Flow with Jets Impinging on a Random Wire Regenerator Matrix, *Proceedings of the 37th Intersociety Energy Conversion Engineering Conference*, The Institute of Electrical and Electronics Engineers, Paper no. 20144, Washington, DC.

Ibrahim, M. B., Rong, W., Simon, T. W., Tew, R., and Gedeon, D. (2003). Microscopic Modeling of Unsteady Convective Heat Transfer in a Stirling Regenerator Matrix, *Proceedings of the 1st International Energy Conversion Engineering Conference*, Portsmouth, VA, August 17–21.

Ibrahim, M., Simon, T., Gedeon, D., and Tew, R. (2004a). Improving the Performance of the Stirling Convertor: Redesign of the Regenerator with Experiments, Computation and Modern Fabrication Techniques, Final Report to the U.S. Department of Energy, no. DE-FC36-00GO10627.

Ibrahim, M. B., Veluri, S., Simon, T., and Gedeon, D. (2004b). CFD Modeling of Surface Roughness in Laminar Flow, Paper no. AIAA-2004-5585, *Proceedings of the 2nd International Energy Conversion Engineering Conference*, Providence, RI, August 16–19.

Ibrahim, M. B., Rong, W., Simon, T., Tew, R., and Gedeon, D. (2004c). Simulations of Flow and Heat Transfer inside Regenerators Made of Stacked Welded Screens Using Periodic Cell Structures, *Proceedings of 2nd International Energy Conversion Engineering Conference*, Paper no. AIAA-2004-5599, Providence, RI.

Ibrahim, M., Simon, T., Mantell, S., Gedeon, D., Qiu, S., Wood, G., and Guidry, D. (2004d). Developing the Next Generation Stirling Engine Regenerator: Designing for Application of Microfabrication Techniques and for Enhanced Reliability and Performance in Space Applications, Phase I Final Report on work done under Radioisotope Power Conversion Technology NRA Contract NAS3-03124, Prepared for NASA Glenn Research Center (published September 2004).

Ibrahim, M., Mittal, M., Jiang, N., and Simon, T. (2005). Validation of Multi-Dimensional Stirling Engine Codes: Modeling of the Heater Head, *Proceedings of 3rd International Energy Conversion Engineering Conference*, Paper no. AIAA-2005-5654, San Francisco, CA.

Ibrahim, M. B., Danila, D., Simon, T., Mantell, S., Sun, L., Gedeon, D., Qiu, S., Wood, J. G., Kelly, K., and McLean, J. (2007). A Microfabricated Segmented-Involute-Foil Regenerator for Enhancing Reliability and Performance of Stirling Engines: Phase II Final Report for the Radioisotope Power Conversion Technology NRA Contract NAS3- 03124, NASA Contractor Report, NASA/CR-2007-215006.

Ibrahim, M. B., Gedeon, D., Wood, G., and McLean, J. (2009a). A Microfabrication of a Segmented-Involute-Foil Regenerator for Enhancing Reliability and Performance of Stirling Engines, Phase III Final Report for the Radioisotope Power Conversion Technology NRA Contract NAS3-03124, NASA Contractor Report, NASA/CR-2009- 215516.

Ibrahim, M., Gedeon, D., and Wood, G. (2009b). Actual-Scale Regenerator Experiments for Improved Design and Manufacturing Practices, Final Report for NASA Grant NNX07AR53G, March 2009.

Idelchick, I. E. (1986). *Handbook of Hydraulic Resistance*, 2nd ed. Hemisphere, New York.

Kagawa, N. (2002). Experimental Study of 3-kW Stirling Engine, *Propulsion and Power*, 18(2): 696–702.

Kagawa, N., Matsuguchi, A., Furutani, S., Takeuchi, T., Araoka, K., and Kurita, T. (2007). Development of a Compact 3-kW Stirling Engine Generator, *Proceedings of the 13th International Stirling Conference*, Tokyo, September 24–26.

Kagawa, N., Sakamoto, M., Nagatomo, S., Komakine, T., Hisoka, S., Sakuma, T., Aral, Y., and Okuda, M. (1988). Development of a 3 kW Stirling Engine for a Residential Heat Pump System, *Proceedings of 4th International Conference on Stirling Engines*, Japan Society of Mechanical Engineers, Tokyo, pp. 1–6.

Kalpakjian, S., and Schmidt, S. R. (2003). *Manufacturing Processes for Engineering Materials*, Pearson Education, Upper Saddle River, NJ.

Kaviany, M. (1995). *Principles of Heat Transfer in Porous Media*, 2nd ed., Mechanical Engineering Series, Springer-Verlag, New York.

Kays, W., and London, A. (1964). *Compact Heat Exchangers*, 2nd ed., McGraw-Hill, New York.

Kays, W. M., and London, A. L. (1984). *Compact Heat Exchanger*, 3rd ed., McGraw-Hill, New York.

Khrustalev, D., and Semenov, S. (2003). Advances in Low Temperature Cryogenic and Miniature Loop Heat Pipes, *Proceedings of the 14th Spacecraft Thermal Control Workshop*, El Segundo, CA.

Kiseev, V. M., Nepomnyashy, A. S., Gruzdova, N. L., and Kim, K. S. (2003). Miniature Loop Heat Pipes for CPU Cooling, *Proceedings of the 7th International Heat Pipe Symposium*, Jeju, Korea.

Kitahama, D., Takizawa, H., Kagawa, N., Matsuguchi, A., and Tsuruno, S. (2003). Performance of New Mesh Sheet for Stirling Engine Regenerator, Paper AIAA 2003-6015, *Proceedings of 1st International Energy Conversion Engineering Conference*, Portsmouth, Virginia, 17–21 August.

Knisely, C. W. (1990). Strouhal Numbers of Rectangular Cylinders at Incidence: A Review and New Data, *J. Fluids Struct.*, 4: 371–393.

Ko, K., and Anand, N. K. (2003). Use of Porous Baffles to Enhance Heat Transfer in a Rectangular Channel, *Int. J. Heat Mass Trans.*, 46(22): 4191–4199.

Konev, S. V., Polasek, F., and Horvat, L. (1987). Investigation of Boiling in Capillary Structures, *Heat Transfer-Sov. Res.*, 19(1): 14–17.

Ku, J. (1999). Operating Characteristics of Loop Heat Pipes, *Proceedings of the 29th International Conference on Environmental Systems*, Denver, CO, July 12–15, SAE Paper no. 1999-01-2007.

Kurzweg, U. H. (1985). Enhanced Heat Conduction in Oscillating Viscous Flows within Parallel-Plate Channels, *J. Fluid Mech.*, 156: 291–300.

Kuwahara, F., Shirota, M., and Nakayama, A. (2001). A Numerical Study of Interfacial Convective Heat Transfer Coefficient in Two-Energy Equation Model for Convection in Porous Media, *Int. J. Heat and Mass Trans.*, 44(6): 1153–1159.

Lage, J. L., Delemos, M. J. S., and Nield, D. A. (2002). Modeling Turbulence in Porous Media, Chapter 8 in *Transport Phenomena in Porous Media II*, edited by Ingham, D. B. and Pop, I., Pergamon, Oxford.

Lange, C. F., Durst, F., and Breuer, M. (1998). Momentum and Heat Transfer from Cylinders in Laminar Cross-Flow at $10^{-4} \leq Re \leq 200$, *Int. J. Heat Mass Trans.*, 41: 3409–3430.

Launder, B. E. (1986). Low Reynolds Number Turbulence Near Walls, UMIST Mechanical Engineering Department, Rept. TFD/86/4, University of Manchester, England, UK.

Launder, B. E., and Spalding, D. B. (1974). The Numerical Computation of Turbulent Flows, *Comp. Methods for Appl. Mech. Eng.*, 3: 269–289.

Luikov, A. V., and Vasiliev, L. L. (1970). Heat and Mass Transfer at Low Temperature (in Russian), Minsk, pp. 5–23.

Malico, I., Zhou, X., and Pereira, J. (2000). Two-Dimensional Numerical Study of Combustion and Pollutants Formation in Porous Burners, *Combust. Sci. Tech.*, 152: 57–79.

Masuoka, T., and Takatsu, Y. (1996). Turbulence Model for Flow through Porous Media, *Int. J. Heat Mass Trans.*, 39(13): 2803–2809.

Masuoka, T., Takatsu, Y., and Inoue, T. (2002). Chaotic Behavior and Transition to Turbulence in Porous Media, *Microscale Thermophysical Engineering*, 6(4), October/November/December.

Matsuguchi, A., Kagawa, N., and Koyama, S. (2009). Improvement of a Compact 3-kW Stirling Engine with Mesh Sheet, *Proceedings of the International Stirling Engine Conference (ISEC)*, Groningen, The Netherlands, November 16–17.

Matsuguchi, A., Aizu, Y., Nishishita, Y., and Kagawa, N. (2008). Performance Analysis of New Matrix Material for the Stirling Engine Regenerator (in Japanese, except for abstract, figure, and table captions), *Proceedings of the Japan Society of Mechanical Stirling*, Vol. 11, 2008.

Matsuguchi, A., Moronaga, T., Furutani, S., Takeuchi, T., and Kagawa, N. (2005). Design and Development of New Matrix for Stirling Engine Regenerator, Paper AIAA 2005-5595, *Proceedings of 3rd International Energy Conversion Engineering Conference*, San Francisco, CA, 15–18 August.

Maydanik, Y. F., and Fershtater, Y. G. (1997). Theoretical Basis and Classification of Loop Heat Pipes and Capillary Pumped Loops, *Proceedings of the 10th International Heat Pipe Conference*, Stuttgart, Germany, September 21–25.

Maydanik, Y. F., Fershtater, Y. G., and Pastukhov, V. G. (1992). Development and Investigation of Two Phase Loops with High Pressure Capillary Pumps for Space Applications, *Proceedings of the 8th International Heat Pipe Conference*, Beijing, China, September 14–18.

Maydanik, Y. F., Vershinin, S. V., Korukov, M. A., and Ochterbeck, J. M. (2005). Miniature Loop Heat Pipes—A Promising Means for Cooling Electronics, *IEEE Trans. Compon. Packag. Technol.*, 28(2): 290–296.

McFadden, G. (2005). Forced Thermal Dispersion within a Representative Stirling Engine Regenerator, M.S. Thesis, Mechanical Engineering Department, University of Minnesota.

Metzger, T., Didierjean, S., and Maillet, D. (2004). Optimal Experimental Estimation of Thermal Dispersion Coefficients in Porous Media, *Int. J. Heat Mass Trans.*, 47(14–16): 3341–3353.

Mitchell, M. P., and Fabris, D. (2003). Improved Flow Patterns in Etched Foil Regenerators, *Cryocoolers 12*, Kluwer Academic/Plenum, New York, pp. 499–505.

Mitchell, M., Gedeon, D., Wood, G., and Ibrahim, M. (2005). Testing Program for Etched Foil Regenerator, Paper AIAA 2005-5515, *Proceedings of 3rd International Energy Conversion Engineering Conference*, San Francisco, CA, August 15–18.

Mitchell, M. P., Gedeon, D., Wood, G., and Ibrahim, M. (2007). Results of Tests of Etched Foil Regenerator Material. *Cryocoolers 14*, Springer U.S. *Proceedings of International Cryocooler Conference*, Boulder, CO.

Miyabe, H., Takahashi, S., and Hamaguchi, K. (1982). An Approach to the Design of Stirling Engine Regenerator Matrix Using Packs of Wire Gauzes, *Proceedings of the 17th Intersociety Energy Conversion Engineering Conference* (IECEC Paper 829306), pp. 1839–1844, Piscataway, NJ, Institute of Electrical and Electronics Engineers.

Moraga, N. O., Rosas, C. E., Bubnovich, V. I., and Tobar, J. R. (2009). Unsteady Fluid Mechanics and Heat Transfer Study in a Double-Tube Air–Combustor Heat Exchanger with Porous Medium, *Int. J. Heat Mass Trans.*, 52(13–14): 3353–3363.

Mudaliar, A. V. (2003). 2-D Unsteady Computational Analysis of Cylinders in Cross-Flow and Heat Transfer, M.S. Thesis, Cleveland State University.

Munson, B., Young, D., and Okiishi, T. (1994). *Fundamentals of Fluid Mechanics*, 2nd ed., John Wiley and Sons, New York.

Nakayama, A. and Kuwahara, F., (2000). Numerical Modeling of Convective Heat Transfer in Porous Media Using Microscopic Structures, *Handbook of Porous Media*, edited by Vafai, K., Marcel Dekker, New York, pp. 441–488.

Nightingale, N. P. (1986). Automotive Stirling Engine, Mod II Design Report, DOE/NASA/0032- 28, NASA CR-175106, MTI86ASE58SRI.

Nikuradse, J. (1933). Strömungsgesetze in rauhen Rohren, VDI-Forschungsch (no. 361).

Niu, Y., Simon, T. W., Ibrahim, M., and Gedeon, D. (2002). Oscillatory Flow and Thermal Field Measurements at the Interface between a Heat Exchanger and a Regenerator of a Stirling Engine, *Proceedings of the 37th Intersociety Energy Conversion Engineering Conference*, Paper no. 20141.

Niu, Y., Simon, T. W., Ibrahim, M., Tew, R., and Gedeon, D. (2003a). Thermal Dispersion of Discrete Jets upon Entrance of a Stirling Engine Regenerator under Oscillatory Flow Conditions, *6th ASME/JSME Thermal Engineering Joint Conference*, Paper no. TED- AJ03-641, Hawaii.

Niu, Y., Simon, T. W., Ibrahim, M., Tew, R., and Gedeon, D. (2003b). Measurements of Unsteady Convective Heat Transfer Rates within a Stirling Regenerator Matrix Subjected to Oscillatory Flow, *Proceedings of the 1st International Energy Conversion Engineering Conference*, Paper no. AIAA-2003-6013.

Niu, Y., Simon, T. W., Ibrahim, M., Tew, R., and Gedeon, D. (2003c). Jet Penetration into a Stirling Engine Regenerator Matrix with Various Regenerator-to-Cooler Spacings, *Proceedings of the 1st International Energy Conversion Engineering Conference*, Paper no. AIAA-2003-6014.

Niu, Y., Simon, T., Gedeon, D., and Ibrahim, M. (2004). On Experimental Evaluation of Eddy Transport and Thermal Dispersion in Stirling Regenerators, *Proceedings of the 2nd International Energy Conversion Engineering Conference*, Paper no. AIAA-2004-5646, Providence, RI.

Niu, Y., McFadden, G., Simon, T., Ibrahim, M., and Wei, R. (2005a). Measurements and Computation of Thermal Dispersion in a Porous Medium, *Proceedings of the 3rd International Energy Conversion Engineering Conference*, Paper no. AIAA-2005-5578, San Franciso, CA.

Niu, Y., McFadden, G., Simon, T., Ibrahim, M., and Rong, W. (2005b). Measurements and Computation of Thermal Dispersion in a Porous Medium, *Proceedings of the 3rd International Energy Conversion Engineering Conference*, AIAA-2005-37923, San Francisco, CA, August 15–18.

Niu, Y., Simon, T., Gedeon, D., and Ibrahim, M. (2006). Direct Measurements of Eddy Transport and Thermal Dispersion in High-Porosity Matrix, *AIAA J. of Thermophysics Heat Trans.*, 20(1), January–March.

Norberg, C. (1993). Flow Around Rectangular Cylinders: Pressure Forces and Wake Frequencies, *J. Wind Engng. Ind. Aerodyn.*, 49: 187–196.

North, M. T., Rosenfeld, J. H., and Shaubach, R. M. (1995). Liquid Film Evaporation from Bidisperse Capillary Wicks in Heat Pipe Evaporators, *Proceedings of the 9th International Heat Pipe Conference*, Albuquerque, NM, May 1–5.

Okajima, A., Nagahisa, A., and Rokugoh, A. (1990). A Numerical Analysis of Flow around Rectangular Cylinders, *JSME Int. J. Ser. II*, 33: 702–711.

Oliveira, A. A. M., and Kaviany, M. (2001). Nonequilibrium in the Transport of Heat and Reactants in Combustion in Porous Media, *Progress in Energy and Combustion Sci.*, 27(5): 523–545.

Organ, A. J. (1997). *The Regenerator and the Stirling Engine*, Wiley, New York.

Organ, A. J. (2000). Two Centuries of Thermal Regenerator, Proceedings of the Institute of Mechanical Engineering, Part C, *J. Mech. Eng. Sci.*, 214(NoC1): 269–288.

Ould-Amer, Y., Chikh, S., Bouhadef, K., and Lauriat, G. (1998). Forced Convection Cooling Enhancement by Use of Porous Materials, *Int. J. Heat Fluid Flow*, 19(3): 251–258.

Pastukhov, V. G., Maydanik, Y. F., and Chernyshova, M. A. (1999). Development and Investigation of Miniature Loop Heat Pipes, *Proceedings of the 29th International Conference on Environmental Systems*, Denver, CO, Paper no. 1999-01-1983.

Paul, B. K., and Terhaar, T. (2000). Comparison of Two Passive Microvalve Designs for Microlamination Architectures, *J. Micromechanics Microeng.*, 10: 15–20.

Pavel, B. I., and Mohamad, A. A. (2004). An Experimental and Numerical Study on Heat Transfer Enhancement for Heat Exchangers Fitted with Porous Media, *Int. J. Heat Mass Trans.*, 47: 4939–4952.

Peacock, J. A., and Stairmand, J. W. (1983). Film Gauge Calibration in Oscillatory Pipe Flow, *J. Phys. E:Sci. Instrum.*, 16: 571–576.

Qiu, S., and Augenblick, J. (2005). Thermal and Structural Analysis of Micro-Fabricated Involute Regenerator, The Space Technology and Applications International Forum, STAIF2005, Albuquerque, NM.

Reutskii, V. G., and Vasiliev, L. L. (1981). Dokl. Akad. Nauk BSSR, 24(11): 1033–1036.

Ritsumei. (2004). www.ritsumei.ac.jp/se/~sugiyama/research/re_5.2e.html. Ritsumeikan University.

Rong, W. (2005). Cleveland State University, Private communication.

Ruhlich, I., and Quack, H. (1999). Investigations on Regenerative Heat Exchangers, *Cryocoolers 10*, edited by R. G. Ross, Jr., Kluwer Academic/Plenum, New York, pp. 265–274.

Sachs, E., Cima, M., Williams, P., Brancazio, D., and Cornie, J. (1992). Three Dimensional Printing: Rapid Tooling and Prototypes Directly from a CAD Model, *Transactions of the ASME*, 114: 481–488.

Samoilenko, L. A., and Preger, E. A. (1966). Investigation of Hydraulic Resistance of Pipelines in the Transient Mode of Flow of Liquids and Gases, *Issled, Vodosnabzhen, Kanalizatsii (Trudy LISI)*: Leningrad. pp. 27–39.

Sandia. (2004). www.sandia.gov/mst/technologies/Meso-Machining.html.

Sathe, S., Peck, R., and Tong, T. (1990). Flame Stabilization and Multimode Heat Transfer in Inert Porous Media, *Combust. Sci. Tech.*, 70: 93–109.

Schlichting, H. (1979). *Boundary-Layer Theory*, McGraw-Hill, New York.

Schlunder, E. U. (1975). Equivalence of One- and Two-Phase Models for Heat Transfer Processes in Packed Beds: One-Dimensional Theory, *Chem. Eng. Sci.*, 30: 449–452.

Semenic, T., and Catton, I. (2006). Heat Removal and Thermophysical Properties of Biporous Evaporators, Paper no. IMECE2006-15928, pp. 377–383; DOI:10.1115/IMECE2006-15928. ASME 2006 International Mechanical Engineering Congress and Exposition (IMECE2006) November 5–10, Chicago, IL, Sponsor: *Heat Transfer Division Heat Transfer*, Volume 2.

Semenic, T., Lin, Y-Yu, and Catton, I. (2008). Thermophysical Properties of Biporous Heat Pipe Evaporators, *ASME J. Heat Trans.*, 130, 022602-1-to-10.

Seume, J. R., Friedman, G., and Simon, T. W. (1992). Fluid Mechanics Experiments in Oscillating Flow Volume I-Report, NASA Contractor Report 189127.

Shah, R. K. (1975). Thermal Entry Length Solutions for the Circular Tube and Parallel Plates, *Proc. Natl. Heat Mass Transfer Conf., 3rd, Indian Inst. Technol.*, Bombay, I, Paper no. HMT-11-75.

Shah, R. K. (1978). A Correlation for Laminar Hydrodynamic Entry Length Solutions for Circular and Noncircular Ducts, *J. Fluids Eng.*, 100: 177–179.

Shah, R. K., and London, A. L. (1978). *Laminar Flow Forced Convection in Ducts*, Academic Press, New York.

Shirey, K., Banks, S., Boyle, R., and Unger, R. (2006). Design and Qualification of the AMS-02 Flight Cryocoolers, *Cryogenics 46* 142–148.

Siegel, R., and Howell, J. R. (2002). *Thermal Radiation Heat Transfer*, 4th ed., Taylor and Francis, Boca Raton, FL.

Simon, T. W., and Seume, J. R. (1988). A Survey of Oscillating Flow in Stirling Engine Heat Exchangers, NASA Contractor Report 182108.

Simon, T. W., Ibrahim, M., Kannaparedy, M., Johnson, T., and Friedman, G. (1992). Transition of Oscillatory Flow in Tubes: An Empirical Model for Application to Stirling Engines, *Proceedings of the 27th Intersociety Energy Conversion Engineering Conference*, Vol. 5, pp. 495–502, Paper no. 929463.

Simon, T. (2003). University of Minnesota, Private communication.

Simon, T., McFadden, G., and Ibrahim, M. (2006). Thermal Dispersion within a Porous Medium Near a Solid Wall, *Proceedings of 13th International Heat Transfer Conference*, Sydney, Australia, Paper no. 1198.99, August 13–18.

Singh, R. (2006). Thermal Control of High-Powered Desktop and Laptop Microprocessors Using Two-Phase and Single-Phase Loop Cooling Systems, Ph.D. thesis, RMIT University, Melbourne, Australia.

Singh, R., Akbarzadeh, A., Dixon, C., and Mochizuki, M. (2004). Experimental Determination of the Physical Properties of a Porous Plastic Wick Useful for Capillary Pumped Loop Applications, *Proceedings of the 13th International Heat Pipe Conference*, Shanghai, China, September 21–25.

Singh, R., Akbarzadeh, A., Dixon, C., Mochizuki, M., Nguyen, T., and Reihl, R. R. (2007). Miniature Loop Heat Pipes with Different Evaporator Configurations for Cooling Compact Electronics, *Proceedings of the 14th International Heat Pipe Conference*, April 22–27, Florianopolis, Brazil, pp. 176–181.

Singh, R., Akbarzadeh, Aliakbar, and Mochizuki, Masataka. (2009). Effect of Wick Characteristics on the Thermal Performance of the Miniature Loop Heat Pipe. *ASME J. Heat Trans.*, 131/082601-1 to 10.

Sohankar, A., Norberg, C., and Davidson, L. (1995). Numerical Simulation of Unsteady Flow Around a Square Two-Dimensional Cylinder, *Proceedings of the 12th Australasian Fluid Mechanics Conference*, Sydney, pp. 517–520.

Sohankar, A., Norberg, C., and Davidson, L. (1996). Numerical Simulation of Unsteady Low-Reynolds Number Flow around a Rectangular Cylinder at Incidence, *Proceedings of the 3rd Int. Colloq. on Bluff Body Aerodynamics and Applications*, Blacksburg, VA, July 28–August 1.

Sohankar, A., Norberg, C., and Davidson L. (1998). Low-Reynolds Number Flow Around a Square Cylinder at Incidence: Study of Blockage, Onset of Vortex Shedding and Outlet Boundary Condition, *Int. J. Num. Meth. in Fluids*, 26: 39–56.

Stephan, K. (1959). Warmeubergang und druckabfall bei nicht ausgebildeter Laminarstromung in Rohren und in ebenen Spalten, *Chem.-Ing.-Tech.*, 31: 773–778.

Stine, W. B., and Diver, R. B. (1994). A Compendium of Dish/Stirling Technology, SAND93- 7026 UC-236.

Sun, L., Mantell, S. C., Gedeon, D., and Simon, T. W. (2004). A Survey of Microfabrication Techniques for Use in Stirling Engine Regenerators, Paper no. AIAA-2004-5647, *Proceedings of the 2nd International Energy Conversion Engineering Conference*, Providence, RI.

Swift, G. W. (1988). Thermacoustic Engines, *J. Acoust. Soc. Am.*, 84(4): 1145–1180.

Takahachi, S., Hamaguchi, K., Miyabe, H., and Fujita, H. (1984). On the Flow Friction and Heat Transfer of the Foamed Metals as the Regenerator Matrix, *Proceedings of the 2nd International Conference on Stirling Engines*, Paper 3–4, Shanghai, The Chinese Society of Naval Architecture and Marine Engineering and the Chinese Society of Engineering Thermophysics.

Takahata, K., and Gianchandani, Y. (2002). Batch Mode Micro-Electro-Discharge Machining, *J. Microelectromechanical Sys.*, 11(2): 102–110.

Takeuchi, T., Kitahama, D., Matsugchi, A., Kagawa, N., and Tsuruno, S. (2004). Performance of New Mesh Sheet for Stirling Engine Regenerator, Paper AIAA 2004-5648, A Collection of Technical Papers, *2nd International Energy Conversion Engineering Conference*, Providence, RI, 16–19 August.

Takizawa, H., Kagawa, N., Matsuguchi, A., and Tsuruno, S. (2002). Performance of New Matrix for Stirling Engine Regenerator, Paper 20057, *Proceedings of 37th Intersociety Energy Conversion Engineering Conference*, Washington, DC, 29–31 July.

Tannehill, J. C., Anderson, D. A., and Platcher, R. H. (1997). *Computational Fluid Mechanics and Heat Transfer*, 2nd ed., Taylor and Francis, Boca Raton, FL.

Taylor, D. R., and Aghili, H. (1984). An Investigation of Oscillating Flow in Tubes, *Proceedings of the 19th Intersociety Energy Conversion Engineering Conference*, Paper 849176, pp. 2033–2036, American Nuclear Society.

Tew, R. C. Jr.; Thieme, L. G., and Miao, D. (1979). Initial Comparison of Single Cylinder Stirling Engine Computer Model Predictions with Test Results, SAE Paper 790327, DOE/NASA/1040-78/30, NASA TM-79044.

Tew, R. C., Simon, T. W., Gedeon, D., Ibrahim, M. B., and Rong, W. (2006). An Initial Non-Equilbrium Porous-Media Model for CFD Simulation of Stirling Regenerators, NASA/TM-2006-214391.

Tew, R. C., Ibrahim, M. B., Danila, D., Simon, T., Mantell, S., Sun, L., Gedeon, D., Kelly, K., McLean, J., Wood, J. G., and Qiu, S. (2007). A Microfabricated Involute-Foil Regenerator for Stirling Engines, NASA/TM-2007-214973.

Thieme, L. G. (1979). Low-Power Baseline Test Results for the GPU-3 Stirling Engine, DOE/NASA/1040-79/6, NASA TM-79103.

Thieme, L. G. (1981). High-Power Baseline and Motoring Test Results for the GPU-3 Stirling Engine, DOE/NASA/51040-31, NASA TM-82646.

Thieme, L. G., and Tew, R. C. Jr. (1978). Baseline Performance of the GPU 3 Stirling Engine, DOE/NASA/1040-78/5, NASA TM-79038.

Tong, L. S., and London, A. L. (1957). Heat Transfer and Flow-Friction Characteristics of Woven-Screen and Cross-Rod Matrices, *Trans. ASME*, pp. 1558–1570.

Tuchinsky, L., and Loutfy, R. (1999). Novel Process for Cellular Materials with Oriented Structure, *Proceedings of the 1st Proceedings of the First International Conference on Metal Foams and Porous Metal Structures* (MetFoam'99), Bremen (Germany).

Urieli, I., and Berchowitz, D. M. (1984). *Stirling Cycle Engine Analysis*, Adam Hilger Ltd., London.

Unger, R. Z., Wiseman, R. B., and Hummon, M. R. (2002). The Advent of Low Cost Cryocoolers, *Cryocoolers 11*, Springer U.S., pp. 79–86.

Vasiliev, L. L. (1993). Open-Type Miniature Heat Pipes, *J. Eng. Physics Thermophysics*, 65(1).

Vasiliev, L. L., Khrustalev, D. K., and Kulakov, A. G. (1991). Highly Efficient Condenser with Porous Element, *21st Int. Conf. on Environmental Systems*, San Francisco, CA.

Walker, G. (1980). *Stirling Engines*, Oxford University Press, New York, July 15–18.

Walker, G., and Vasishta, V. (1971). Heat-Transfer and Flow-Friction Characteristics of Dense-Mesh-Wire-Screen Stirling Cycle Regenerators, In: K. D. Timmerhaus, Editor, *Advances in Cryogenic Engineering*, Vol. 16, pp. 302–311, Plenum Press, New York.

Wang, J., and Catton, I. (2004). Vaporization Heat Transfer in Biporous Wicks of Heat Pipe Evaporators, *Proceedings of the International Heat Pipe Conference*, Shanghai, China, September 21–25.

Ward, J. C. (1964). Turbulent Flow in Porous Media, *J. Hydraulics Division*, 90(HY5): 1–12.

Watson, E. J. (1983). Diffusion in Oscillatory Pipe Flow, *J. Fluid Mech.*, 133: 233–244.

West, C. D. (1986). *Principles and Applications of Stirling Engines*, Van Nostrand Reinhold, New York.

Wilcox, D. C. (1998). *Turbulence Modeling for CFD*, DCW Industries, La Canada, CA.

Williamson, C. H. K. (1996). Vortex Dynamics in the Cylinder Wake, *Annu. Rev. Fluid Mech.*, 28: 477–539.

Wilson, S. D., Dyson, R. W., Tew, R. C., and Demko, R. (2005). Experimental and Computational Analysis of Unidirectional Flow through Stirling Engine Heater Head, *3rd International Energy Conversion Engineering Conference*, Paper no. AIAA-2005-5539, San Francisco, CA.

Wong, W. A., Wood, J. G., and Wilson, K. (2008). Advanced Stirling Convertor (ASC)—From Technology Development to Future Flight Product, NASA/TM—2008-215282.

Wood, J. G., Carroll, C., and Penswick, L. B. (2005). Advanced 80 W Stirling Convertor Development Progress, *Space Technology Applications International Forum (STAIF)* paper.

Yeh, C. C., Liu, B. H., and Chen, Y. M. (2008). A Study of Loop Heat Pipe with Biporous Wicks, *Heat Mass Transfer*, 44: 1537–1547.

Zhao, T. S., and Cheng, P. (1996). Oscillatory Pressure Drops through a Woven-Screen Packed Column Subjected to a Cyclic Flow, *Cryogenics 36*, pp. 333–341.

Zhang, J., and Dalton, C. (1998). A Three Dimensional Simulation of Steady Approach Flow Past a Circular Cylinder at Low Reynolds Number, *Int. J. Num. Meth. in Fluids*, 26: 1003–1022.

Zhdanok, S. A., Dobrego, K. V., and Futko, S. I. (2000). Effect of Porous Media Transparency on Spherical and Cylindrical Filtrational Combustion Heaters Performance, *Int. J. Heat Mass Trans.*, 43: 3469–3480.

Index